Springer Theses

Recognizing Outstanding Ph.D. Research

Aims and Scope

The series "Springer Theses" brings together a selection of the very best Ph.D. theses from around the world and across the physical sciences. Nominated and endorsed by two recognized specialists, each published volume has been selected for its scientific excellence and the high impact of its contents for the pertinent field of research. For greater accessibility to non-specialists, the published versions include an extended introduction, as well as a foreword by the student's supervisor explaining the special relevance of the work for the field. As a whole, the series will provide a valuable resource both for newcomers to the research fields described, and for other scientists seeking detailed background information on special questions. Finally, it provides an accredited documentation of the valuable contributions made by today's younger generation of scientists.

Theses may be nominated for publication in this series by heads of department at internationally leading universities or institutes and should fulfill all of the following criteria

- They must be written in good English.
- The topic should fall within the confines of Chemistry, Physics, Earth Sciences, Engineering and related interdisciplinary fields such as Materials, Nanoscience, Chemical Engineering, Complex Systems and Biophysics.
- The work reported in the thesis must represent a significant scientific advance.
- If the thesis includes previously published material, permission to reproduce this must be gained from the respective copyright holder (a maximum 30% of the thesis should be a verbatim reproduction from the author's previous publications).
- They must have been examined and passed during the 12 months prior to nomination.
- Each thesis should include a foreword by the supervisor outlining the significance of its content.
- The theses should have a clearly defined structure including an introduction accessible to scientists not expert in that particular field.

Indexed by zbMATH.

More information about this series at http://www.springer.com/series/8790

Alfred Zong

Emergent States in Photoinduced Charge-Density-Wave Transitions

Doctoral Thesis accepted by Massachusetts Institute of Technology, MA, USA

Springer

Alfred Zong
University of California
Berkeley, CA, USA

ISSN 2190-5053 ISSN 2190-5061 (electronic)
Springer Theses
ISBN 978-3-030-81750-3 ISBN 978-3-030-81751-0 (eBook)
https://doi.org/10.1007/978-3-030-81751-0

This Springer imprint is published by the registered company Springer Nature Switzerland AG
The registered company address is: Gewerbestrasse 11, 6330 Cham, Switzerland

Supervisor's Foreword

Phase transitions lie in the heart of quantum material research. Knowledge of their underlying mechanisms helps us tailor material properties and guides our search for novel states of matter. With recent advances in femtosecond laser technology, ultrafast light pulses emerge as a versatile tool in the pursuit of phase engineering and control. Photoinduced transitions may display distinct features compared to their equilibrium counterpart and often lead to exotic orders that are not accessible in equilibrium, such as high-temperature superconducting behavior, topological transition, and a wide variety of metastable states.

In this book, Dr. Alfred Zong investigated photoinduced transitions by examining compounds that host a charge density wave (CDW). In equilibrium, a charge density wave is a spatially modulated condensate of electrons accompanied by a periodic lattice distortion when certain metallic crystals are cooled below a critical temperature. CDWs provide an excellent platform to study transitions in a highly nonequilibrium setting for several reasons. First, across a CDW transition, both electrons and lattice ions undergo dramatic changes, which can be monitored by a suite of ultrafast pump-probe techniques. Here, Dr. Zong employed three time-resolved instruments: electron diffraction, angle-resolved photoemission, and optical spectroscopy. They give access to the lattice and electron dynamics that, when taken together, yield profound insights into the nonequilibrium pathway of a light-induced CDW transition. Second, across many classes of quantum materials, the CDW phase is found in close proximity or in coexistence with a number of other ground states, including superconductivity, antiferromagnetism, and a correlated insulating state of a Mott insulator or an excitonic insulator. By perturbing the CDW with a femtosecond light pulse, one could potentially manipulate these related phases of matter and study how competing or coupled phases evolve out of equilibrium. Third, CDWs in many compounds are known to be susceptible to perturbation, which can yield interesting dynamical states, such as a sliding condensate under a DC voltage and electronic mode locking when an additional AC voltage is applied. Hence, it is likely that laser excitation could also give rise to internal deformation of the CDW and some type of metastability, which may find applications in ultrafast devices.

Dr. Zong made several important contributions towards our understanding of how a symmetry-broken state responds to a perturbation by ultrashort laser light, as exemplified in different CDW compounds. Upon photoexcitation, the transient restoration of the broken symmetry is found to exhibit dynamical slowing-down, reminiscent of critical slowing-down in equilibrium transitions. The subsequent relaxation back to the ground state, however, differs from an equilibrium picture as there are distinct recovery timescales for the order parameter amplitude and its phase coherence. The key to explain this discrepancy is photoinduced topological defects in the CDW, which could persist long after photoexcitation and disrupt the long-range coherence. In some cases, Dr. Zong showed that these light-induced defects are essentially metastable and last over days or weeks, resulting in chiral domains that could be reversibly flipped by either another light pulse or by thermal annealing. An interesting scenario arises when there is a phase competition between two nearly degenerate CDW phases, out of which only the slightly dominant phase exists in equilibrium while the subdominant one is suppressed. Dr. Zong observed that a light pulse transiently weakens the dominant order, and at the same time, the subdominant CDW emerges. The interplay between these two CDW orders helps us learn important characteristics of phase competition in far-from-equilibrium systems, where light-induced defects and order parameter fluctuations play important roles.

To summarize, this book by Dr. Zong provides many key insights into nonequilibrium symmetry-breaking transitions, developed using the platform of charge density waves. The principals discussed in the book can be generalized to other broken symmetry states, such as superconductivity and magnetic ordering, bringing us one step closer towards manipulating phases of matter using a laser pulse.

Cambridge, MA, USA
June 2021

Nuh Gedik

Preface

The past five years has witnessed tremendous progress in time-resolved techniques, ranging from atomic-resolution attosecond microscopy to large-scale free electron lasers. At the same time, new classes of quantum materials – broadly characterized by their unusual properties arising from special symmetry, dimensionality, correlation, or topology – are either synthesized or artificially assembled, leading to discoveries that constantly challenge the existing frameworks in condensed matter physics. This book sits at the confluence of the two topics: The overarching theme is to leverage ultrafast pulses to understand and manipulate various properties of quantum materials.

Our materials of choice are charge-density-wave (CDW) compounds, first discussed by Fröhlich and Peierls in the 1950s and intensely studied in the 1970s to 1980s for their peculiar transport properties. They represent a prototypical broken-symmetry state, akin to a low-energy version of crystallization, where electron density and ionic positions modulate with a periodicity different from the underlying lattice. Recently, there has been a reviving interest in CDWs because they are often found in materials that host other exotic ground states, for example, unconventional superconductors and excitonic insulators. Decoding the electron-phonon and electron-electron interactions in CDWs is thought to offer some key insights into the complex phase diagram of these compounds.

From the perspective of ultrafast sciences, CDWs provide a versatile playground to investigate two fundamental questions: (i) how symmetry-breaking phase transitions proceed in an out-of-equilibrium setting, and (ii) how light pulses selectively suppress or enhance proximal phases of matter. The first question entails the study of both the amplitude and the phase of a complex order parameter, which may be readily applied to other symmetry-breaking transitions. Collective excitations of a CDW add another lens to study the nonequilibrium dynamics: With the help of femtosecond pulses, one can observe these excitations in real time. The second question touches upon a central characteristic of many quantum materials, where competing or coexisting orders are important factors to consider when we study anomalous material properties. Using light pulses to traverse the boundary between neighboring phases not only yields insights into the driving force behind the equilibrium phase diagram but also holds promises for engineering functional properties at an ultrafast timescale.

Two CDW systems are examined in this book: rare-earth tritellurides ($R\mathrm{Te}_3$) and octahedrally coordinated tantalum disulfide ($1T$-TaS_2). Though both of them share a layered, quasi-two-dimensional structure, they differ in several important aspects. For example, while $R\mathrm{Te}_3$ is considered a simple metal well-described by a single-particle band structure, strong correlation effects exist in $1T$-TaS_2 and it becomes a Mott insulator at low temperature. In terms of their ground state CDW distortion, while $R\mathrm{Te}_3$ has a stripe or checkerboard pattern with no chirality, $1T$-TaS_2 has a planar chiral structure. Furthermore, the CDW is incommensurate in $R\mathrm{Te}_3$, but can fall into many morphologies ranging from completely incommensurate to completely commensurate in $1T$-TaS_2. These materials cover a wide spectrum of phenomenology, allowing us to study the interaction between CDW and light under different conditions.

The book is organized as follows. Chapter 1 offers a bird's-eye view of ultrafast studies of quantum materials, including a discussion of the state-of-the-art time-resolved probes and my take on their future directions. The scientific review provides a context for the results presented in Chaps. 4, 5, 6, 7, and 8 and highlights additional experiments undertaken during my graduate study. Chapter 2 gives a formal introduction to CDWs, laying the theoretical foundation for subsequent chapters. It also details the properties of the two CDW compounds examined in this book. Chapter 3 is devoted to ultrafast electron diffraction, an important probe used throughout the book. The chapter reviews recent progress of this nascent technique, followed by details of instrumentation, sample fabrication, and data analysis.

Chapters 4, 5, 6, and 7 focus on $R\mathrm{Te}_3$. In Chap. 4, we will study the relaxation dynamics of a complex order parameter following photoexcitation, where the CDW amplitude and phase coherence are decoupled in their temporal evolutions due to photoinduced topological defects. In Chap. 5, we shift the focus from the relaxation to the melting of the broken-symmetry state. I will provide evidence for dynamical slowing-down of the photoinduced phase transition, which parallels critical slowing-down in an equilibrium transition. In Chap. 6, I will present the discovery of a light-induced CDW in LaTe_3, which is suppressed in equilibrium due to phase competition. The ultrafast dynamics of the competing orders is examined in Chap. 7. Despite the competing relationship, they develop simultaneously in an out-of-equilibrium regime, mediated by photoinduced fluctuations.

While the above phenomena in $R\mathrm{Te}_3$ last only a few picoseconds, Chap. 8 shows an example of a photoinduced metastable state in $1T$-TaS_2. At room temperature, a single pulse of light can reversibly switch the planar chirality of the nearly commensurate CDW in $1T$-TaS_2 . We will use a suite of probes to characterize the resulting metastable configuration.

This book showcases many facets of the interaction between a CDW and a femtosecond light pulse, which give rise to a variety of emergent states out of equilibrium. The results presented here serve to deepen our understanding of photoinduced phase transitions, paving the way towards the design and control of other broken-symmetry states with light.

Parts of this thesis have been published in the following journal articles:

1. Zong, A., Shen, X., Kogar, A., Ye, L., Marks, C., Chowdhury, D., Rohwer, T., Freelon, B., Weathersby, S., Li, R., Yang, J., Checkelsky, J., Wang, X., Gedik, N.: Ultrafast manipulation of mirror domain walls in a charge density wave. Sci. Adv. **4**, eaau5501 (2018)
2. Zong, A., Kogar, A., Bie, Y.-Q., Rohwer, T., Lee, C., Baldini, E., Ergeçen, E., Yilmaz, M.B., Freelon, B., Sie, E.J., Zhou, H., Straquadine, J., Walmsley, P., Dolgirev, P.E., Rozhkov, A.V., Fisher, I.R., Jarillo-Herrero, P., Fine, B.V., Gedik, N.: Evidence for topological defects in a photoinduced phase transition. Nat. Phys. **15**, 27–31 (2019)
3. Zong, A., Dolgirev, P.E., Kogar, A., Ergeçen, E., Yilmaz, M.B., Bie, Y.-Q., Rohwer, T., Tung, I.-C., Straquadine, J., Wang, X., Yang, Y., Shen, X., Li, R., Yang, J., Park, S., Hoffmann, M.C., Ofori-Okai, B.K., Kozina, M.E., Wen, H., Wang, X., Fisher, I.R., Jarillo-Herrero, P., Gedik, N.: Dynamical slowing-down in an ultrafast photoinduced phase transition. Phys. Rev. Lett. **123**, 097601 (2019)
4. Kogar, A., Zong, A., Dolgirev, P.E., Shen, X., Straquadine, J., Bie, Y.-Q., Wang, X., Rohwer, T., Tung, I.-C., Yang, Y., Li, R., Yang, J., Weathersby, S., Park, S., Kozina, M.E., Sie, E.J., Wen, H., Jarillo-Herrero, P., Fisher, I.R., Wang, X., Gedik, N.: Light-induced charge density wave in $LaTe_3$. Nat. Phys. **16**, 159–163 (2020)
5. Dolgirev, P.E., Rozhkov, A.V., Zong, A., Kogar, A., Gedik, N., Fine, B.V.: Amplitude dynamics of the charge density wave in $LaTe_3$: theoretical description of pump-probe experiments. Phys. Rev. B **101**, 054203 (2020)
6. Dolgirev, P.E., Michael, M.H., Zong, A., Gedik, N., Demler, E.: Self-similar dynamics of order parameter fluctuations in pump-probe experiments. Phys. Rev. B **101**, 174306 (2020)
7. Fichera, B.T., Kogar, A., Ye, L., Gökce, B., Zong, A., Checkelsky, J.G., Gedik, N.: Second harmonic generation as a probe of broken mirror symmetry. Phys. Rev. B **101**, 241106(R) (2020)
8. Zong, A., Kogar, A., Gedik, N.: Phase competition and light-induced ordering in charge density waves. Proc. SPIE **11684** (Invited Paper), 1168412 (2021)

Acknowledgments

First and foremost, I am extremely fortunate to have Nuh Gedik as my thesis advisor. Nuh holds the magic of presenting deep concepts in a simple, mesmerizing, yet rigorous manner. During our first meeting, Nuh seeded my interest in ultrafast sciences with a simple sketch of $\Delta R/R$ traces and Floquet-Bloch bands. Ever since, he has shaped my research outlook and scientific imagination. I learned from him how to think bold – he constantly pushes technical boundaries and treads uncharted territory – and at the same time, how to turn dreams into reality with meticulous attention to details. I also learned from him how to deal with failure – be it manuscript rejection or a leaky vacuum chamber. During my second year, there was a period when every single setup in the lab malfunctioned, not to mention the recurring episodes of flood throughout our group history. Nuh always handled these demoralizing setbacks with great composure, reminding us this is nothing but an integral part of experimental science. I cannot thank him more for granting me tremendous freedom to pursue a variety of experiments both on campus and at national labs. The liberty to collaborate with so many top-notch groups has really broadened my scientific horizon and forged long-lasting friendship. Nuh never micromanages, yet he is always available for a chat and is generous with honest critiques, both inside and outside physics. He truly cares about every member in the lab and throws his full support to groom us as independent scientists.

Nuh's other magic is to assemble a diverse group of gifted experimentalists, and I have the real privilege to learn from each of them. I owe my optical skills to Timm Rohwer. Listening to his mini lectures during optical alignment bestows fun and enlightenment upon a tedious job. Whenever some equipment stops working, Timm jumps straight in and can always find a fix. He is one of the very few people who defy the *Second Law of Optimization*, which states "attempts to improve a working laser only serve to deteriorate the performance in a monotonic fashion." Byron Freelon pioneered the design and construction of the keV ultrafast electron diffraction setup, which forms the backbone of many projects in this book. He never gets annoyed by my endless questions, from CAD drawings to vacuum lines. Anshul Kogar was my longtime officemate and comrade for the UED projects, and we spent countless hours together both in the lab and during beamtimes. We can literally chat or debate about physics for hours (apologies to our neighbors), and he is a constant source of inspiration for new ideas and perspectives, as exemplified by his blog, This Condensed Life: https://thiscondensedlife.wordpress.com/. Anshul has an extremely personable character, gluing different members of the group together. I am deeply inspired by his diverse interest in science and his ever-present optimism, even during a time when every single component of the UED fell apart. Yifan Su is the newest member of the UED subgroup and I have had the fortune to mentor him in the past 2 years. He is a quick learner, full of curiosity, and eager to get hands dirty – I have no doubt that UED will be in good care.

Outside the UED subgroup, Fahad Mahmood and Inna Vishik are the first people who helped me navigate the rhythm of the lab. Like a big brother and sister, they kept no secret about the essential skills of grad school, which may include first-hand information of free food and a stockpile of Red Bulls. Edbert Sie is an all-round scientist, mentor, and role model, who seems to know everything and always goes one extra mile when helping with whatever I asked him, even after he moved to the other side of the country. Edoardo Baldini not only has infinite knowledge about a vast array of materials and ultrafast spectroscopies, he never hesitates to share his expertise with others. I have carefully kept his handwritten notes out of our discussions, which will serve me for a long time. I am also grateful for the opportunity to work with Suyang Xu, who taught me the importance of embracing new ideas and to delve deep behind the façade. His sharp questions drove me to re-examine my prior knowledge and basic concepts, often giving rise to new insights. Baiqing Lyu is an indispensable buddy for the past 2 years. We collaborated on a number of projects and we traveled together across the country for experiments. Besides a huge repertoire of interesting material systems, I also admire his tenacity and disciplined work ethics, both inside and outside the lab. My limited knowledge of attosecond science is attributed to Doron Azoury. Chatting with him about future scientific directions generates the rare adrenaline of discovering something truly exciting. Changmin Lee, Dongsung Choi, and generations of the tr-arpes group have dedicated tremendous effort to maintaining the most complex instrument in the lab. Behind each photoemission spectrum presented in this book is hours of bolt tightening, mirror tweaking, and code debugging, among many others. The optical pump-probe setups in the lab (and the room itself) would not exist without the hard work of Zhanybek Alpichshev, Emre Ergeçen, Mehmet Yilmaz, and Batyr Ilyas. The second harmonic generation beamline was revived by Bryan Fichera, who added his unique touch on the LabVIEW control and the python analysis: https://bfichera.github.io/shgpy/. I am also delighted to work alongside Carina Belvin, not just in the lab, but also as members of the physREFS: https://physrefs.mit.edu/ team. Besides being a brilliant physicist, she truly cares about the welfare of other graduate students. Her service to a number of student organizations in the department undoubtedly brightens the lives of many. Over the years, many other members of the group have made my stay in the lab full of fond memories. To Daniel Pilon, Özge Özel, Clifford Allington, Nikesh Koirala, Azel Murzabekova, Guy Marcus, Andrew Xu, Manan Bajaj, and Nolan Peard: I appreciate your company and eagerness to help at all times.

The close-knit CMX community at MIT is another highlight of my graduate study. Monica Wolf is the head of the family and she takes meticulous care of each "kid" on the second floor of Bldg 13. The physical proximity of the CMX groups has forged many long-lasting collaborations – including but not limited to long-term loans of cryogens, Keithley's, and leak detectors. Thanks to Pablo Jarillo-Herrero who makes his lab a shared facility, I had the chance to work closely with Ya-Qing Bie and Xirui Wang and to learn the technique of 2D material assembly. These experiments would not be possible without Dahlia Klein, who makes sure the glovebox is well maintained despite the high-traffic usage. Linda Ye from the group of Joseph Checkelsky is the go-to person when I have a question on transport experiments. She is always happy to share her beautiful crystals and the magic trick of gluing wires on tiny samples. I also had the privilege to work on a number of Raman experiments with Jiarui Li from the group of Riccardo Comin. In addition, Riccardo and Maxim Metlitski are the other members of my thesis committee. I am grateful for their effort in perusing this book and their thoughtful feedback. Outside the lab – whether it is on a badminton court, during a road trip, at a hotpot table, or in the Student Center – I am lucky to be surrounded by a group of CMX friends – Qiong Ma, Zhihai Zhu, Qi Song, Yuan Cao, and Yafang Yang – who have made my graduate life memorable and colorful.

A major part of the book is focused on rare-earth tritellurides; the high-quality crystals were grown by Joshua Straquadine and Philip Walmsley from the group of Ian Fisher at Stanford. I am also indebted to Ian for introducing me to the world of quantum materials through his iconic class AP 204. Ian is always encouraging when a new experimental idea is proposed and chatting with Ian never fails to reveal fresh insights hidden in the data. I am also grateful for the close collaboration with the SLAC UED team led by Xijie Wang, who is dedicated to pushing the boundaries of instrumentation on multiple fronts. The SLAC team – Renkai Li, Jie Yang, Stephen Weathersby, Suji Park, Alexander Reid, Michael Kozina, and Duan Luo – is always generous with sharing their expertise on all aspects of UED. My special gratitude goes to Xiaozhe Shen, who seems to master every detail of the instrument and together we probably broke the record of the longest sleepless measurement at the beamline. At SLAC UED, I am also privileged to have worked closely with Haidan Wen, Qi Zhang, Kyle Hwangbo, and I-Cheng Tung. Through this multi-institution collaboration, I am additionally thankful for the opportunity to conduct experiments at Argonne under the guidance of Haidan and his group: Qian Li, Samuel Marks, and Jiawei Zhang. Haidan taught me how to perform synchrotron X-ray diffraction, which is further augmented by both temporal and spatial resolutions.

Over the years, I appreciate the opportunity to collaborate with many other talented experimentalists: Ismail El Baggari from the group of Lena Kourkoutis at Cornell University; Alex McLeod from the group of Dmitri Basov at Columbia University; Makoto Hashimoto and Donghui Lu at SSRL; and Yongtao Cui at UC Riverside. Furthermore, I want to express my heartfelt gratitude to numerous people who spent considerable effort on growing high-quality crystals and fabricating samples of demanding dimension: Cong Su and Haozhe Wang from the group of Jing Kong at MIT; Di Lu, Seung Sae Hong and Bai Yang Wang from the group of Harold Hwang at Stanford University; Carolyn Marks and Cigdem Keskinbora at Harvard University; Kyle Seyler and Genevieve Clark from the group of Xiaodong Xu at University of Washington; Hengdi Zhao from the group of Gang Cao at University of Colorado Boulder; Kai Du from the group of Sang-Wook Cheong at Rutgers University; and Dong Wu from the group of Nanling Wang at Peking University.

During my graduate study, I am honored to have worked with a number of brilliant theorists, who provided much-needed insights into many surprising findings. I truly enjoyed discussing with Boris Fine and Alexander Rozhkov, who are both sharp when dealing with experimental data and creative when coming up with theoretical scenarios. Much of my theoretical knowledge on pump-probe dynamics was taught by Pavel Dolgirev from the group of Eugene Demler. Talking to Pavel requires me to summon all my brain power and the dead neurons are very well spent. He forces me to think hard about what exactly I have measured, which often forms the basis for another exciting prediction. Right here at MIT, I am thankful for many intellectually stimulating discussions across departments, with special thanks to Debanjan Chowdhury, Patrick Lee, Tian Xie, Zhiwei Ding, and Jiawei Zhou.

A significant source of inspiration during my PhD comes from my undergraduate advisor, Zhi-Xun Shen, and my mentors in his group: Yu He, Shuolong Yang, and Sudi Chen. They set an extremely high standard for both a good scientist and a caring brother. Their unconditional support has walked me through the darkest days while their grilling questions are constantly pushing me to strive for a better self. Talking to them reminds me of the purest sense of curiosity that brought me to this field, and I cannot be more blessed than having their wisdom to guide me along the way.

In the process of publishing this PhD thesis, I am indebted to Sam Harrison, Dinesh Vinayagam, and Pearly Percy for their editorial help. I would like to give special thanks to Emma Berger for her loyal readership and thorough examination of the manuscript.

This book is not possible without the unwavering support from my parents. On the surface, my mother does not fully approve my decision to pursue a PhD in physics while giving up the six-digit salary in the Silicon Valley. Deep in her heart, she is immensely proud and poured all her love when I struggled through the graduate career. My father is much more relaxed, and our chats have largely remained scientific and technical. Together, they are my best role models, and from them, I learned the guiding virtues that shaped my world view: equanimity (静), diligence (勤), prudence (思), integrity (诚), and love (爱). You are my heroes.

This book was written during an unprecedented time as the COVID-19 pandemic has paralyzed the entire globe. The pain is further aggravated by the highly polarizing politics within the USA, recent events on racial injustice, as well as an escalating geopolitical war between the two superpowers in the world. I feel extremely privileged to have the luxury of burying myself under physics despite the heart-wrenching turbulence outside. I hope this book may provide a temporary escape for the readers, and we as scientists could make the world a better place in our own little way.

Berkeley, CA, USA
August 2021

宗国 Alfred Zong

Contents

Acronyms

1D	One-dimensional
2D	Two-dimensional
2PPE	Two-photon photoemission
3D	Three-dimensional
AFM	Antiferromagnetism
AHE	Anomalous Hall effect
AM	Amplitude mode
ARPES	Angle-resolved photoemission spectroscopy
BCS	Bardeen-Cooper-Schrieffer
BZ	Brillouin zone
CAP	Coherent acoustic phonon
CCD	Charge-coupled device
CDW	Charge density wave
CEP	Carrier-envelope phase
CF	ConFlat
cgs	Centimeter-gram-second
CMR	Colossal magnetoresistance
DECP	Displacive excitation of coherent phonon
DFPT	Density-functional perturbation theory
DLRO	Diagonal long-range order
DW	Domain wall

EELS	Electron energy loss spectroscopy
E_F	Fermi level
EI	Excitonic insulator
EMCCD	Electron-multiplying charge-coupled device
EPC	Electron-phonon coupling

FEL	Free-electron laser
FFT	Fast Fourier transform
FIB	Focused ion beam
FM	Ferromagnetism
FSN	Fermi surface nesting
FTIR	Fourier-transform infrared spectroscopy
FWHM	Full width at half maximum

HHG	High-harmonic generation
HTSC	High-T_c superconductor

IMT	Insulator-to-metal transition
ISRS	Impulsive stimulated Raman scattering

JPR	Josephson plasma resonance

LAPE	Laser-assisted photoemission
LEED	Low-energy electron diffraction
LIA	Lock-in amplifier
LRO	Long-range order

MBE	Molecular beam epitaxy
MCP	Microchannel plate
MFM	Magnetic force microscopy
MIM	Microwave impedance microscopy
MIR	Mid-infrared
ML	Monolayer

NCCDW	Nearly-commensurate charge density wave
NIR	Near-infrared
NMR	Nuclear magnetic resonance

OAPM	Off-axis parabolic mirror
OD	Outer diameter
ODLRO	Off-diagonal long-range order
OFHC	Oxygen-free high thermal conductivity
OPA	Optical parametric amplifier
PDMS	polydimethylsiloxane
PIPT	Photoinduced phase transition
PLD	Pulsed laser deposition
QSL	Quantum spin liquid
RA-SHG	Rotational anisotropy second harmonic generation
RF	Radio frequency
RIXS	Resonant inelastic X-ray scattering
r.l.u.	Reciprocal lattice unit
ROI	Region of interest
SAD	Selected area diffraction
SC	Superconductivity
SDW	Spin density wave
SHG	Second harmonic generation
SI	International System of Units
SNOM	Scanning near-field optical microscopy
SOC	Spin-orbit coupling
STEM	Scanning transmission electron microscopy
STM	Scanning tunneling microscopy
TB	Tight binding
TDFS	Transient depletion field screening
TDS	Thermal diffuse scattering
TDTS	Time-domain terahertz spectroscopy
TEM	Transmission electron microscopy
TFISH	Terahertz field-induced second harmonic
TMD	Transition metal dichalcogenide
TO	Transverse optical
TOS	Transient optical spectroscopy
tr	Time-resolved
TTM	Two-temperature model

UED	Ultrafast electron diffraction
UEM	Ultrafast electron microscopy
UHV	Ultra-high vacuum
UV	Ultraviolet
vdW	van der Waals
XRD	X-ray diffraction
XUV	Extreme ultraviolet

Chapter 1
Ultrafast Sciences in Quantum Materials

Abstract Ultrashort laser pulses, with duration in the femto-to-picosecond regime, usher in an era of time-domain study of quantum materials, in which many-body interactions can lead to remarkable properties that are beyond the single-particle description of solids by the band theory. Not only do ultrafast pulses reveal microscopic dynamics and collective excitations with unprecedented temporal resolution, the corresponding high peak field also enables a suite of nonlinear spectroscopies that are otherwise inaccessible. More recently, an intense light pulse has also been shown to induce a variety of nonequilibrium states, from superconductivity to ferroelectricity, whose origins are subject of intense investigations. This chapter offers a review of ultrafast studies in quantum materials, and recent progress is discussed from the perspective of both ultrafast probing and control of condensed matter systems. The review is concluded with a description of advanced femtosecond techniques used in this dissertation and an outlook of next-generation tools.

An ultrafast movie of photoinduced evolution of an electronic band. Individual frames were taken using time- and angle-resolved photoemission spectroscopy on $ErTe_3$.

A. Zong, *Emergent States in Photoinduced Charge-Density-Wave Transitions*, Springer Theses, https://doi.org/10.1007/978-3-030-81751-0_1

The field of ultrafast sciences dates back to the discovery of laser mode-locking in the early 1960s [1], shortly after the invention of laser itself. Mode-locking refers to the phenomenon that different oscillating modes in a laser cavity maintain equal frequency spacings with a fixed phase relationship to one another (Fig. 1.1b). Under these conditions, the laser output as a function of time consists of a comb of short pulses, whose duration is limited by the bandwidth via the Fourier transform relation. The progress in this field was accelerated by the advent of Ti:sapphire lasers that produced sub-100-fs pulses with self-mode-locking [2], a spontaneous process similar to synchronized flashing of fireflies, also common to other ensembles of coupled nonlinear oscillators [3, 4]. The technology of producing ultrashort pulses is complemented by the capability of packing tremendous amount of energy within each pulse, achieved using chirped pulse amplification [5], an important discovery leading to the Nobel Prize in Physics in 2018. Nowadays, it is routinely possible to attain high-repetition-rate ($\geq$100 kHz), high-energy ($\geq$1 mJ), and ultrashort ($\leq$100 fs) pulses at near-infrared (NIR) wavelengths. These pulses can, in turn, be used to create ultrafast light sources across decades of wavelengths via a variety of nonlinear processes, such as extreme ultraviolet (XUV) to soft X-ray pulses using high-harmonic generation (HHG) (Fig. 1.1c), mid-infrared (MIR) pulses using supercontinuum generation and parametric amplification, and few-cycle terahertz pulses using optical rectification (Fig. 1.1d).

Over the past decade, these ultrafast light pulses, ranging from sub-femtosecond to picosecond in duration, have played an instrumental role in our understanding of quantum materials. Broadly speaking, these materials are systems where symmetry, topology, dimensionality, and strong correlations lead to a host of exotic ground states and manifest as some macroscopic observables [10–13]. Such quantum materials include, for example, unconventional superconductors, quantum spin liquids, strange metals, and topological matter, where intricate interactions between charge, spin, orbital, and lattice degrees of freedom give rise to a complex phase diagram that is beyond the prediction of the best theories to date [14].

From a bird's eye view, there are two central questions in the study of quantum materials. First, *what mechanisms drive the formation of a particular ground state?* Second, *how do we manipulate equilibrium states of matter to achieve a desired property?* In the following, I will offer a few perspectives on how intense, ultrashort laser pulses help address these questions. I will also highlight specific contributions from works detailed in this dissertation.

1.1 New Insights into Correlated Ground States

Just like a typical linear response measurement, watching how a ground state responds to an ultrafast laser pulse reveals valuable information about the equilibrium state itself. Besides the apparent application of detecting unoccupied states by, for example, two-photon photoemission (2PPE) or time- and angle-resolved photoemission spectroscopy (tr-ARPES) [6] (Fig. 1.2a), perturbation by a light pulse offers two main advantages. First, a high peak electric field, routinely larger than

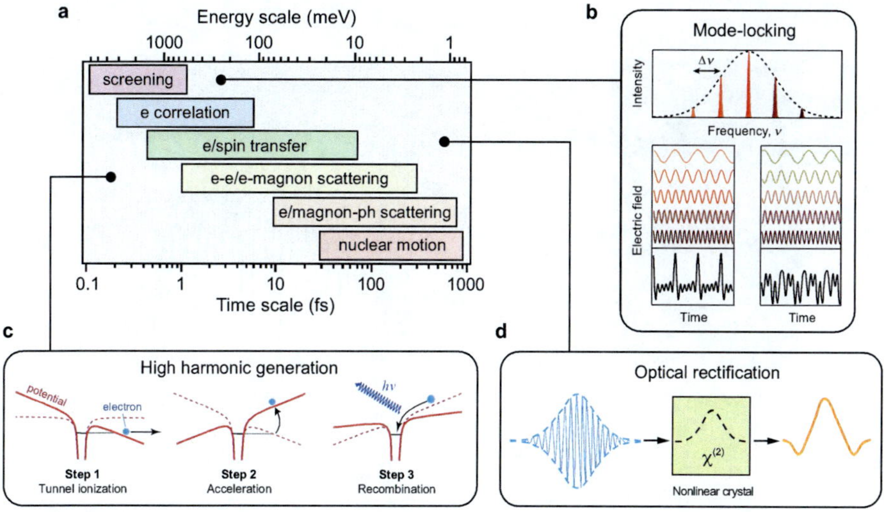

Fig. 1.1 Time and energy scales in solids probed by ultrafast laser pulses. (**a**) Elementary excitations and decay processes in solids. The characteristic time (τ) and energy (Γ) follow the uncertainty relation, $\tau\Gamma = \hbar = 658\,\text{meV}\cdot\text{fs}$ [6]. (**b**) Illustration of laser mode-locking. The emission power spectrum includes multiple axial resonator modes spaced by frequency difference $\Delta\nu$ (*top*), which is greatly exaggerated. When successive modes have fixed phase difference (mode-locked), the time-domain profile consists of a train of sharp peaks (*bottom left*). If the phase differences are random, the peaks disappear (*bottom right*). (**c**) Classical picture of high-harmonic generation (HHG). The strong electric field in the incident laser pulse modulates the atomic potential, causing ionization, acceleration, and recombination of bound electrons. (**d**) Schematic of optical rectification for generating THz pulses. In a nonlinear crystal, the induced second-order polarization has a time-dependence proportional to the envelope of the incident pulse. This time-varying polarization sources the THz radiation. (Adapted with permission from Ref. [6], Elsevier and Ref. [7], Wiley (**a**); Ref. [8], IOP Publishing (**c**); Ref. [9], SPIE (**d**))

0.01 V/Å across a wide range of wavelengths, induces many nonlinear effects that sensitively depend on the symmetry, topology, and correlations in a material. Second, the temporal resolution, from sub-femtosecond to few picoseconds, matches the typical timescales of dynamical processes in solids (Fig. 1.1a), such as carrier screening ($\lesssim$1 fs), charge or spin transfer (sub-fs to 10s fs), electron–electron scattering (few fs to 100s fs), and electron–phonon scattering (10s fs to 1 ps) [7]. These two characteristics of an ultrafast pulse provide us with a new lens to investigate ground states of matter, as illustrated below.

1.1.1 Nonlinear Spectroscopy

Due to the high electric field in an ultrashort laser pulse, one needs to consider material response beyond the linear regime. For example, the time-dependent

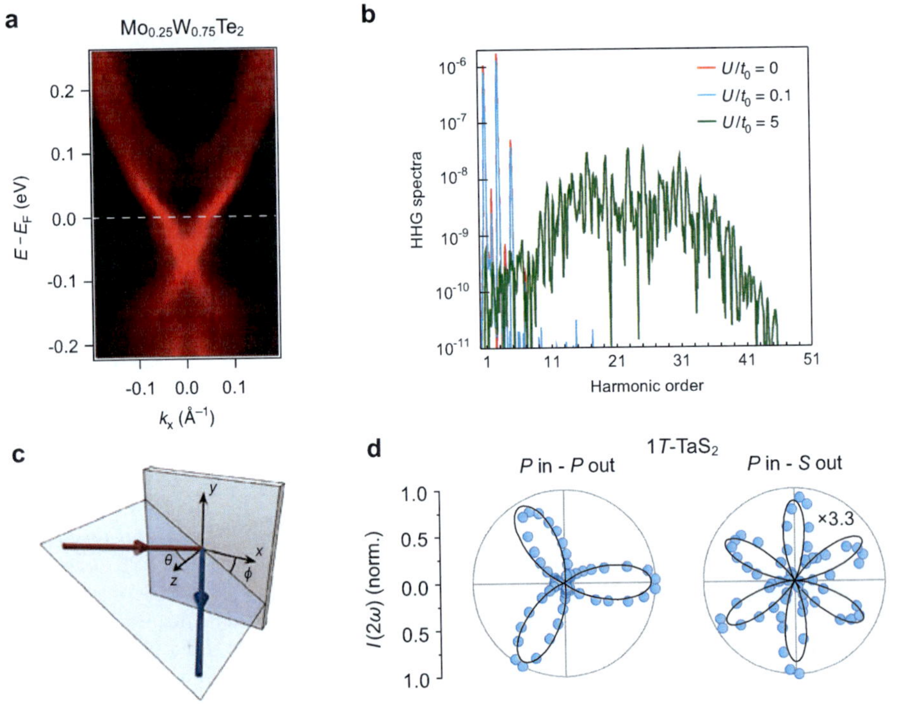

Fig. 1.2 **Nonlinear spectroscopy and detection of unoccupied states.** (**a**) Electronic dispersion of Weyl semimetal $Mo_{0.25}W_{0.75}Te_2$ above and below the Fermi energy, measured by tr-ARPES (1.48 eV pump photon and 5.92 eV probe photon). Spectra were taken at $\sim$8 K with a fixed momentum k_y near the predicted positions of the Weyl points. (**b**) Theoretical high-harmonic spectra emitted by a strongly correlated system using a one-dimensional Fermi–Hubbard model with half-filling. Distinct features are observed for different combinations of on-site repulsion U and nearest-neighbor hopping t_0. Calculation assumed 32.9 THz incident photon with a 0.1 V/Å field amplitude and a hopping rate $t_0 = 0.52$ eV. (**c**) Rotational anisotropy second harmonic generation (RA-SHG) as a sensitive probe of crystalline symmetry. A NIR photon is incident at angles (ϕ, θ) on the sample surface, and its second harmonic intensity is detected for various incoming and outgoing photon polarizations. (**d**) Second harmonic intensity as a function of ϕ for $1T$-TaS_2 at 356 K, where the material is in an incommensurate CDW phase. Intensities are normalized to the maximum value in the P_{in}-P_{out} polarization channel. Solid curves are fits to the surface point group C_{3v}. (Adapted from Ref. [15] (**a**); adapted with permission from Ref. [16], Springer Nature Ltd (**b**)).

electric polarization $\mathbf{P}(t)$ induced in a solid would depend on higher powers of the electric field $\mathbf{E}(t)$ in the incident light:

$$P_i(t) = \chi^{(1)}_{ij} E_j(t) + \chi^{(2)}_{ijk} E_j(t)E_k(t) + \chi^{(3)}_{ijkl} E_j(t)E_k(t)E_l(t) + \cdots , \tag{1.1}$$

where $\chi^{(n)}$ is the n-th-order susceptibility tensor, and cgs units and the Einstein summation convention are adopted. If the incident electric field oscillates as $\Re\left[e^{i\omega t}\right]$, the $\chi^{(n)}$ term corresponds to the generation of the n-th harmonic of the incident photon because the electric dipole oscillates as $\Re\left[e^{in\omega t}\right]$. The detection of emitted harmonics provides a measurement of these susceptibility tensors $\chi^{(n)}$, which are in turn tied to the space group symmetry of the material.

In particular, focusing on the $\chi^{(2)}$ term, the technique of rotational anisotropy second harmonic generation (RA-SHG) has proven to be an ultra-sensitive probe of lattice, magnetic, and electronic symmetries of a crystal, including its bulk, surface, and interfaces with other materials [17] (Fig. 1.2c, d). Its ability to detect subtle symmetry changes that often elude other probes has greatly contributed to our understanding of many strongly correlated systems, such as cuprate high-T_c superconductors (HTSCs) [18] and putative quantum spin liquids (QSLs) [19]. In Chap. 8, I will also show an example where RA-SHG can be used to differentiate the mirror domains of a light-induced state in $1T$-TaS_2 [20]. In this case, mirror-symmetry breaking does not occur in the original crystal lattice, but instead in the superlattice that forms due to a charge density wave (CDW) transition.

In addition to uncovering symmetries, nonlinear spectroscopy is of great importance in characterizing the topology of electronic wavefunctions, such as Berry phase and Berry curvature dipole [21–24]. In the study of QSLs, it may even provide a viable path to observe spin fractionalization [25]. For sufficiently high fields ($\gtrsim$0.1–1 V/Å), nonlinear response can reach well beyond the $\chi^{(2)}$ term, giving rise to the phenomena of HHG in solids, a topic of intense study over the last decade [26]. These experiments typically require materials with a large bandgap and an incident photon in the MIR to THz regime to avoid carrier excitation and sample damage. Nonetheless, metallic systems such as graphene have been demonstrated to exhibit HHG without degradation [27]. Though the mechanism behind these high harmonics remains unresolved, it is generally believed that the process involves a collective motion of band electrons driven by the strong field, such as dynamical Bloch oscillations. Such intra- or inter-band motion of electrons is intimately related to both the band structure and their correlations, which affect the strength, cutoff, and general shape of the HHG spectra. These spectra could enable us to reconstruct the valence potential and electron density in a material at the picometer spatial scale [28], and probe correlation effects and many-body dynamics at the femtosecond timescale [16] (Fig. 1.2b). Despite its infancy, solid state HHG has emerged as a versatile tool to examine the complex atomic environment in quantum materials both in and out of equilibrium.

1.1.2 Coherent Oscillatory Dynamics

In a symmetry-broken phase, elementary excitations of the order parameter offer another lens to characterize the ground state. However, these excitations are incoherent under equilibrium conditions, and one could only study their space- and time-averaged properties, such as the Debye–Waller intensity reduction of a Bragg peak due to thermal excitation of phonons. This situation changes upon a photoexcitation event instigated by a femtosecond laser pulse, which could launch the coherent dynamics of an order parameter. The most commonly observed coherent motion is that of an optical or acoustic phonon, in which lattice ions over a macroscopic sample volume move in perfect synchrony. Typically, coherent optical phonons are generated by impulsive stimulated Raman scattering (ISRS) [29], displacive excitation (DECP) [30], or transient depletion field screening (TDFS) [31]. On the other hand, coherent acoustic phonons (CAPs) may result from five different sources [32, 33]: electron–phonon deformation potential, thermoelasticity, the inverse piezoelectric effect, electrostriction, and magnetostriction. An example of CAP is shown in Fig. 1.3a, detected by ultrafast electron diffraction (UED), a technique featured in Chap. 3.

The synchronized motion is not confined to lattice phonon modes. In Chaps. 4 and 8, I will show two examples of coherent amplitude modes excited in two different CDW materials [36, 37]. They manifest as an oscillatory feature in time-resolved optical spectroscopy and diffraction, and their oscillation amplitude and frequency encode information about the local environment in an excited medium. By using two pump pulses, a strong photoexcitation event followed by a weak one, one could further track the time evolution of these oscillations and gain insights into the dynamics of the order parameter throughout a nonequilibrium phase transition [38–40].

When the symmetry-broken states involve intertwined degrees of freedom, photoexcited coherent dynamics are particularly helpful in identifying the driving force that stabilizes the ground state [34, 41]. Consider, for example, a recent study on the Verwey transition in Fe_3O_4 [34], which is known for its complexity due to the interplay between structural rearrangement and charge–orbital ordering. In thermal equilibrium, any signature of its order parameter excitation is obscured by a broad conductivity continuum near the transition temperature (T_V). By contrast, coherent oscillations can be observed in ultrafast time-domain terahertz spectroscopy (TDTS), which are attributed to the elementary excitations of a trimeron order [42]. These coherent modes decrease in energy as the effective lattice temperature after laser excitation increases toward T_V, suggesting the critical role of trimeron through the Verwey transition (Fig. 1.3b). Such a frequency reduction in a coherent mode of the order parameter is a general feature in broken-symmetry states after photoexcitation, first demonstrated in early studies of CDW systems [43–45]. As I will elaborate in Chap. 5, this softening can be interpreted as a flattening of the free energy landscape near the transition temperature.

Finally, when different degrees of freedom are measured by their corresponding time-resolved probes, coherent features in the temporal evolution provide a natural link between observables in different techniques. For example, while the self-energy of electronic quasiparticles can be measured by ARPES, the lattice configuration can be determined by X-ray diffraction (XRD). If both the electronic energy and ionic position are seen to oscillate at the same phonon frequency after photoexcitation, one could then obtain a model-independent measure of the mode-specific electron–phonon coupling strength [35] (Fig. 1.3c, d). This procedure, which only relies on the coherent response in the time domain, could be applied to other coupled orders, establishing an experimental paradigm for measuring fundamental physical quantities.

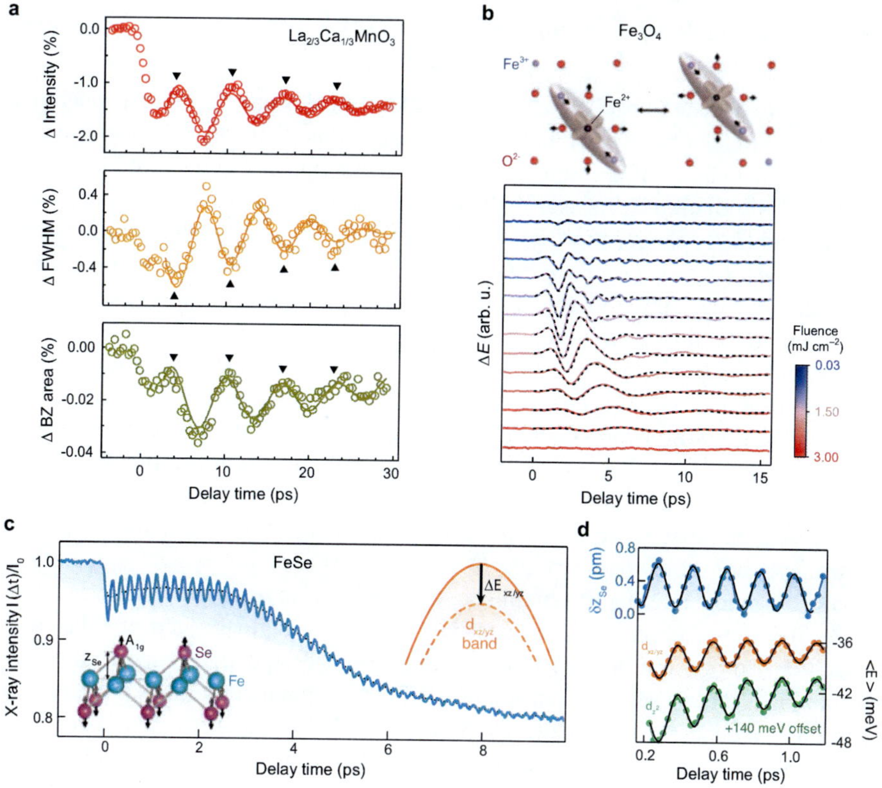

Fig. 1.3 Coherent oscillatory dynamics. (**a**) Coherent acoustic phonon (CAP) in a 20-nm-thick freestanding manganite film, excited by a 1038-nm NIR pulse. The 0.16-THz CAP is captured by the Bragg peaks in ultrafast electron diffraction, where peak intensity, width, and projected Brillouin zone area all show a phase-locked oscillatory feature. Data were taken at room temperature with 26-keV electrons. (**b**) *Top*: schematic of the sliding motion of a trimeron (cigar-shaped rod) below the Verwey transition temperature (T_V) of Fe_3O_4. The polaron coherently tunnels between an Fe^{2+} and an adjacent Fe^{3+} site. *Bottom*: This motion leads to collective oscillations seen in the transmitted electric field (ΔE), measured by ultrafast THz spectroscopy. The oscillatory modes soften with increasing absorbed fluence of the 1.55-eV pump that raises the effective lattice temperature toward T_V, suggesting their critical role in the Verwey transition. Sample was kept at 7 K. (**c**) Coherent optical phonon (*inset, left*) excited in a 60 unit cell FeSe film on a Nb-$SrTiO_3$ substrate, showing up as oscillations in the (004) Bragg peak intensity in tr-XRD. The lattice motion also modulates the electronic band energies (*inset, right*). Sample was kept at 20 K. (**d**) Se displacement (δz_{Se}) calculated from the X-ray intensity modulation, and the corresponding energy shift ($\langle E \rangle$) measured by tr-ARPES. Combining δz with $\langle E \rangle$ allows a model-free estimate of the electron–phonon coupling strength for this phonon. (Adapted with permission from Ref. [34], Springer Nature Ltd (**b**); Ref. [35], AAAS (**c**, **d**))

1.1.3 Timescale of Elementary Processes

The ability to measure observables on ultrafast timescales has greatly expanded our knowledge of electrons, spins, and lattice dynamics on their intrinsic timescales. I often use the analogy of a simple pendulum: While one can deduce the period of small-angle oscillation by measuring its length, a time-domain measurement would allow us to directly clock its movement after we nudge it slightly out of equilibrium. In solids, the dynamics of quasiparticles and collective excitations carry a signature timescale, making it possible to disentangle the contributions of individual degrees of freedom to the formation of a ground state.

A simple example of a separation of timescales is captured in the two-temperature model (TTM), one for conduction electrons and the other for lattice ions, which offers a phenomenological description of the ultrafast dynamics in a metallic system [47, 48]. Upon photoexcitation, light electrons are first excited to a high-energy state and fast electron–electron scatterings establish a temperature scale within the electronic subsystem. The excited electrons then relax through electron–phonon collisions, leading to a delayed rise of the lattice temperature. Though the model appears too simplistic, one could make some reasonable estimate of the electron–phonon coupling constant ($\lambda_{\text{el-ph}}$) through the carrier relaxation time [49], a procedure applied in early time-resolved studies of superconductors [48]. In Fig. 1.4a, we see an extension of the TTM, which incorporates an additional degree of freedom to model the dynamics in a CDW material [50]. The main idea of this extension, including other so-called N-temperature models, is to ascribe a temperature to each subsystem under consideration. The assumption of a thermal state for individual subsystems may not be valid, and in Chap. 7, we will discuss a theoretical model that explicitly considers nonthermal effects, including a proliferation of order parameter fluctuations after photoexcitation [51].

An accurate measurement of microscopic timescales is particularly important in the study of phase transitions. For example, transient optical spectroscopy (TOS) on cuprate HTSCs provides a detailed account of how quasiparticles recombine and condense into a Cooper pair [52, 53]. Examining the quasiparticle relaxation time yields rich information on the momentum dependence of the superconducting gap [54]. In Chap. 4, I will look into the relaxation dynamics in a CDW material, from which we can learn how the amplitude and phase of the complex order parameter reestablish after photoexcitation.

With improved instrumental temporal resolution over the past decade, one could look beyond the relaxation dynamics, which usually take place beyond hundreds of femtoseconds to few picoseconds. Instead, one may focus on the initial system response after photoexcitation. This shift in perspective has proven essential in unveiling the driving force of many interesting ground states such as CDWs, putative excitonic insulators (EIs), and Mott insulators. The working hypothesis is that the initial response time of a system should reflect the dominant elementary process that stabilizes the ground state, such as phonon softening, exciton condensation, or electron localization due to Coulomb repulsion (Fig. 1.4d). This scheme has been

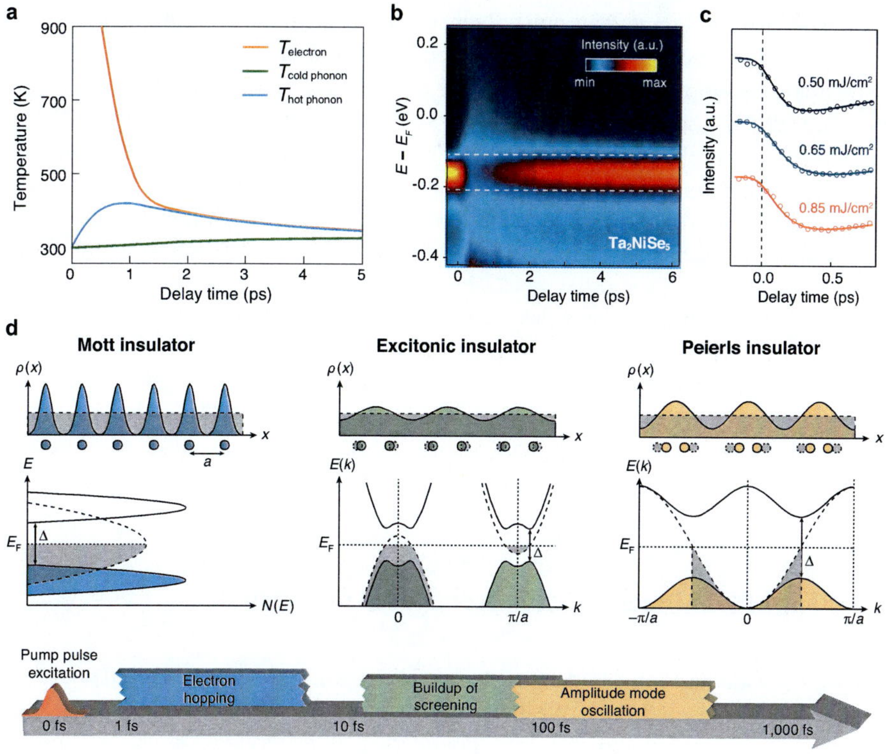

Fig. 1.4 Timescales of initial system response after photoexcitation. (**a**) Example of a three-temperature model that captures the temperature evolutions in a system initially kept at 300 K. Phonons are divided into two categories: "hot phonons" that couple strongly to electrons and "cold phonons" that represent the rest of the lattice modes. Energy transfer between electrons and cold phonons is neglected in the calculation. (**b**) Photoemission intensity of a putative excitonic insulator (EI), Ta_2NiSe_5, as a function of energy and pump–probe delay. Intensity was integrated over ± 0.05 Å^{-1} in the k_x–k_z plane around the Γ point. Data was acquired at 14 K, using 6.2-eV probe photons and 1.55-eV pump photons with an absorbed pump fluence of 0.85 mJ/cm^2. (**c**) Evolution of photoemission intensity in the valence band, where the energy integration window is shown in the dashed box in (**b**). The initial decay proceeds over a phononic timescale of 0.3–0.4 ps, hinting at a structural origin of the spectral gap in Ta_2NiSe_5 as opposed to a dominant EI contribution. (**d**) Three classes of insulators that may accompany a CDW ground state. Schematic of real-space electronic density (ρ) and ionic position (*first row*), energy distribution (*second row*), and characteristic timescales of the initial responses to impulsive NIR photoexcitation (*third row*). Dashed lines with gray shading represent the metallic states before the transition into the gapped ground state. (Adapted with permission from Ref. [46], Springer Nature Ltd (**d**))

applied, for example, to differentiate various types of CDWs [46], to identify the elementary excitation of a Mott insulator [55], or to pin down the electronic origin of a controversial insulator-to-metal transition (IMT) [56]. Recently, our study of the initial response time in a putative EI, Ta_2NiSe_5, also casts doubt on its EI nature and instead reveals a dominant structural origin of its ground state [57] (Fig. 1.4b, c). In Chap. 5, we will take a closer look at this initial response and study its dependence on the incident pulse fluence. This systematic measurement reveals surprising critical behavior in a nonequilibrium phase transition.

It should be emphasized that if multiple degrees of freedom are of interest, it is often not easy to track and compare their timescales simultaneously. This calls for a time-resolved probe that can capture the response of each subsystem together so that any comparison is made under identical photoexcitation and sample conditions. An example is offered by UED, a technique in which one can distinguish scattering from valence electrons and nuclei cores by examining different Bragg orders [58]. Specifically, while lower order peaks have a predominant contribution from the valence electrons, higher order peaks are determined by the nuclei positions. This unique characteristic has been leveraged to study the IMT transition in VO_2 [59, 60] and the CDW transition in $1T$-TaSeTe [61]. In Chap. 7, I will offer another example where the concurrent visualization of multiple orders in UED yields important information about how competing order parameters evolve in an out-of-equilibrium context.

1.2 Light Engineering of Novel States of Matter

The past decade has witnessed an explosion in our ability to control and manipulate quantum materials [12, 14]. We have an increasingly diverse toolbox of tuning parameters, which help to push the boundaries of different extreme conditions never imagined before. For example, in the condensed matter community, it is now possible to explore material properties at a temperature well under 1 mK using a dilution refrigerator in combination with adiabatic nuclear cooling [62], or at a pressure up to 1 TPa using a diamond anvil cell [63]. At the time of writing, a static magnetic field up to 45 T is also available [64], whereas a pulsed field can reach 182 T [65]. Progress in attaining more and more extreme conditions is complemented by the rapid developments in material synthesis and device fabrication, which add extra dimensions to engineering new states of matter. Besides chemical doping in bulk crystal growth, techniques such as pulsed laser deposition (PLD) and molecular beam epitaxy (MBE) can systematically introduce lattice strain in thin films down to a single atomic layer. With improved technologies to exfoliate, transfer, and pattern quasi-two-dimensional (quasi-2D) materials [66], it is now possible to gate, stack, and twist various kinds of van der Waals (vdW) heterostructures to engineer and explore exotic properties such as superconductivity [67, 68].

The ability to generate ultrafast light pulses with a wide range of wavelengths, pulse durations, and pulse energies provides a unique path to manipulate material properties. Typically, one expects that the laser pulse injects a tremendous amount of energy into a material, hence increasing its temperature and entropy. This situation is analogous to thermal heating and the system becomes more disordered as ground states are melted. However, with tailored laser pulses, it is possible to realize ordered states out of equilibrium. This section is devoted to delineating some important mechanisms of these light-induced states while providing the context for some of my own discoveries in Chaps. 6, 7, and 8.

1.2.1 Nonlinear Phononics

In solids, phonons have a characteristic energy scale from 10 to 100 meV, corresponding to an oscillation frequency in the THz to MIR regime. For infrared-active modes, a light pulse can directly couple to the lattice vibrational coordinate and drive ionic motion [69]. We can estimate the ionic displacement in this driven lattice motion based on the induced polarization density, **P**, which takes the following form in SI units

$$\mathbf{P}(\omega) = \epsilon_0 \chi(\omega) \mathbf{E}(\omega), \tag{1.2}$$

where ϵ_0 is the vacuum permittivity, and $\chi(\omega)$ is the complex susceptibility along the direction of the incident electric field **E** with an angular frequency ω. The value of $\chi(\omega)$ can be obtained from the complex optical conductivity σ via the relation [70]

$$|\epsilon_0 \chi(\omega)| = \frac{1}{\omega} |\sigma(\omega)|, \tag{1.3}$$

and the value of $\sigma = \sigma_1 + i\sigma_2$ can be readily obtained from TDTS or Fourier transform infrared spectroscopy (FTIR). Given **P**, the atomic displacement d can then be approximated as

$$d = \frac{|\mathbf{P}|}{nQ}, \tag{1.4}$$

where Q is the ionic charge and n is the density of dipoles, obtained by examining the specific phonon mode driven by the light pulse. As a concrete example, in a study of an underdoped cuprate superconductor $YBa_2Cu_3O_{6+\delta}$, Kaiser and coworkers resonantly drove a 20-THz B_{1u} phonon that involves an oscillating apical oxygen ion along the c-axis [71]. Under a pulse fluence of 4 mJ/cm^2, which corresponds to a peak electric field around 0.03 V/Å, it is estimated that oxygen ions could move up to 10 pm, which is approximately 5% of the equilibrium Cu-O distance. To get a sense of this displacement amplitude, let us consider the prototypical ferroelectric transition of $BaTiO_3$, which is accompanied by a cubic-to-tetragonal structural transition. The largest ionic displacement during the transition is a displacement of the oxygen ion along the c-axis by 10 pm, measured relative to the barium ion at the origin [72]. This displacement is comparable to the light-induced movement of apical oxygens in $YBa_2Cu_3O_{6+\delta}$, so clearly, resonant phonons driven by an

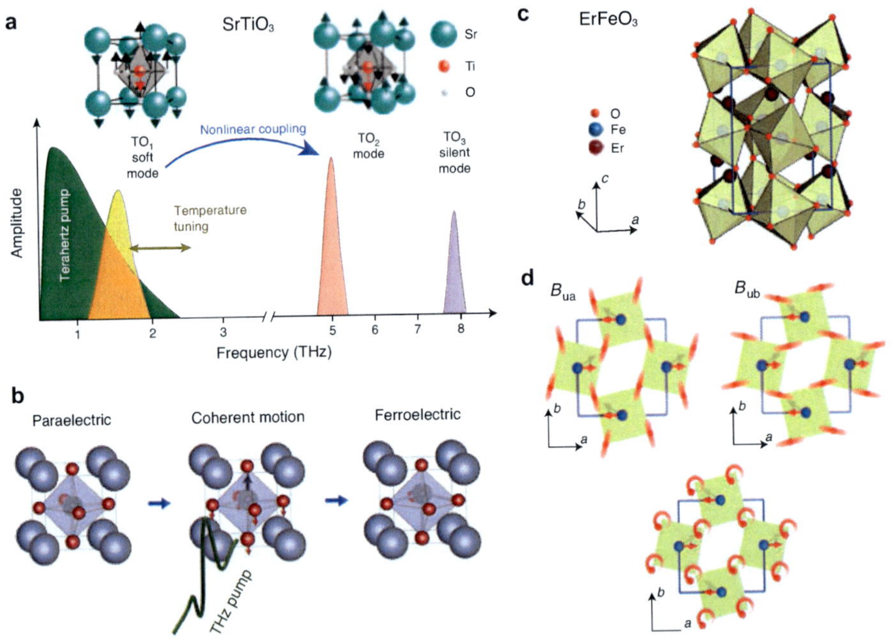

Fig. 1.5 Coherent phonons driven by THz or MIR pulses. (**a**) Schematic of THz-driven phonons up-conversion in $SrTiO_3$. The THz pump has a spectrum (green) that overlaps with the soft transverse optical phonon (TO_1, yellow), whose energy changes with temperature. Due to phonon anharmonicity, the resonantly excited TO_1 mode can couple to higher-frequency modes TO_2 and TO_3, which can be observed as coherent oscillations in a Bragg peak using tr-XRD.The two lowest-frequency zone-center TO eigenmodes are displayed at the top. (**b**) Coherent driving of the soft mode in $SrTiO_3$ by an intense THz pulse could induce a paraelectric-to-ferroelectric transition, where the Ti^{4+} ion in the center is vertically displaced relative to the oxygen octahedron, hence giving rise to a dipole moment. This transition can be detected by THz field-induced second harmonic (TFISH) generation spectroscopy. (**c**) Crystal structure of an antiferromagnetic insulator, $ErFeO_3$, featuring an orthorhombically distorted perovskite structure. (**d**) Linearly polarized 20-THz MIR pulses are able to drive the in-plane B_{ua} and B_{ub} phonons (*top*). They result in an elliptical lattice polarization (*bottom*), mimicking the application of a magnetic field and leading to the excitation of spin precession carried by the iron ions. (Adapted with permission from Ref. [73], Springer Nature Ltd (**a**); Ref. [74], AAAS (**b**); Ref. [75], Springer Nature Ltd (**c**, **d**))

intense laser pulse allow us to access a highly nonlinear regime, with the potential to dramatically alter material properties.

Selective phonon pumping can be directly visualized by watching the atomic motion using tr-XRD. In the study of a prototypical transition metal oxide $SrTiO_3$, Kozina and coworkers applied an intense THz pulse, whose spectrum range of 0.2–2.5 THz overlaps with the soft zone-center transverse optical (TO) phonon, an infrared-active mode [73]. The ensuing atomic motion not only tracks the temporal profile of the driving field, it also exhibits several high frequency phonons due to the anharmonic phonon–phonon coupling that is particularly pronounced in this nonlinear regime (Fig. 1.5a). Beyond this proof-of-principal demonstration, the material $SrTiO_3$ and its soft TO phonon are interesting in their own right: unlike its ferroelectric cousins such as $BaTiO_3$ and $PbTiO_3$, the soft phonon in $SrTiO_3$ never induces a ferroelectric transition due to quantum fluctuations [76]. It has been found that driving a $SrTiO_3$ crystal with strong THz or MIR pulses would induce a transient or metastable ferroelectric state [74, 77] (Fig. 1.5b), demonstrating an ultrafast pathway toward a symmetry-breaking transition out of equilibrium.

This targeted driving of specific phonon modes greatly broadens the scope of quantum phase control via optical excitation and has created a wide spectrum of transient phenomena such as epitaxy strain-induced IMT [78], ultrafast piezomagnetism [79], and phonon engineering of an effective magnetic field [75] (Fig. 1.5c, d). Perhaps the most famous and somewhat controversial application of nonlinear phononics is the observation of light-induced superconductivity in underdoped $YBa_2Cu_3O_{6+\delta}$ [71, 80–82] and K_3C_{60} [83, 84], far beyond their respective equilibrium superconducting transition temperature. In $YBa_2Cu_3O_{6+\delta}$, it was thought that the light-driven motion of the apical oxygen in the bilayer Cu-O planes enhances the $d_{x^2-y^2}$ character of the in-plane electronic structure, thus favoring a superconducting transition [81, 82]. Similarly, in K_3C_{60}, the excitation of the stretching and compressing modes of the hexagons and pentagons in the fullerene molecule is thought to assist the phonon-mediated pairing [83]. It should be noted that recent experiments on $YBa_2Cu_3O_{6+\delta}$ [85] and K_3C_{60} [86] have called these interpretations into questions, and it remains an active research area to theoretically understand these fascinating superconducting-like signatures in a highly excited state.

1.2.2 Photoinduced Metastable States

When electrons are excited by an intense pulse, the free energy landscape in the phase space may undergo a dramatic transformation such that the system is trapped in a local but not global minimum. This leads to the emergence of a metastable structure that may persist long after the pulse arrival. The exact mechanism for entering a local minimum is often unresolved and differs from one system to another. Possible candidates include transient heating or fast quenching [87], carrier screening [88], photodoping [89], defect creation [90], and plastic lattice distortion [74, 91]. In Sect. 1.2.3, we will further discuss the role of intertwined orders in reaching the local minima of an excited state.

When the ground state order is fragile and somewhat unstable, it is plausible that an ultrafast pulse could disturb the subtle balance of competing energy scales and lead to the appearance of a metastable state [92]. For example, this is the case for $1T$-TaS_2, which I will formally introduce in Chap. 2. Its ground state below 183 K has a commensurate CDW, which is accompanied by the formation of a Mott gap in its Ta 5d band. This Mott state is rather fragile and can be easily collapsed by pressure, charge doping, isovalent substitution, or a CDW domain wall [93, 94]. Upon the incidence of a single femtosecond NIR pulse, it was found that $1T$-TaS_2 enters a "hidden" metallic state that is characterized by strongly modified single-particle and collective mode spectra [90] (Fig. 1.6a). Subsequent works by scanning tunneling microscopy (STM) found that an electrical pulse from the STM tip could achieve a similar effect, creating a metastable state of $1T$-TaS_2 that features a mosaic of disordered CDW domains spanning a few nanometers [95, 96]. Interestingly, when the laser pulse fluence exceeds a threshold value of approximately 3.5 mJ/cm^2, the commensurate CDW state no longer enters the "hidden" metallic state but is instead transformed into an amorphous phase characterized by a hyperuniform pattern of localized charges [87]. This shows the versatility of ultrafast pulses in sculpting the metastable state of a material.

While the ground state of $1T$-TaS_2 is naturally unstable against perturbations, one can also design the ground state of a material through advanced material synthesis to make it susceptible to a photoinduced phase transition. For example, in an epitaxially strained thin film of $La_{2/3}Ca_{1/3}MnO_3$, a train of femtosecond 1.5-eV pulses has been shown to induce a ferromagnetic (FM) metallic patch against an antiferromagnetic (AFM) insulating background at 80 K [98] (Fig. 1.6b). These metallic patches persist indefinitely at this temperature, and their emergence may be explained by the transient relief of epitaxial strain and the associated reduction in the Jahn–Teller distortion of the MnO_6 octahedra [97]. Progress in thin film growth further allows the possibility of creating layered heterostructures that are intrinsically metastable. This is the case seen in an atomic-scale $PbTiO_3$/$SrTiO_3$ superlattice grown on a $DyScO_3$ substrate. The ground state of the system is reached by a fine balance of polarization, electric, gradient, and elastic energies, resulting in ferroelectric–ferroelastic domains and polarization vortices. After the illumination by femtosecond pulses with photon energy exceeding the $PbTiO_3$ bandgap, these domains and vortices disappear and a pristine three-dimensional (3D) supercrystal emerges [88] (Fig. 1.6c, d). Again, the supercrystalline phase is metastable, estimated to last for at least a year under ambient conditions.

It is worth emphasizing that the photoinduced metastable states in all these examples require the pulse fluence to exceed a minimum threshold. In the case of $1T$-TaS_2 and $PbTiO_3$/$SrTiO_3$, it was further found that the threshold fluence positively correlates with the pulse duration [88, 90], indicating that a thermal heating scenario is unable to account for the phenomenology. In fact, raising the sample temperature will result in the return to the original ground state [87, 88, 90, 97]. The fully reversible behavior thus opens the possibility of ultrafast switching to and from a metastable state: The forward path is provided by a femtosecond pulse with a high peak electric filed, while the backward path could be implemented by a longer pulse or a pulse train, whose primary contribution is laser-induced heating. In Chap. 8, I will discuss another unusual photoinduced metastable structure in $1T$-TaS_2, this time at room temperature. The ultrafast switching is again reversible, and it involves the creation and annihilation of CDW mirror domains.

In addition to controlling charge, orbital, and lattice, degrees of freedom, ultrafast pulses are instrumental in charting a path toward metastable spin states in magnetic systems [99]. At the fundamental time limit imposed by the Landauer principle [100], Schlauderer and coworkers demonstrated the fastest and least-dissipative all-

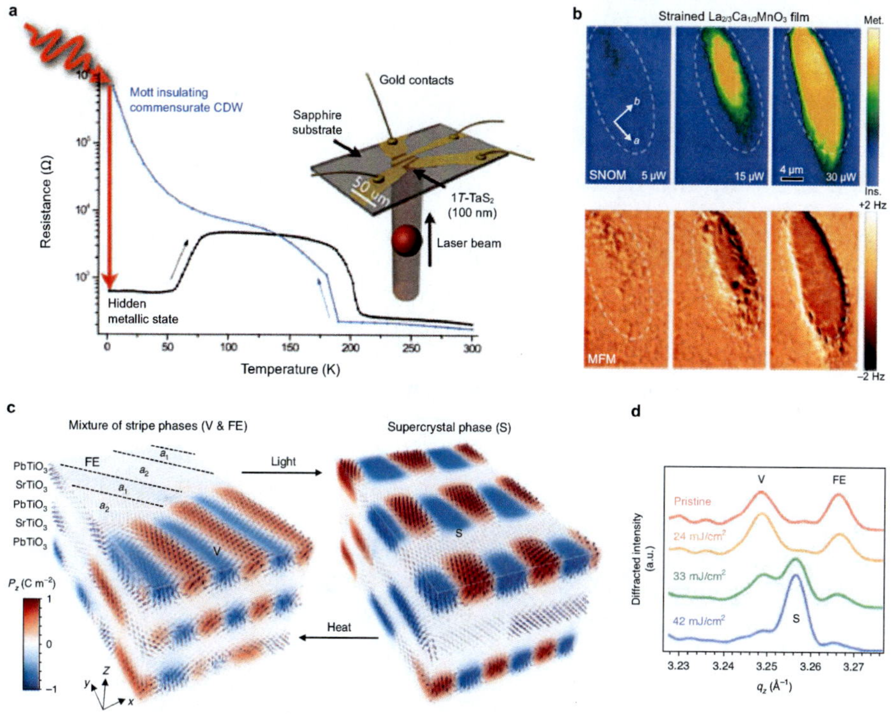

Fig. 1.6 Metastable states induced by ultrafast pulses. (**a**) Resistivity reduction (red arrow) of $1T$-TaS_2 at 1.5 K by a 35-fs, 800-nm laser pulse. Blue curve is the resistance measured on cooling; the jump around 183 K indicates a transition from the nearly commensurate to the commensurate CDW phase, which is also a Mott insulator. Black curve is the resistance in the warming branch, showing a recovery to the usual, high-temperature state. (**b**) Nano-imaging of a photoinduced ferromagnetic metallic patch within epitaxially strained $La_{2/3}Ca_{1/3}MnO_3$ film at 70 K, which is an antiferromagnetic insulator in equilibrium. Top and bottom rows are simultaneous scanning near-field optical microscopy (SNOM) and magnetic force microscopy (MFM) images of the photoinduced metallicity and magnetic moments, respectively. Images were taken after a train of 1.5-eV pulses at incremental power levels (labeled) were incident onto the area demarcated by the dashed oval. (**c**) Illustration of a light-induced supercrystal in a $PbTiO_3$/$SrTiO_3$ superlattice grown on a $DyScO_3$ substrate, calculated by a phase-field model. The equilibrium state contains a mixture of in-plane ferroelectric–ferroelastic a_1/a_2 domains (FE) and polar vortices (V). Arrows indicate local polar displacements. Red and blue colors indicate the z-component of the polarization (P_z). White regions correspond to in-plane polarization. Shining a single sub-picosecond optical pulse or pulse trains results in a metastable supercrystal phase (S), which contains long-range orders of ferroelectric, ferroelastic, and polar vortices in 3D. Thermal annealing reverts the process. (**c**) Static XRD intensity showing the peaks corresponding to the V, FE, and S phases after single shots of 400-nm pulses were applied on the sample with indicated fluences. (Adapted with permission from Ref. [90], AAAS (**a**); Ref. [97], Springer Nature Ltd (**b**); Ref. [88], Springer Nature Ltd (**c, d**))

coherent spin switching to a long-lived local-minimum in a model antiferromagnet, $TmFeO_3$ [101]. A more permanent, all-optical steering of net magnetic moments has also been demonstrated in a cobalt-substituted yttrium iron garnet [102]. Specifically, resonant pumping of select d–d transitions in the cobalt ions has enabled ultrafast reading and writing of magnetic domains, paving the way toward ever faster and less dissipative memory devices.

1.2.3 Tipping the Balance Between Intertwined Orders

A fascinating aspect of quantum materials lies in the plethora of possible intertwined ground states, giving rise to many complex phase diagrams yet to be explained. One needs to look no further than the cuprate HTSCs [103, 104], where superconductivity coexists or borders various forms of CDW, spin density wave (SDW), and electron-nematic orders, not to mention the mysterious pseudogap and the strange metal behavior in the non-superconducting state (Fig. 1.7a). Another canonical example is found in manganites [105, 106], where colossal magnetoresistance (CMR) was discovered and a large number of spin-, charge-, and orbital-ordered phases have since been identified. Whenever two states lie in close proximity in the phase diagram, they may either compete or co-develop. In both cases, perturbing one order by an ultrafast pulse would inevitably modify the coupled order, possibly leading to new states of matter out of equilibrium.

Competing Phases

One heavily scrutinized example of phase competition is again found in cuprate HTSCs, where charge order is evidenced to compete with superconductivity in equilibrium [104, 110, 111]. This competition is most pronounced near the $p = 1/8$ hole doping, where the superconducting dome exhibits a dip, and the amplitude and correlation length of the charge order are the largest [111]. This decrease in the superconducting T_c is the most dramatic in the $La_{2-x}Ba_xCuO_4$ family, where T_c is suppressed almost entirely by the stripe phase that consists of charged rivers separating regions of an oppositely phased AFM order [112–114]. In addition, under high magnetic field where superconductivity is suppressed, X-ray scattering revealed the sharpening and amplitude restoration of the charge order peak, which could even have long coherence in the direction perpendicular to the Cu-O plane [115–117].

This competition has been leveraged in several ultrafast experiments that reported signatures of light-induced superconductivity while transiently melting the charge order, for example, in $La_{1.675}Eu_{0.2}Sr_{0.125}CuO_4$ [108] and $La_{2-x}Ba_xCuO_4$ [118] (Fig. 1.7b). In the $YBa_2Cu_3O_{6+\delta}$ family, the picture seems less clear. The initial reports of optically enhanced coherent transport was found with $\delta = 0.45$ (hole doping $p = 0.08$) and 0.5 ($p = 0.09$) [71, 80, 119], and femtosecond X-ray scattering showed a transiently suppressed charge order at $\delta = 0.6$ ($p = 0.11$) [120]. However, a recent experiment on the $\delta = 0.67$ ($p = 1/8$) compound has instead found evidence of light-enhanced charge order, where both its amplitude

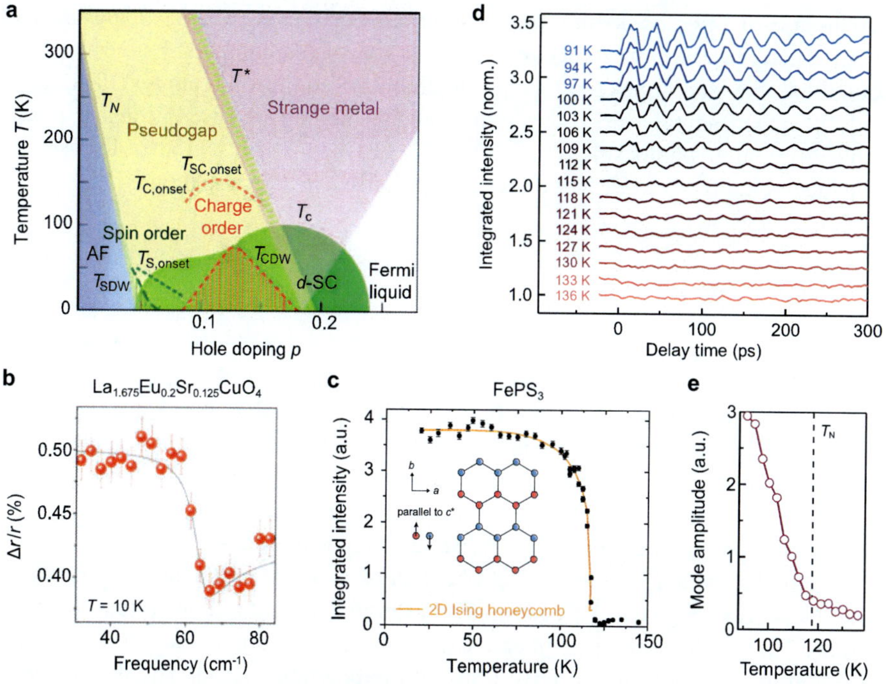

Fig. 1.7 **Intertwined orders out of equilibrium.** (**a**) Phase diagram of cuprate HTSCs as of 2015, featuring various types of spin and charge orders in addition to d-wave superconductivity. The superconducting dome (green) suffers a dip around $p = 1/8$ as a result of phase competition with the charge order. (**b**) Light-induced superconductivity in stripe-ordered cuprate $La_{1.675}Eu_{0.2}Sr_{0.125}CuO_4$. At 10 K, the sample is non-superconducting as T_c is suppressed by the stripe order. After excitation with MIR pulses at 16 μm, a Josephson plasma resonance (JPR) appears as an edge at 60 cm^{-1} in the transient c-axis reflectance 5 ps after the photoexcitation, measured by time-resolved terahertz spectroscopy. The JPR edge may be interpreted as a signature of 3D superconductivity in cuprates, which arises at the expense of melting the competing stripe phase. (**c**) Temperature dependence of the integrated intensity of the $(\frac{3}{2}, \frac{1}{2}, 0.34)$ magnetic Bragg peak of the layered zigzag antiferromagnet, $FePS_3$, measured by single-crystal neutron diffraction. It indicates a Néel temperature $T_N = 117.52$ K after fitting the data to the 2D Ising model in a honeycomb lattice. In-plane spin arrangement is shown in the inset [107], where only Fe atoms are drawn. (**d**) Evolution of integrated intensity of $(\bar{3}\bar{3}1)$ Bragg peak in $FePS_3$ after excitation by a 260-nm light pulse, measured by UED. Intensities are normalized to values before photoexcitation and traces are vertically shifted by 0.15 for clarity. A coherent acoustic phonon (CAP) at 36 GHz was excited. (**e**) Amplitude of the CAP as a function of temperature, showing a dramatic enhancement below T_N due to transient melting of the spin order and strong spin–lattice coupling. (Adapted with permission from Ref. [104], Springer Nature Ltd (**a**); Ref. [108], AAAS (**b**); Ref. [109], APS (**c**))

and coherence grow transiently when it is photoexcited below the superconducting T_c [121].

In Chaps. 6 and 7, I will present our discovery of a light-induced CDW due to the competition between two orthogonally oriented density waves in equilibrium. Unlike the previous experiments in cuprates, we were able to observe the time evolution of the two competing orders simultaneously using a single experimental setup. This gives us a more accurate comparison of the competing states and we found that they could develop together in an out-of-equilibrium context. The large body of experimental work is accompanied by theoretical descriptions of phase competition after photoexcitation [122, 123]. In Chap. 7, we will also discuss our own theoretical framework based on a non-perturbative effective model [51].

Coupled Orders

In many quantum materials, different degrees of freedom undergo a cooperative change that leads to a particular ground state. In Sect. 1.2.2, we have already seen several examples, such as the opening of a Mott gap in the commensurate CDW phase of $1T$-TaS_2 (more on this in Sect. 2.4.3) and the concomitant change in the Jahn–Teller distortion, spin orientation, and charge gap in manganites. The closely coupled nature among microscopic subsystems motivates us to use ultrafast light to perturb one specific order therefore inducing changes in others.

A rich playground for this type of phenomena is found in the interplay between spin and lattice degrees of freedom. For example, Kim and coworkers observed a transient SDW order after photoexciting the parent compound of a pnictide HTSC above its equilibrium SDW transition temperature [124]. The dynamics of the transient spin order follows a coherently excited phonon with a frequency as high as 5.5 THz, and it was suggested that the lattice vibration drives the incipient SDW fluctuation into the transient order. In materials with strong magnetostriction, light-induced ultrafast demagnetization is accompanied by significant changes in the lattice motion [125–127]. For example, in a 2D zigzag antiferromagnet $FePS_3$ [107, 109], we observed a dramatic enhancement of coherent acoustic phonons below T_N, which are induced by a transient melting of the AFM order (Fig. 1.7c–e). A mystery in these photoinduced demagnetization phenomena concerns the question of where the spin angular momentum goes. Using tr-XRD, Dornes and coworkers demonstrated the ultrafast Einstein–de Haas effect [128]: The angular momentum lost from the spins is transferred to the lattice on sub-picosecond timescales, again confirming the intimate relation between the two subsystems.

1.2.4 Coherent Light–Matter Interaction

Not only does a light pulse provide a sudden change in the free energy landscape, it could also hybridize with the electronic state to attain a coherent quantum superposition of photons and electrons. According to the Floquet theorem [129], if a time-dependent Hamiltonian has a period T in time, its energy eigenstates will be spaced by $2\pi/T$. The consequence of this theorem in a crystalline system under a periodic drive by light is the creation of Floquet–Bloch states [130], which have been directly observed with tr-ARPES on the surface of a topological insulator

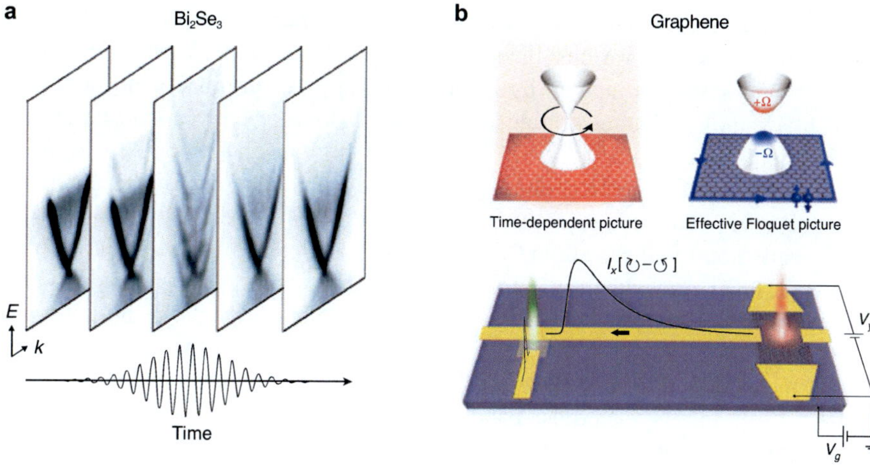

Fig. 1.8 Light-induced Floquet bands. (**a**) Driving a 3D topological insulator, Bi_2Se_3, with an intense 160-meV MIR pulse results in the formation of Floquet–Bloch bands detected by tr-ARPES. These Floquet–Bloch bands of the surface state manifest as replicas of the original Dirac cone. Spectra were taken at 30 K with a probe photon energy of 6.3 eV. (**b**) Coherent interaction between the Dirac cone in graphene and circularly polarized light (*left*) opens a topological gap in the effective Floquet band dispersion (*right*). The gap is characterized by the presence of Berry curvature (Ω). Topologically protected transport develops if the Fermi level (E_F) lies inside the gap, leading to an anomalous Hall effect (AHE). The anomalous Hall currents were detected using the device shown at the bottom. When graphene is excited by a 500-fs, 191-meV circularly polarized MIR pulse (red beam), the generated helicity-dependent current I_x[↻ − ↺] is probed at various time delays in the photoconductive switch, which is activated by a second laser pulse (green beam). Anomalous currents were measured as a function of longitudinal voltage V_y and backgate voltage V_g, the latter of which controls the E_F. (Adapted with permission from Ref. [132], Springer Nature Ltd (**a**); Ref. [133], Springer Nature Ltd (**b**))

[131, 132]. The signature of these Floquet–Bloch states is the appearance of replicas of the original band along the energy axis spaced by the driving photon energy $\hbar\omega$ (Fig. 1.8a), where the intensity of the n-th side band scales as

$$I_n \propto J_n(\beta)^2, \quad \text{where } \beta \propto \frac{E}{\omega^2}. \tag{1.5}$$

Here, J_n is the Bessel function of the first kind, and E is the in-plane component of the electric field in the light pulse. Though these Floquet–Bloch side bands may look similar to those created by laser-assisted photoemission (LAPE), one may readily differentiate the two effects by examining the momentum dependence of the side band intensity and the avoided crossings between the replicas [131, 132]. A key experimental detail is the usage of a MIR pump pulse with energy below the bulk bandgap of the material. This is to minimize incoherent carrier excitation, which could easily mask the weak side bands.

The demonstration of Floquet–Bloch states has spurred the field of Floquet engineering, in which various theoretical proposals have been put forward to realize exotic states of matter by photon-dressing the original band structure [12]. For example, Floquet driving may result in topologically nontrivial states, such as topological insulators [134], fractional Chern insulators [135], and Weyl semimetals [136]. One interesting proposal of inducing an anomalous Hall current in graphene [137] has been recently realized, where McIver and coworkers opened a topologically nontrivial gap at the Dirac point with a circularly polarized MIR pulse [133] (Fig. 1.8b). Beyond the realm of manipulating topological phases, Floquet engineering has been proposed to modify the coupling strength of microscopic parameters in a many-body system, for example, the spin–exchange interaction, which can be altered by photo-assisted hopping [138] or by drive-induced orbital change [139]. In 2D materials, band engineering through Floquet driving has particular relevance in the field of valleytronics [140]. For example, light pulses could act like a pseudo-magnetic field to lift the degeneracy of the two valleys in monolayer WSe_2 and WS_2, analogous to the AC Stark effect [141, 142]. One can target a selected valley by changing the helicity of the light pulse due to valley-dependent selection rules, inviting future applications based on ultrafast and coherent manipulations of the valley degree of freedom.

1.3 Advanced Time-Resolved Techniques

The vast majority of ultrafast experiments are conducted using the pump–probe scheme. This is similar to stroboscopic photography pioneered by Harold Edgerton, who produced the revolutionary *Milk Drop Coronet* photograph in 1957 [143] (Fig. 1.9). There are remarkable parallels between a pump–probe experiment and the milk drop image. The pump laser pulse (milk drop) perturbs a material system in equilibrium (stationary and flat milk surface), launching dynamics in the electrons, lattice, and possibly spin (formation of ripples and the coronet). A probe pulse, be it another electromagnetic pulse or an electron bunch, interacts with the system in motion (flash light), and the post-interaction probe pulse is captured by various imaging systems or analyzers (film). By varying the time delay between the pump and the probe pulse, one can capture the dynamics at different stages. We could accumulate better statistics by repeatedly perturbing the system at a fixed delay with a high repetition rate, assuming that the dynamics stays the same from one pulse to another and that the system re-equilibrates before the next pulse arrives.

As mentioned at the beginning of this chapter, the advent of mode-locked lasers catapulted the field of ultrafast sciences into full sprint. In this section, I will first introduce the techniques I co-developed and used in the past few years. I will conclude by speculating the future directions of time-resolved techniques that could help address some fundamental questions in our field.

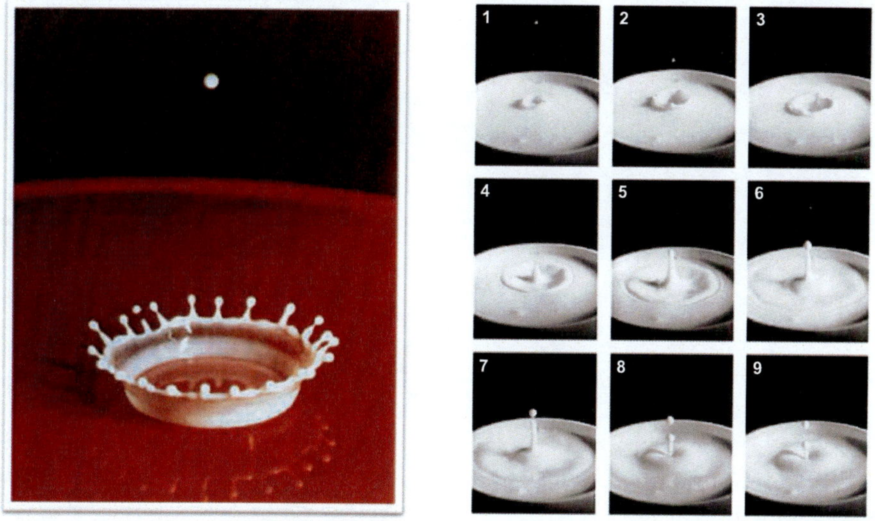

Fig. 1.9 Milk Drop Coronet. Iconic stroboscopic photographs by Harold Edgerton. By varying the time delay between the falling milk drop and the flash light, different moments of the milk surface dynamics can be captured. (Adapted with permission from refs. [143, 144], MIT Museum)

1.3.1 Probes Involved in This Dissertation

MeV and keV Ultrafast Electron Diffraction

UED is the primary tool used in this dissertation and instrumentation details are explained in Chap. 3. It was pioneered by the group of Ahmed H. Zewail with initial conception in 1991 [145, 146]. Zewail was awarded the Nobel Prize in Chemistry in 1999 for his groundbreaking work in applying femtosecond techniques to study chemical reactions. For solids, one may consider UED as a high-speed video camera that records a movie of the atomic motion in a crystal. In a typical UED setup, the pump is an ultrashort light pulse, while the probe is an electron pulse, whose energy could range from tens of keV to a few MeV, depending on the electron generation mechanism (Fig. 1.10b). At each pump–probe delay, a snapshot of electron diffraction is taken, which encodes rich information about the lattice such as structural transformations, strain, defects, and coherent phonon modes [147].

XUV Time-Resolved ARPES

While UED primarily tracks the time evolution of lattice ions, tr-ARPES provides information on the electron dynamics by directly probing the quasiparticle self-energy [148]. The specific system I co-developed with my colleagues uses probe

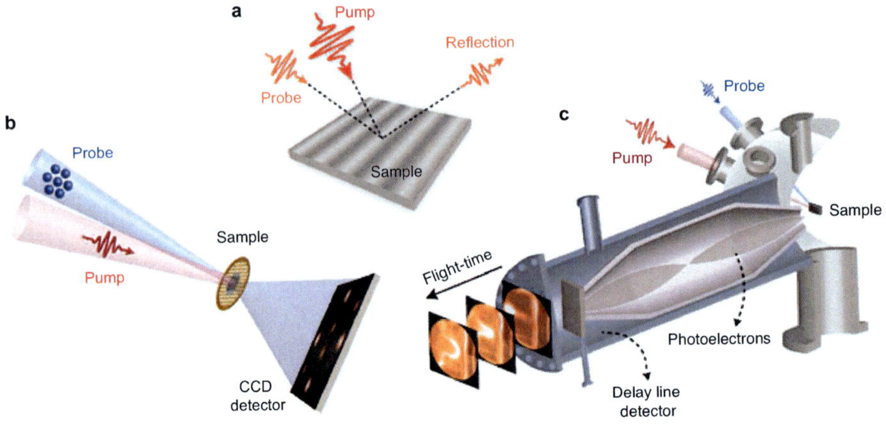

Fig. 1.10 Schematics of time-resolved probes used in the dissertation. (**a**) Transient optical spectroscopy (TOS) in the reflection geometry. The probe is a white light supercontinuum from 500 to 700 nm. (**b**) Ultrafast electron diffraction (UED) in the transmission geometry. Thin samples ($\leq$100 nm) are mounted on copper grids (shown here) or silicon nitride windows. CCD, charge-coupled device. (**c**) Time- and angle-resolved photoemission spectroscopy (tr-ARPES) using a time-of-flight detector. Constant-energy photoemission intensity maps are shown for a CDW material, $ErTe_3$, measured at 15 K

photons with 10.75 eV energy (115 nm), which covers the first Brillouin zone of most quantum materials [149]. It operates with an XUV photon flux of 2.4×10^{10} photons/s at a 250 kHz repetition rate, achieving an energy resolution as good as 16 meV and a time resolution of 230 fs. This time–bandwidth product is approximately two times the lower bound imposed by the Fourier transform limit of a Gaussian pulse [150],

$$\Delta\tau[\text{fs}] \cdot \Delta E[\text{meV}] \geq 1825[\text{fs} \cdot \text{meV}], \tag{1.6}$$

where $\Delta\tau$ is the pulse width and ΔE is the energy linewidth, both measured at full width at half maximum (FWHM).

The laser source for this XUV tr-ARPES system is a commercial Yb:KGW regenerative amplifier system (PHAROS SP-10-600-PP, Light Conversion), whose 1.19-eV (1038-nm) fundamental beam is split into pump and probe branches right after the output. The pump branch is passed into an optical parametric amplifier (OPA, Light Conversion ORPHEUS-LYRA), which is capable of generating a tunable range from UV to MIR. The probe branch is first frequency-tripled to 3.58 eV (346 nm) and then focused onto an ~100–μm spot at the entrance of a hollow-core fiber filled with xenon gas (XUUS, KMLabs), generating the 9th harmonic at 10.75 eV (115 nm). It is worth highlighting that the energy of the 3.58-eV pulse is rather small (3 μJ at 250 kHz repetition rate), which is to be contrasted to HHG-based setups that usually require much larger pulse energies (from hundreds of μJ to a few mJ) [149].

The resulting XUV pulse is passed through a custom-built off-plane Czerny–Turner monochromator (McPherson OP-XCT) with a 500 grooves/mm gold-coated grating (Richardson Gratings, Newport) [151]. This special design can minimize pulse width broadening and enhance throughput efficiency. After passing through a 0.5-mm-thick LiF window (Korth Kristalle GmbH) and the exit slits of the monochromator, the probe pulse is focused onto the sample with a gold-coated toroidal mirror. The pump pulse is independently focused to the sample, with a $\sim 10^{\circ}$ angle relative to the probe arm. The sample is held in an ultrahigh vacuum (UHV) with a base pressure better than 10^{-10} torr at room temperature, achieved using a differential pumping scheme along the XUV beamline after the fiber. Photoelectrons are collected by a time-of-flight detector (Scienta ARTOF 10k), which makes simultaneous measurements of the energy and in-plane momenta possible (Fig. 1.10c). The typical angular range of the detector is $\pm 15^{\circ}$, covering a momentum range of ± 0.33 Å^{-1} at the Fermi level (E_F), assuming a typical work function of 4.5 eV [152] and an untilted sample plane relative to the detector.

Overall, the high repetition rate, tunable pump wavelength, high energy, and high time resolutions make this tr-ARPES setup uniquely purposed for investigating a wide range of quantum materials. In particular, compared to systems based on a hemispherical electron analyzer, our time-of-flight detector enables the efficient measurement of Fermi surface evolution after photoexcitation. Two such examples will be shown in Chaps. 4 and 6, where we will focus on the gap dynamics of a CDW family.

Transient Optical Spectroscopy

Transient optical spectroscopy (TOS) encompasses a large number of pump–probe techniques that measure the electron, spin, and lattice dynamics [48] (Fig. 1.10a). For photons in the UV to NIR wavelength, the relatively straightforward data acquisition scheme with photodiodes and lock-in amplifiers enables the detection of tiny changes ($\sim 10^{-7}$) after photoexcitation. The ease of setup and the superior sensitivity compared to UED and tr-ARPES make TOS an attractive alternative to study ultrafast dynamics.

The setup used in this dissertation measures transient optical reflectivity and transmission across a white light supercontinuum from 500 to 700 nm.[1] The driving laser is a commercial Ti:sapphire regenerative amplifier system (Wyvern 500/1000, KMLabs), which outputs 70-fs, 1.59-eV (780-nm) pulses at a 30 kHz repetition rate. The pulse is split into pump and probe branches, and the latter is focused onto a sapphire crystal for the white light generation. After being reflected from or transmitted through the sample, the probe pulse is directed to a monochromator.

---

[1]The experimental station in our group was first developed by Edbert Sie, which was later reconfigured and improved by Emre Ergeçen and Mehmet Yilmaz.

A photodiode is used for lock-in detection at each selected wavelength from the monochromator.

The transient change of optical reflectivity (ΔR) or transmission (ΔT) in this wavelength range (500 to 700 nm) is indicative of photoinduced modifications of inter-band transitions. In a simplified picture, ΔR or ΔT describes the transient population of excited carriers in a semiconductor, or the change of electronic temperature in a metal [48]. The exact interpretation is much more nuanced, depending on factors such as band renormalization, optical transition matrix elements, Coulomb interaction, and excitons. Nevertheless, this technique offers a complementary perspective to the electron and lattice dynamics observed in UED and tr-ARPES. In Chap. 4, we will show that TOS is able to detect a coherent excitation of the CDW order parameter. In Chap. 5, transient reflectivity measurements reveal a clear signature of critical behavior in a nonequilibrium phase transition. Both discoveries were not possible without the high sensitivity and temporal resolution of the TOS setup.

1.3.2 Future Directions

In this section, I will provide my musings and speculations on the next-generation time-resolved techniques and how they can advance our understanding and control of quantum materials.[2] The categories below are by no means exhaustive, but I do believe that achieving even a subset of them already represents a great leap in our community.

Multimodal Probes with Sophisticated Sample Environment

The meaning of "multimodal" is two-fold. First, multiple time-resolved techniques may be used to investigate a single material, in a bid to disentangle the intertwined dynamics of multiple degrees of freedom (Fig. 1.11a). A challenge in this approach is to maintain identical sample conditions across different techniques, which often necessitates in-situ sample growth, transfer, and measurement. This is similar to the construction of the "vacuum highway" in recent years, which connects chambers of, for example, MBE, PLD, ARPES, and STM. Coupling several ultrafast light sources into such an integrated system would be a new milestone.

The development of multiple probing techniques should also take into consideration the sample environment, such as its temperature range and the possibility

[2]Some points in this section echo the perspective of a group of junior scientists, who gathered in January 2020 at The Johns Hopkins University under the organization of N. P. Armitage to brainstorm the future of the correlated electron problem [14]. I am honored to participate in this exercise, getting truly inspired by the many vigorous discussions and debates.

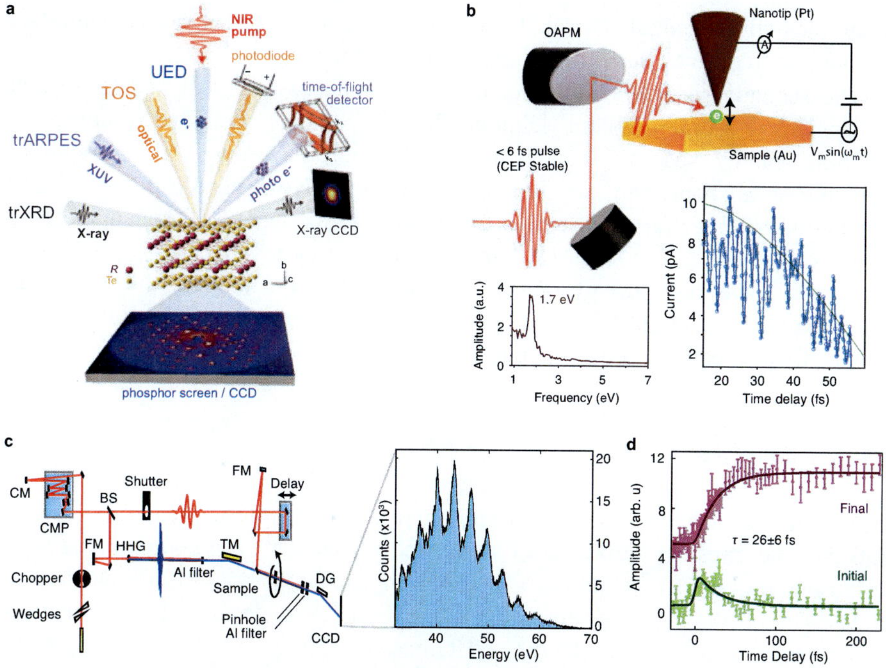

Fig. 1.11 Frontiers of ultrafast spectroscopies and microscopies. (**a**) Multimodal approach of time-resolved investigations. Different probes are sensitive to different degrees of freedom. Together, they reveal a comprehensive picture of the microscopic dynamics. (**b**) Attosecond coherent field-driven STM. Carrier-envelope phase (CEP)-stable pulses of 810-nm wavelength are focused at a Pt/Ir tip by an off-axis parabolic mirror (OAPM) (*top*). Field-driven tunneling current from a gold nanorod displays fast oscillations at long time delay after photoexcitation, which are interpreted as localized surface plasmon resonances (*bottom right*). Fourier analysis shows an oscillation frequency of 1.7 eV (*bottom left*). (**c**) Table-top attosecond XUV spectroscopy. The schematic of the setup (*left*) shows a sub-5-fs 760-nm pump arm and a broadband XUV probe arm, where a typical high-harmonic spectrum is shown. *CM* collimating mirror, *CMP* chirped mirror compressor, *BS* beam splitter, *FM* focusing mirror, *DG* diffraction grating. (**d**) Characteristic timescale of photoinduced insulator-to-metal transition (IMT) in VO_2 (26 ± 6 fs) revealed by absorption changes from the vanadium $M_{2,3}$ edge. Purple and green traces represent changes in relative spectral weight around the $M_{2,3}$ edge of the original insulating phase and the quasi-equilibrium metallic state, respectively. (Adapted with permission from Ref. [162], AAAS (**b**); Ref. [56], NAS (**c**, **d**))

of applying tunable magnetic field and pressure. Certain probes are intrinsically incompatible with specific environments. For example, standard ARPES does not work with a magnetic field or a pair of diamond anvil cells. Nevertheless, developments are underway to circumvent these restrictions, with reports of ARPES studies on current-flowing samples [153], electrostatically gated heterostructures [154], and crystals under an adjustable uniaxial strain [155–157]. A perennial issue regarding sample environment in typical ultrafast experiments is the pump laser-induced steady-state heating. This problem limits most experiments to temperatures above the boiling point of ^{4}He (4.2 K). It awaits to be seen whether studying ultrafast dynamics at ultralow temperatures may become a reality.

The second aspect of "multimodal" entails the possibility of adding several dimensions to a time-resolved technique, hence achieving combined resolution in the time, energy, space, and momentum axis. There have been developments to combine certain conjugate axes, such as time-resolved X-ray nano-diffraction [158] that has both momentum and spatial resolutions, or time-resolved resonant inelastic X-ray scattering (RIXS) [159, 160] and electron energy loss spectroscopy (EELS) [161] that have both energy and time resolutions. Nonetheless, the resolution product of $\Delta x \Delta p$ or $\Delta t \Delta E$ is far beyond the Heisenberg limit, so future work is needed to drastically improve these technical benchmarks.

Attosecond Science

Sub-femtosecond pulses or pulse trains usually come from the HHG process that involves the transient ionization of atoms and subsequent recombination between electrons and ions [163] (Fig. 1.1c). It is not immediately obvious that the attosecond timescale and the corresponding 10s–100s eV energy scale are relevant for most processes in solids. Nevertheless, HHG-based techniques offer two important pieces of information: (i) element specificity through XUV/soft-X-ray absorption spectroscopy, and (ii) phase sensitivity through a holographic detection scheme in photoemission. Regarding the first point, table-top or free-electron laser-based absorption spectroscopies have been used to track element-specific evolution of local bonding, magnetization, and lattice structure [56, 164] (Fig. 1.11c, d). Without the constraint of the uncertainty principle, both time and energy resolutions can be optimal (as to fs in time and ~10 meV in energy). As for the second point, using successive harmonics in an HHG pulse train, the attosecond community has successfully imaged the phase of a complex wavefunction in atomic orbitals by leveraging the interference between different quantum pathways of photoemission [165, 166]. A similar holographic detection scheme may be deployed to solid systems, which could enable phase-sensitive photoemission measurements and allow the investigation into, for example, the sign of the superconducting gap in correlated superconductors.

The generation of attosecond pulses has also been realized in electrons [167, 168], opening new opportunities to study fundamental processes in ultrafast diffraction with unprecedented timing precision. For example, using single-crystalline silicon as a demonstration, Morimoto and Baum showed that there is no time difference between the scatterings into Bragg spots at different orders. Specifically, electron-crystal scattering with 70-keV electrons occurs with no delay beyond ~10 as, which puts an upper bound on the interaction time between high-energy electrons and the lattice. With much larger momentum coverage compared to XUV or X-ray pulses, these electron pulses would serve as a versatile complement to other attosecond spectroscopies.

Ultrafast Microcopy and Single-Shot Measurement

The ability to observe elementary processes at their intrinsic length- and timescales calls for the development of ultrafast microscopy. The demonstration of time-resolved STM using THz pulses [169, 170] represents a significant step toward that ideal at the atomic scale. Recently, the temporal resolution of STM was further pushed into the few-fs regime, revealing a 1.7 eV plasmon mode in a gold nanorod [162] (Fig. 1.11b). At the mesoscopic length scale, a range of time-resolved tools have also been developed [48], such as ultrafast electron microscopy (UEM), time-resolved X-ray diffraction microscopy, wide-field optical microscopy, and scanning near-field optical microscopy (SNOM).

A fundamental problem of time-resolved microscopy is the stochastic nature of the material response to each pump pulse. In a typical stroboscopic setup, statistics are accumulated by repeatedly perturbing the material in equilibrium. While the response in the momentum space may be identical from pulse to pulse, the real-space arrangement can be drastically different unless there is a coherent response such as molecular vibrations [171] and acoustic phonons [172]. In this regard, scanning probes are often not suitable for studying the ultrafast spatial dependence, and efforts should be directed toward single-shot full-field imaging with electrons or photons [173–175]. These single-shot experiments have an added advantage of studying irreversible photoinduced transitions, greatly expanding the type of processes available to existing ultrafast techniques.

Accelerator-Based Light and Electron Sources

The single-shot scheme discussed in the preceding paragraph is only possible with a high flux of the probe beam. For a single light pulse, roughly 10^{14} photons in the XUV/soft-X-ray wavelength are needed for imaging in a few nanometer range [175]. For an electron pulse, as its scattering cross section is much greater (10^5 to 10^6) than X-rays [176], the flux needed for a reasonable statistic is about 10^8 electrons per pulse. At the time of writing, such a bright photon source is only possible from free-electron lasers (FELs), where coherent electromagnetic radiation is produced as relativistic electrons oscillate when they travel through the periodic magnetic field in an undulator [177]. The high photon flux also enables nonlinear spectroscopy in the soft X-ray range, such as second harmonic generation [178]. This type of nonlinear experiment allows us to conduct ultrafast core-level spectroscopy that is particularly sensitive to element-specific changes at interfaces, which may be buried or in contact with a liquid or gas.

As for electrons, to minimize space charging when packing a large number of them into a tiny spacetime volume, we need to use a highly relativistic beam, whose kinetic energy well exceeds the electron rest mass (0.511 MeV). The acceleration of electrons to such a high energy is achieved using a radio-frequency (RF) photoinjector, the same type of technology for accelerating electrons in an FEL.

The potential of accelerator-based ultrafast sources goes beyond producing brilliant probe beams. On the pump side, coherent undulator radiation can also be utilized to generate high-electric-field (0.01 V/Å), high-repetition-rate (100 kHz) THz pulses with a continuously tunable photon frequency (0.1–1.1 THz) [179, 180]. These intense THz pulses are expected to drive a variety of collective excitations in the ground state, such as the Higgs mode of cuprate HTSCs [181, 182], yielding valuable insights into the formation of respective order parameters across the phase transition. One could envision that intense, wavelength-tunable light sources would become an indispensable workhorse for future nonlinear phononics and spectroscopies.

References

1. Smith, P.W.: Mode-locking of lasers. Proc. IEEE **58**, 1342–1357 (1970)
2. Spence, D.E., Kean, P.N., Sibbett, W.: 60-fsec pulse generation from a self-mode-locked Ti:sapphire laser. Opt. Lett. **16**, 42–44 (1991)
3. Wiesenfeld, K., Colet, P., Strogatz, S.H.: Synchronization transitions in a disordered Josephson series array. Phys. Rev. Lett. **76**, 404–407 (1996)
4. Richard, M., Wouters, M., Dang, L.S.: Spontaneous coherence within a gas of exciton-polaritons in telluride microcavities. In: Slavcheva, G., Roussignol, P. (eds.) Optical Generation and Control of Quantum Coherence in Semiconductor Nanostructures, pp. 265–291. Springer, Berlin/Heidelberg (2010)
5. Strickland, D., Mourou, G.: Compression of amplified chirped optical pulses. Opt. Commun. **56**, 219–221 (1985)
6. Petek, H., Ogawa, S.: Femtosecond time-resolved two-photon photoemission studies of electron dynamics in metals. Prog. Surf. Sci. **56**, 239–310 (1997)
7. Bovensiepen, U., Kirchmann, P.S.: Elementary relaxation processes investigated by femtosecond photoelectron spectroscopy of two-dimensional materials. Laser Photonics Rev. **6**, 589–606 (2012)
8. Pfeifer, T., Spielmann, C., Gerber, G.: Femtosecond x-ray science. Rep. Prog. Phys. **69**, 443–505 (2006)
9. O'Sullivan, C.M., Murphy, J.A.: Electro-optic crystals (optical rectification). In: Field Guide to Terahertz Sources, Detectors, and Optics, chap. 13, pp. 13–14. SPIE, Bellingham (2012)
10. Lu, D., Vishik, I.M., Yi, M., Chen, Y., Moore, R.G., et al.: Angle-resolved photoemission studies of quantum materials. Annu. Rev. Condens. Matter Phys. **3**, 129–167 (2012)
11. Keimer, B., Moore, J.E.: The physics of quantum materials. Nat. Phys. **13**, 1045–1055 (2017)
12. Basov, D.N., Averitt, R.D., Hsieh, D.: Towards properties on demand in quantum materials. Nat. Mater. **16**, 1077–1088 (2017)
13. Sobota, J.A., He, Y., Shen, Z.-X.: Angle-resolved photoemission studies of quantum materials. Rev. Mod. Phys. **93**, 025006 (2021)
14. Alexandradinata, A., Armitage, N.P., Baydin, A., Bi, W., Cao, Y., et al.: The future of the correlated electron problem. arXiv 2010.00584 (2020)
15. Belopolski, I., Sanchez, D.S., Ishida, Y., Pan, X., Yu, P., et al.: Discovery of a new type of topological Weyl fermion semimetal state in $Mo_xW_{1-x}Te_2$. Nat. Commun. **7**, 13643 (2016)
16. Silva, R.E.F., Blinov, I.V., Rubtsov, A.N., Smirnova, O., Ivanov, M.: High-harmonic spectroscopy of ultrafast many-body dynamics in strongly correlated systems. Nat. Photon. **12**, 266–270 (2018)

17. Torchinsky, D.H., Hsieh, D.: Rotational anisotropy nonlinear harmonic generation. In: Kumar, C.S.S.R. (ed.) Magnetic Characterization Techniques for Nanomaterials, pp. 1–49. Springer, Berlin/Heidelberg (2017)
18. Zhao, L., Belvin, C.A., Liang, R., Bonn, D.A., Hardy, W.N., et al.: A global inversion-symmetry-broken phase inside the pseudogap region of $YBa_2Cu_3O_y$. Nat. Phys. **13**, 250–254 (2017)
19. Laurita, N.J., Ron, A., Han, J.W., Scheie, A., Sheckelton, J.P., et al.: Evidence for a parity broken monoclinic ground state in the $S = 1/2$ Kagomé antiferromagnet herbertsmithite. arXiv 1910.13606 (2019)
20. Fichera, B.T., Kogar, A., Ye, L., Gökce, B., Zong, A., et al.: Second harmonic generation as a probe of broken mirror symmetry. Phys. Rev. B **101**, 241106 (2020)
21. Sipe, J.E., Shkrebtii, A.I.: Second-order optical response in semiconductors. Phys. Rev. B **61**, 5337–5352 (2000)
22. Moore, J.E., Orenstein, J.: Confinement-induced Berry phase and helicity-dependent photocurrents. Phys. Rev. Lett. **105**, 026805 (2010)
23. Sodemann, I., Fu, L.: Quantum nonlinear Hall effect induced by Berry curvature dipole in time-reversal invariant materials. Phys. Rev. Lett. **115**, 216806 (2015)
24. Morimoto, T., Nagaosa, N.: Topological nature of nonlinear optical effects in solids. Sci. Adv. **2**, e1501524 (2016)
25. Wan, Y., Armitage, N.P.: Resolving continua of fractional excitations by spinon echo in THz 2D coherent spectroscopy. Phys. Rev. Lett. **122**, 257401 (2019)
26. Ghimire, S., Reis, D.A.: High-harmonic generation from solids. Nat. Phys. **15**, 10–16 (2019)
27. Yoshikawa, N., Tamaya, T., Tanaka, K.: High-harmonic generation in graphene enhanced by elliptically polarized light excitation. Science **356**, 736–738 (2017)
28. Lakhotia, H., Kim, H.Y., Zhan, M., Hu, S., Meng, S., et al.: Laser picoscopy of valence electrons in solids. Nature **583**, 55–59 (2020)
29. Stevens, T.E., Kuhl, J., Merlin, R.: Coherent phonon generation and the two stimulated Raman tensors. Phys. Rev. B **65**, 144304 (2002)
30. Zeiger, H.J., Vidal, J., Cheng, T.K., Ippen, E.P., Dresselhaus, G., et al.: Theory for displacive excitation of coherent phonons. Phys. Rev. B **45**, 768–778 (1992)
31. Fö, Dekorsy, T.: Coherent phonons in bulk and low-dimensional semiconductors. In: Lanzani, G., Cerullo, G., De Silvestri, S. (eds.) Coherent Vibrational Dynamics, pp. 129–172. CRC Press, Boca Raton (2007)
32. Ruello, P., Gusev, V.E.: Physical mechanisms of coherent acoustic phonons generation by ultrafast laser action. Ultrasonics **56**, 21–35 (2015)
33. Korff Schmising, C.V., Harpoeth, A., Zhavoronkov, N., Ansari, Z., Aku-Leh, C., et al.: Ultrafast magnetostriction and phonon-mediated stress in a photoexcited ferromagnet. Phys. Rev. B **78**, 060404 (2008)
34. Baldini, E., Belvin, C.A., Rodriguez-Vega, M., Ozel, I.O., Legut, D., et al.: Discovery of the soft electronic modes of the trimeron order in magnetite. Nat. Phys. **16**, 541–545 (2020)
35. Gerber, S., Yang, S.-L., Zhu, D., Soifer, H., Sobota, J.A., et al.: Femtosecond electron-phonon lock-in by photoemission and x-ray free-electron laser. Science **357**, 71–75 (2017)
36. Zong, A., Kogar, A., Bie, Y.-Q., Rohwer, T., Lee, C., et al.: Evidence for topological defects in a photoinduced phase transition. Nat. Phys. **15**, 27–31 (2019)
37. Zong, A., Shen, X., Kogar, A., Ye, L., Marks, C., et al.: Ultrafast manipulation of mirror domain walls in a charge density wave. Sci. Adv. **4**, eaau5501 (2018)
38. Yusupov, R., Mertelj, T., Kabanov, V.V., Brazovskii, S., Kusar, P., et al.: Coherent dynamics of macroscopic electronic order through a symmetry breaking transition. Nat. Phys. **6**, 681–684 (2010)
39. Mertelj, T., Kusar, P., Kabanov, V.V., Giraldo-Gallo, P., Fisher, I.R., et al.: Incoherent topological defect recombination dynamics in $TbTe_3$. Phys. Rev. Lett. **110**, 156401 (2013)
40. Ravnik, J., Vaskivskyi, I., Mertelj, T., Mihailovic, D.: Real-time observation of the coherent transition to a metastable emergent state in $1T$-TaS_2. Phys. Rev. B **97**, 075304 (2018)

41. Ron, A., Chaudhary, S., Zhang, G., Ning, H., Zoghlin, E., et al.: Ultrafast enhancement of ferromagnetic spin exchange induced by ligand-to-metal charge transfer. Phys. Rev. Lett. **125**, 197203 (2020)
42. Senn, M.S., Wright, J.P., Attfield, J.P.: Charge order and three-site distortions in the Verwey structure of magnetite. Nature **481**, 173–176 (2012)
43. Demsar, J., Biljaković, K., Mihailovic, D.: Single particle and collective excitations in the one-dimensional charge density wave solid $K_{0.3}MoO_3$ probed in real time by femtosecond spectroscopy. Phys. Rev. Lett. **83**, 800–803 (1999)
44. Demsar, J., Forró, L., Berger, H., Mihailovic, D.: Femtosecond snapshots of gap-forming charge-density-wave correlations in quasi-two-dimensional dichalcogenides $1T$-TaS_2 and $2H$-$TaSe_2$. Phys. Rev. B **66**, 041101 (2002)
45. Yusupov, R.V., Mertelj, T., Chu, J.H., Fisher, I.R., Mihailovic, D.: Single-particle and collective mode couplings associated with 1- and 2-directional electronic ordering in metallic RTe_3 (R=Ho, Dy, Tb). Phys. Rev. Lett. **101**, 24602 (2008)
46. Hellmann, S., Rohwer, T., Kalläne, M., Hanff, K., Sohrt, C., et al.: Time-domain classification of charge-density-wave insulators. Nat. Commun. **3**, 1069 (2012)
47. Anisimov, S.I., Kapeliovich, B.L., Perel'Man, T.L.: Electron emission from metal surfaces exposed to ultrashort laser pulses. Zh. Eksp. Teor. Fiz. **66**, 776–781 (1974)
48. Prasankumar, R.P., Taylor, A.J.: Optical Techniques for Solid-State Materials Characterization. CRC Press, Boca Raton (2011)
49. Allen, P.B.: Theory of thermal relaxation of electrons in metals. Phys. Rev. Lett. **59**, 1460–1463 (1987)
50. Dolgirev, P.E., Rozhkov, A.V., Zong, A., Kogar, A., Gedik, N., et al.: Amplitude dynamics of charge density wave in $LaTe_3$: theoretical description of pump-probe experiments. Phys. Rev. B **101**, 054203 (2020)
51. Dolgirev, P.E., Michael, M.H., Zong, A., Gedik, N., Demler, E.: Self-similar dynamics of order parameter fluctuations in pump-probe experiments. Phys. Rev. B **101**, 174306 (2020)
52. Rothwarf, A., Taylor, B.N.: Measurement of recombination lifetimes in superconductors. Phys. Rev. Lett. **19**, 27–30 (1967)
53. Gedik, N., Blake, P., Spitzer, R.C., Orenstein, J., Liang, R., et al.: Single-quasiparticle stability and quasiparticle-pair decay in $YBa_2Cu_3O_{6.5}$. Phys. Rev. B **70**, 014504 (2004)
54. Kabanov, V.V., Demsar, J., Podobnik, B., Mihailovic, D.: Quasiparticle relaxation dynamics in superconductors with different gap structures: theory and experiments on $YBa_2Cu_3O_{7-\delta}$. Phys. Rev. B **59**, 1497–1506 (1999)
55. Ligges, M., Avigo, I., Golež, D., Strand, H.U.R., Beyazit, Y., et al.: Ultrafast doublon dynamics in photoexcited $1T$-TaS_2. Phys. Rev. Lett. **120**, 166401 (2018)
56. Jager, M.F., Ott, C., Kraus, P.M., Kaplan, C.J., Pouse, W., et al.: Tracking the insulator-to-metal phase transition in VO_2 with few-femtosecond extreme UV transient absorption spectroscopy. Proc. Natl. Acad. Sci. U. S. A. **114**, 9558–9563 (2017)
57. Baldini, E., Zong, A., Choi, D., Lee, C., Michael, M.H., et al.: The spontaneous symmetry breaking in Ta_2NiSe_5 is structural in nature. arXiv 2007.02909 (2020)
58. Zheng, J.C., Zhu, Y., Wu, L., Davenport, J.W.: On the sensitivity of electron and X-ray scattering factors to valence charge distributions. J. Appl. Cryst. **38**, 648–656 (2005)
59. Morrison, V.R., Chatelain, R.P., Tiwari, K.L., Hendaoui, A., Bruhács, A., et al.: A photoinduced metal-like phase of monoclinic VO_2 revealed by ultrafast electron diffraction. Science **346**, 445–448 (2014)
60. Otto, M.R., René de Cotret, L.P., Valverde-Chavez, D.A., Tiwari, K.L., Émond, N., et al.: How optical excitation controls the structure and properties of vanadium dioxide. Proc. Natl. Acad. Sci. U. S. A. **116**, 450–455 (2019)
61. Li, J., Li, J., Sun, K., Wu, L., Li, R., et al.: Concurrent probing of electron-lattice dephasing induced by photoexcitation in $1T$-TaSeTe using ultrafast electron diffraction. Phys. Rev. B **101**, 100304 (2020)
62. Prakash, O., Kumar, A., Thamizhavel, A., Ramakrishnan, S.: Evidence for bulk superconductivity in pure bismuth single crystals at ambient pressure. Science **355**, 52–55 (2017)

63. Dubrovinskaia, N., Dubrovinsky, L., Solopova, N.A., Abakumov, A., Turner, S., et al.: Terapascal static pressure generation with ultrahigh yield strength nanodiamond. Sci. Adv. **2**, e1600341 (2016)
64. Battesti, R., Beard, J., Böser, S., Bruyant, N., Budker, D., et al.: High magnetic fields for fundamental physics. Phys. Rep. **765–766**, 1–39 (2018)
65. Nakamura, D., Yamashita, T., Ishikawa, H., Hiroi, Z., Takeyama, S.: Extremely wide $\frac{1}{3}$ magnetic plateau of volborthite $Cu_3V_2O_7(OH)_2 \cdot 2H_2O$ in ultrahigh magnetic fields up to 182 T. Phys. Rev. B **98**, 020404 (2018)
66. Novoselov, K.S., Mishchenko, A., Carvalho, A., Castro Neto, A.H.: 2D materials and van der Waals heterostructures. Science **353**, aac9439 (2016)
67. Cao, Y., Fatemi, V., Fang, S., Watanabe, K., Taniguchi, T., et al.: Unconventional superconductivity in magic-angle graphene superlattices. Nature **556**, 43–50 (2018)
68. Yu, Y., Ma, L., Cai, P., Zhong, R., Ye, C., et al.: High-temperature superconductivity in monolayer $Bi_2Sr_2CaCu_2O_{8+\delta}$. Nature **575**, 156–163 (2019)
69. Subedi, A., Cavalleri, A., Georges, A.: Theory of nonlinear phononics for coherent light control of solids. Phys. Rev. B **89**, 220301 (2014)
70. Dressel, M., Grüner, G.: Electrodynamics of Solids: Optical Properties of Electrons in Matter. Cambridge University Press, Cambridge (2002)
71. Kaiser, S., Hunt, C.R., Nicoletti, D., Hu, W., Gierz, I., et al.: Optically induced coherent transport far above T_c in underdoped $YBa_2Cu_3O_{6+\delta}$. Phys. Rev. B **89**, 184516 (2014)
72. Harada, J., Pedersen, T., Barnea, Z.: X-ray and neutron diffraction study of tetragonal barium titanate. Acta Cryst. **A26**, 336–344 (1970)
73. Kozina, M., Fechner, M., Marsik, P., van Driel, T., Glownia, J.M., et al.: Terahertz-driven phonon upconversion in $SrTiO_3$. Nat. Phys. **15**, 387–392 (2019)
74. Li, X., Qiu, T., Zhang, J., Baldini, E., Lu, J., et al.: Terahertz field-induced ferroelectricity in quantum paraelectric $SrTiO_3$. Science **364**, 1079–1082 (2019)
75. Nova, T.F., Cartella, A., Cantaluppi, A., Först, M., Bossini, D., et al.: An effective magnetic field from optically driven phonons. Nat. Phys. **13**, 132–136 (2017)
76. Müller, K.A., Burkard, H.: $SrTiO_3$: an intrinsic quantum paraelectric below 4 K. Phys. Rev. B **19**, 3593–3602 (1979)
77. Nova, T.F., Disa, A.S., Fechner, M., Cavalleri, A.: Metastable ferroelectricity in optically strained $SrTiO_3$. Science **364**, 1075–1079 (2019)
78. Caviglia, A.D., Scherwitzl, R., Popovich, P., Hu, W., Bromberger, H., et al.: Ultrafast strain engineering in complex oxide heterostructures. Phys. Rev. Lett. **108**, 136801 (2012)
79. Disa, A.S., Fechner, M., Nova, T.F., Liu, B., Först, M., et al.: Polarizing an antiferromagnet by optical engineering of the crystal field. Nat. Phys. **16**, 937–941 (2020)
80. Hu, W., Kaiser, S., Nicoletti, D., Hunt, C.R., Gierz, I., et al.: Optically enhanced coherent transport in $YBa_2Cu_3O_{6.5}$ by ultrafast redistribution of interlayer coupling. Nat. Mater. **13**, 705–711 (2014)
81. Mankowsky, R., Subedi, A., Först, M., Mariager, S.O., Chollet, M., et al.: Nonlinear lattice dynamics as a basis for enhanced superconductivity in $YBa_2Cu_3O_{6.5}$. Nature **516**, 71–73 (2014)
82. Liu, B., Först, M., Fechner, M., Nicoletti, D., Porras, J., et al.: Pump frequency resonances for light-induced incipient superconductivity in $YBa_2Cu_3O_{6.5}$. Phys. Rev. X **10**, 011053 (2020)
83. Mitrano, M., Cantaluppi, A., Nicoletti, D., Kaiser, S., Perucchi, A., et al.: Possible light-induced superconductivity in K_3C_{60} at high temperature. Nature **530**, 461–464 (2016)
84. Cantaluppi, A., Buzzi, M., Jotzu, G., Nicoletti, D., Mitrano, M., et al.: Pressure tuning of light-induced superconductivity in K_3C_{60}. Nat. Phys. **14**, 837–841 (2018)
85. Zhang, S.J., Wang, Z.X., Xiang, H., Yao, X., Liu, Q.M., et al.: Photoinduced nonequilibrium response in underdoped $YBa_2Cu_3O_{6+x}$ probed by time-resolved terahertz spectroscopy. Phys. Rev. X **10**, 011056 (2020)
86. Budden, M., Gebert, T., Buzzi, M., Jotzu, G., Wang, E., et al.: Evidence for metastable photo-induced superconductivity in K_3C_{60}. Nat. Phys. **17**, 611–618 (2021)

87. Gerasimenko, Y.A., Vaskivskyi, I., Litskevich, M., Ravnik, J., Vodeb, J., et al.: Quantum jamming transition to a correlated electron glass in $1T$-TaS_2. Nat. Mater. **18**, 1078–1083 (2019)
88. Stoica, V.A., Laanait, N., Dai, C., Hong, Z., Yuan, Y., et al.: Optical creation of a supercrystal with three-dimensional nanoscale periodicity. Nat. Mater. **18**, 377–383 (2019)
89. Gedik, N., Yang, D.-S., Logvenov, G., Bozovic, I., Zewail, A.H.: Nonequilibrium phase transitions in cuprates observed by ultrafast electron crystallography. Science **316**, 425–429 (2007)
90. Stojchevska, L., Vaskivskyi, I., Mertelj, T., Kusar, P., Svetin, D., et al.: Ultrafast switching to a stable hidden quantum state in an electronic crystal. Science **344**, 177–180 (2014)
91. Sie, E.J., Nyby, C.M., Pemmaraju, C.D., Park, S.J., Shen, X., et al.: An ultrafast symmetry switch in a Weyl semimetal. Nature **565**, 61–66 (2019)
92. Zhang, J., Averitt, R.D.: Dynamics and control in complex transition metal oxides. Annu. Rev. Mater. Res. **44**, 19–43 (2014)
93. Qiao, S., Li, X., Wang, N., Ruan, W., Ye, C., et al.: Mottness collapse in $1T$-$TaS_{2-x}Se_x$ transition-metal dichalcogenide: an interplay between localized and itinerant orbitals. Phys. Rev. X **7**, 041054 (2017)
94. Aishwarya, A., Howard, S., Padhi, B., Wang, L., Cheong, S.-W., et al.: Visualizing 1D zigzag Wigner crystallization at domain walls in the Mott insulator TaS_2. arXiv 1906.11983 (2019)
95. Ma, L., Ye, C., Yu, Y., Lu, X.F., Niu, X., et al.: A metallic mosaic phase and the origin of Mott-insulating state in $1T$-TaS_2. Nat. Commun. **7**, 10956 (2016)
96. Cho, D., Cheon, S., Kim, K.-S., Lee, S.-H., Cho, Y.-H., et al.: Nanoscale manipulation of the Mott insulating state coupled to charge order in $1T$-TaS_2. Nat. Commun. **7**, 10453 (2016)
97. McLeod, A.S., Zhang, J., Gu, M.Q., Jin, F., Zhang, G., et al.: Multi-messenger nanoprobes of hidden magnetism in a strained manganite. Nat. Mater. **19**, 397–404 (2020)
98. Zhang, J., Tan, X., Liu, M., Teitelbaum, S.W., Post, K.W., et al.: Cooperative photoinduced metastable phase control in strained manganite films. Nat. Mater. **15**, 956–960 (2016)
99. Kirilyuk, A., Kimel, A.V., Rasing, T.: Ultrafast optical manipulation of magnetic order. Rev. Mod. Phys. **82**, 2731–2784 (2010)
100. Landauer, R.: Irreversibility and heat generation in the computing process. IBM J. Res. Develop. **5**, 183–191 (1961)
101. Schlauderer, S., Lange, C., Baierl, S., Ebnet, T., Schmid, C.P., et al.: Temporal and spectral fingerprints of ultrafast all-coherent spin switching. Nature **569**, 383–387 (2019)
102. Stupakiewicz, A., Szerenos, K., Afanasiev, D., Kirilyuk, A., Kimel, A.V.: Ultrafast nonthermal photo-magnetic recording in a transparent medium. Nature **542**, 71–74 (2017)
103. Bednorz, J.G., Müller, K.A.: Possible high T_c superconductivity in the Ba-La-Cu-O system. Z. Phys. B **64**, 189–193 (1986)
104. Keimer, B., Kivelson, S.A., Norman, M.R., Uchida, S., Zaanen, J.: From quantum matter to high-temperature superconductivity in copper oxides. Nature **518**, 179–186 (2015)
105. Urushibara, A., Moritomo, Y., Arima, T., Asamitsu, A., Kido, G., et al.: Insulator-metal transition and giant magnetoresistance in $La_{1-x}Sr_xMnO_3$. Phys. Rev. B **51**, 14103–14109 (1995)
106. Schiffer, P., Ramirez, A.P., Bao, W., Cheong, S.-W.: Low temperature magnetoresistance and the magnetic phase diagram of $La_{1-x}Ca_xMnO_3$. Phys. Rev. Lett. **75**, 3336–3339 (1995)
107. Lançon, D., Walker, H.C., Ressouche, E., Ouladdiaf, B., Rule, K.C., et al.: Magnetic structure and magnon dynamics of the quasi-two-dimensional antiferromagnet $FePS_3$. Phys. Rev. B **94**, 214407 (2016)
108. Fausti, D., Tobey, R.I., Dean, N., Kaiser, S., Dienst, A., et al.: Light-induced superconductivity in a stripe-ordered cuprate. Science **331**, 189–191 (2011)
109. Rule, K.C., McIntyre, G.J., Kennedy, S.J., Hicks, T.J.: Single-crystal and powder neutron diffraction experiments on $FePS_3$: search for the magnetic structure. Phys. Rev. B **76**, 134402 (2007)
110. Comin, R., Damascelli, A.: Resonant X-ray scattering studies of charge order in cuprates. Annu. Rev. Condens. Matter Phys. **7**, 369–405 (2016)

111. Frano, A., Blanco-Canosa, S., Keimer, B., Birgeneau, R.J.: Charge ordering in superconducting copper oxides. J. Phys.: Condens. Matter **32**, 374005 (2020)
112. Fujita, M., Goka, H., Yamada, K., Matsuda, M.: Competition between charge- and spin-density-wave order and superconductivity in $La_{1.875}Ba_{0.125-x}Sr_xCuO_4$. Phys. Rev. Lett. **88**, 167008 (2002)
113. Abbamonte, P., Rusydi, A., Smadici, S., Gu, G.D., Sawatzky, G.A., et al.: Spatially modulated 'Mottness' in $La_{2-x}Ba_xCuO_4$. Nat. Phys. **1**, 155–158 (2005)
114. Hücker, M., Zimmermann, M.v., Gu, G.D., Xu, Z.J., Wen, J.S., et al.: Stripe order in superconducting $La_{2-x}Ba_xCuO_4$. Phys. Rev. B **83**, 104506 (2011)
115. sChang, J., Blackburn, E., Holmes, A.T., Christensen, N.B., Larsen, J., et al.: Direct observation of competition between superconductivity and charge density wave order in $YBa_2Cu_3O_{6.67}$. Nat. Phys. **8**, 871–876 (2012)
116. Blanco-Canosa, S., Frano, A., Loew, T., Lu, Y., Porras, J., et al.: Momentum-dependent charge correlations in $YBa_2Cu_3O_{6+\delta}$ superconductors probed by resonant x-ray scattering: evidence for three competing phases. Phys. Rev. Lett. **110**, 187001 (2013)
117. Gerber, S., Jang, H., Nojiri, H., Matsuzawa, S., Yasumura, H., et al.: Three-dimensional charge density wave order in $YBa_2Cu_3O_{6.67}$ at high magnetic fields. Science **350**, 949–952 (2015)
118. Nicoletti, D., Casandruc, E., Laplace, Y., Khanna, V., Hunt, C.R., et al.: Optically induced superconductivity in striped $La_{2-x}Ba_xCuO_4$ by polarization-selective excitation in the near infrared. Phys. Rev. B **90**, 100503 (2014)
119. Liang, R., Bonn, D.A., Hardy, W.N.: Evaluation of CuO_2 plane hole doping in $YBa_2Cu_3O_{6+x}$ single crystals. Phys. Rev. B **73**, 180505 (2006)
120. Först, M., Frano, A., Kaiser, S., Mankowsky, R., Hunt, C.R., et al.: Femtosecond x rays link melting of charge-density wave correlations and light-enhanced coherent transport in $YBa_2Cu_3O_{6.6}$. Phys. Rev. B **90**, 184514 (2014)
121. Wandel, S., Boschini, F., Neto, E.H.D.S., Shen, L., Na, M.X., et al.: Light-enhanced charge density wave coherence in a high-temperature superconductor. arXiv 2003.04224 (2020)
122. Patel, A.A., Eberlein, A.: Light-induced enhancement of superconductivity via melting of competing bond-density wave order in underdoped cuprates. Phys. Rev. B **93**, 195139 (2016)
123. Sun, Z., Millis, A.J.: Transient trapping into metastable states in systems with competing orders. Phys. Rev. X **10**, 021028 (2020)
124. Kim, K.W., Pashkin, A., Schäfer, H., Beyer, M., Porer, M., et al.: Ultrafast transient generation of spin-density-wave order in the normal state of $BaFe_2As_2$ driven by coherent lattice vibrations. Nat. Mater. **11**, 497–501 (2012)
125. Henighan, T., Trigo, M., Bonetti, S., Granitzka, P., Higley, D., et al.: Generation mechanism of terahertz coherent acoustic phonons in Fe. Phys. Rev. B **93**, 220301 (2016)
126. Jal, E., López-Flores, V., Pontius, N., Ferté, T., Bergeard, N., et al.: Structural dynamics during laser-induced ultrafast demagnetization. Phys. Rev. B **95**, 184422 (2017)
127. Reid, A.H., Shen, X., Maldonado, P., Chase, T., Jal, E., et al.: Beyond a phenomenological description of magnetostriction. Nat. Commun. **9**, 388 (2018)
128. Dornes, C., Acremann, Y., Savoini, M., Kubli, M., Neugebauer, M.J., et al.: The ultrafast Einstein-de Haas effect. Nature **565**, 209–212 (2019)
129. Galitskiĭ, V.M., Goreslavskiĭ, S.P., Elesin, V.F.: Electric and magnetic properties of a semiconductor in the field of a strong electromagnetic wave. Zh. Eksp. Teor. Fiz. **57**, 207–217 (1969)
130. Faisal, F.H.M., Kamiński, J.Z.: Floquet-Bloch theory of high-harmonic generation in periodic structures. Phys. Rev. A **56**, 748–762 (1997)
131. Wang, Y.H., Steinberg, H., Jarillo-Herrero, P., Gedik, N.: Observation of Floquet-Bloch states on the surface of a topological insulator. Science **342**, 453–457 (2013)
132. Mahmood, F., Chan, C.-K., Alpichshev, Z., Gardner, D., Lee, Y., et al.: Selective scattering between Floquet-Bloch and Volkov states in a topological insulator. Nat. Phys. **12**, 306–310 (2016)

133. McIver, J.W., Schulte, B., Stein, F.-U., Matsuyama, T., Jotzu, G., et al.: Light-induced anomalous Hall effect in graphene. Nat. Phys. **16**, 38–41 (2020)
134. Lindner, N.H., Refael, G., Galitski, V.: Floquet topological insulator in semiconductor quantum wells. Nat. Phys. **7**, 490–495 (2011)
135. Grushin, A.G., Gómez-León, Á., Neupert, T.: Floquet fractional Chern insulators. Phys. Rev. Lett. **112**, 156801 (2014)
136. Hübener, H., Sentef, M.A., De Giovannini, U., Kemper, A.F., Rubio, A.: Creating stable Floquet-Weyl semimetals by laser-driving of 3D Dirac materials. Nat. Commun. **8**, 13940 (2017)
137. Oka, T., Aoki, H.: Photovoltaic Hall effect in graphene. Phys. Rev. B **79**, 081406 (2009)
138. Mikhaylovskiy, R.V., Hendry, E., Secchi, A., Mentink, J.H., Eckstein, M., et al.: Ultrafast optical modification of exchange interactions in iron oxides. Nat. Commun. **6**, 8190 (2015)
139. Chaudhary, S., Hsieh, D., Refael, G.: Orbital Floquet engineering of exchange interactions in magnetic materials. Phys. Rev. B **100**, 220403 (2019)
140. Schaibley, J.R., Yu, H., Clark, G., Rivera, P., Ross, J.S., et al.: Valleytronics in 2D materials. Nat. Rev. Mater. **1**, 16055 (2016)
141. Kim, J., Hong, X., Jin, C., Shi, S.-F., Chang, C.-Y.S., et al.: Ultrafast generation of pseudo-magnetic field for valley excitons in WSe_2 monolayers. Science **346**, 1205–1208 (2014)
142. Sie, E.J., McIver, J.W., Lee, Y.-H., Fu, L., Kong, J., et al.: Valley-selective optical Stark effect in monolayer WS_2. Nat. Mater. **14**, 290–294 (2015)
143. Edgerton, H.E.: Milk Drop Coronet. https://edgerton-digital-collections.org (1957). Museum number: HEE-SC-08811 ©2010 MIT, courtesy of MIT Museum, Cambridge. MIT Museum, Edgerton Digital Collections. Accessed: 14 Aug 2020
144. Edgerton, H.E.: Milk Drop in Milk (sequence). https://edgerton-digital-collections.org (1935). Museum number: HEE-NC-35007–35015 ©2010 MIT, courtesy of MIT Museum, Cambridge. MIT Museum, Edgerton Digital Collections. Accessed: 14 Aug 2020
145. Zewail, A.H.: Femtosecond transition-state dynamics. Faraday Discuss. Chem. Soc. **91**, 207–237 (1991)
146. Williamson, J.C., Zewail, A.H.: Structural femtochemistry: experimental methodology. Proc. Natl. Acad. Sci. U. S. A. **88**, 5021–5025 (1991)
147. Yang, D.S., Gedik, N., Zewail, A.H.: Ultrafast electron crystallography. 1. Nonequilibrium dynamics of nanometer-scale structures. J. Phys. Chem. C **111**, 4889–4919 (2007)
148. Damascelli, A., Hussain, Z., Shen, Z.-X.: Angle-resolved photoemission studies of the cuprate superconductors. Rev. Mod. Phys. **75**, 473–541 (2003)
149. Lee, C., Rohwer, T., Sie, E.J., Zong, A., Baldini, E., et al.: High resolution time- and angle-resolved photoemission spectroscopy with 11 eV laser pulses. Rev. Sci. Instrum. **91**, 043102 (2020)
150. Koch, T.L., Bowers, J.E.: Nature of wavelength chirping in directly modulated semiconductor lasers. Electron. Lett. **20**, 1038 (1984)
151. Frassetto, F., Cacho, C., Froud, C.A., Turcu, I.C.E., Villoresi, P., et al.: Single-grating monochromator for extreme-ultraviolet ultrashort pulses. Opt. Express **19**, 19169–19181 (2011)
152. He, Y., Vishik, I.M., Yi, M., Yang, S., Liu, Z., et al.: Invited Article: high resolution angle resolved photoemission with tabletop 11 eV laser. Rev. Sci. Instrum. **87**, 011301 (2016)
153. Kaminski, A., Rosenkranz, S., Norman, M.R., Randeria, M., Li, Z.Z., et al.: Destroying coherence in high-temperature superconductors with current flow. Phys. Rev. X **6**, 031040 (2016)
154. Nguyen, P.V., Teutsch, N.C., Wilson, N.P., Kahn, J., Xia, X., et al.: Visualizing electrostatic gating effects in two-dimensional heterostructures. Nature **572**, 220–223 (2019)
155. Riccò, S., Kim, M., Tamai, A., McKeown Walker, S., Bruno, F.Y., et al.: In situ strain tuning of the metal-insulator-transition of Ca_2RuO_4 in angle-resolved photoemission experiments. Nat. Commun. **9**, 4535 (2018)

156. Flötotto, D., Bai, Y., Chan, Y.-H., Chen, P., Wang, X., et al.: In situ strain tuning of the Dirac surface states in Bi_2Se_3 films. Nano Lett. **18**, 5628–5632 (2018)
157. Pfau, H., Chen, S.D., Yi, M., Hashimoto, M., Rotundu, C.R., et al.: Momentum dependence of the nematic order parameter in iron-based superconductors. Phys. Rev. Lett. **123**, 066402 (2019)
158. Wen, H., Cherukara, M.J., Holt, M.V.: Time-resolved x-ray microscopy for materials science. Annu. Rev. Mater. Res. **49**, 389–415 (2019)
159. Dean, M.P.M., Cao, Y., Liu, X., Wall, S., Zhu, D., et al.: Ultrafast energy- and momentum-resolved dynamics of magnetic correlations in the photo-doped Mott insulator Sr_2IrO_4. Nat. Mater. **15**, 601–605 (2016)
160. Mitrano, M., Lee, S., Husain, A.A., Delacretaz, L., Zhu, M., et al.: Ultrafast time-resolved x-ray scattering reveals diffusive charge order dynamics in $La_{2-x}Ba_xCuO_4$. Sci. Adv. **5**, eaax3346 (2019)
161. Feist, A., Bach, N., Rubiano da Silva, N., Danz, T., Möller, M., et al.: Ultrafast transmission electron microscopy using a laser-driven field emitter: femtosecond resolution with a high coherence electron beam. Ultramicroscopy **176**, 63–73 (2017)
162. Garg, M., Kern, K.: Attosecond coherent manipulation of electrons in tunneling microscopy. Science **367**, 411–415 (2020)
163. Krausz, F., Ivanov, M.: Attosecond physics. Rev. Mod. Phys. **81**, 163–234 (2009)
164. Geneaux, R., Marroux, H.J.B., Guggenmos, A., Neumark, D.M., Leone, S.R.: Transient absorption spectroscopy using high harmonic generation: a review of ultrafast X-ray dynamics in molecules and solids. Philos. Trans. R. Soc. A **377**, 20170463 (2019)
165. Villeneuve, D.M., Hockett, P., Vrakking, M.J.J., Niikura, H.: Coherent imaging of an attosecond electron wave packet. Science **356**, 1150–1153 (2017)
166. Huismans, Y., Rouzee, A., Gijsbertsen, A., Jungmann, J.H., Smolkowska, A.S., et al.: Time-resolved holography with photoelectrons. Science **331**, 61–64 (2011)
167. Morimoto, Y., Baum, P.: Diffraction and microscopy with attosecond electron pulse trains. Nat. Phys. **14**, 252–256 (2018)
168. Morimoto, Y., Baum, P.: Attosecond control of electron beams at dielectric and absorbing membranes. Phys. Rev. A **97**, 033815 (2018)
169. Cocker, T.L., Jelic, V., Gupta, M., Molesky, S.J., Burgess, J.A.J., et al.: An ultrafast terahertz scanning tunnelling microscope. Nat. Photon. **7**, 620–625 (2013)
170. Tian, Y., Yang, F., Guo, C., Jiang, Y.: Recent advances in ultrafast time-resolved scanning tunneling microscopy. Surf. Rev. Lett. **25**, 1841003 (2018)
171. Cocker, T.L., Peller, D., Yu, P., Repp, J., Huber, R.: Tracking the ultrafast motion of a single molecule by femtosecond orbital imaging. Nature **539**, 263–267 (2016)
172. Cremons, D.R., Plemmons, D.A., Flannigan, D.J.: Femtosecond electron imaging of defect-modulated phonon dynamics. Nat. Commun. **7**, 11230 (2016)
173. Mo, M.Z., Shen, X., Chen, Z., Li, R.K., Dunning, M., et al.: Single-shot mega-electronvolt ultrafast electron diffraction for structure dynamic studies of warm dense matter. Rev. Sci. Instrum. **87**, 11D810 (2016)
174. Liang, J., Wang, L.V.: Single-shot ultrafast optical imaging. Optica **5**, 1113–1127 (2018)
175. Helk, T., Zürch, M., Spielmann, C.: Perspective: towards single shot time-resolved microscopy using short wavelength table-top light sources. Struct. Dyn. **6**, 010902 (2019)
176. Glaeser, R.M.: Electron crystallography of biological macromolecules. Annu. Rev. Phys. Chem. **36**, 243–275 (1985)
177. Kondratenko, A.M., Saldin, E.L.: Generation of coherent radiation by a relativistic electron beam in an ondulator. Part. Accel. **10**, 207–216 (1980)
178. Lam, R.K., Raj, S.L., Pascal, T.A., Pemmaraju, C.D., Foglia, L., et al.: Soft x-ray second harmonic generation as an interfacial probe. Phys. Rev. Lett. **120**, 023901 (2018)
179. Green, B., Kovalev, S., Asgekar, V., Geloni, G., Lehnert, U., et al.: High-field high-repetition-rate sources for the coherent THz control of matter. Sci. Rep. **6**, 22256 (2016)

180. Salén, P., Basini, M., Bonetti, S., Hebling, J., Krasilnikov, M., et al.: Matter manipulation with extreme terahertz light: progress in the enabling THz technology. Phys. Rep. **836–837**, 1–74 (2019)
181. Katsumi, K., Tsuji, N., Hamada, Y.I., Matsunaga, R., Schneeloch, J., et al.: Higgs mode in the d-wave superconductor $Bi_2Sr_2CaCu_2O_{8+x}$ driven by an intense terahertz pulse. Phys. Rev. Lett. **120**, 117001 (2018)
182. Chu, H., Kim, M.-J., Katsumi, K., Kovalev, S., Dawson, R.D., et al.: Phase-resolved Higgs response in superconducting cuprates. Nat. Commun. **11**, 1793 (2020)

Chapter 2
Charge Density Waves

Abstract Upon cooling, a charge density wave (CDW) spontaneously develops in certain metallic crystals. It is characterized by a spatially modulated condensate of electrons and a periodic distortion of lattice ions, both of which break the original translational symmetry of the crystal. CDW compounds provide an excellent model system to study the interaction between an ultrafast light pulse and a broken-symmetry state because the complex CDW order parameter—involving an amplitude and a phase—can be well monitored by several time-resolved probes, which have access to both structural and electronic degrees of freedom. This chapter lays the theoretical foundation of CDW formation, describes the characteristics of its complex order parameter, and connects the theoretical pictures with experimental observables in diffraction and photoemission measurements. The discussions are concluded by an introduction of two CDW materials, rare-earth tritellurides, and tantalum disulfide, which are investigated in detail in this dissertation.

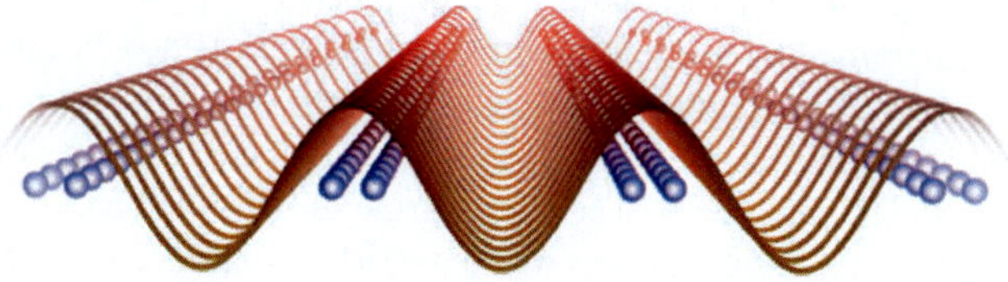

Artist's impression of a 2D unidirectional charge density wave. Spheres represent lattice ions, while sinusoidal curves represent electron densities

A. Zong, *Emergent States in Photoinduced Charge-Density-Wave Transitions*, Springer Theses, https://doi.org/10.1007/978-3-030-81751-0_2

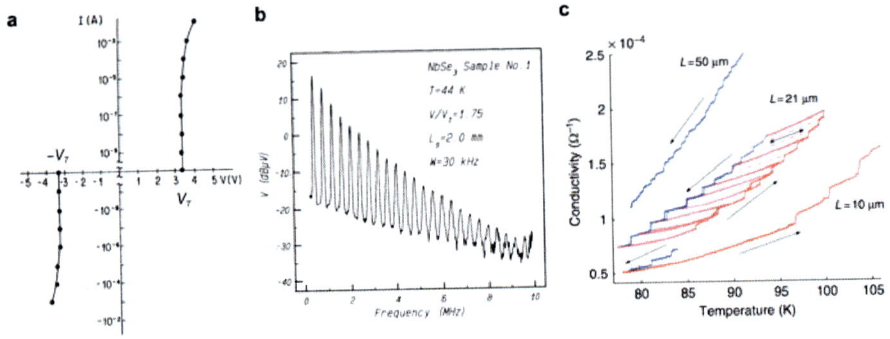

Fig. 2.1 Nonlinear transport properties of CDW systems. (**a**) Current–voltage curve parallel to the *b*-axis of a quasi-1D compound, $K_{0.3}MoO_3$, measured in its incommensurate CDW phase. Beyond threshold values at $\pm V_T$, the condensate is driven into a sliding state, resulting in an orders-of-magnitude increase in the current. (**b**) Spectrum of voltage oscillations in response to an applied dc current in a quasi-1D material, $NbSe_3$, measured in its sliding CDW state. The classical model of these oscillations involves a moving CDW condensate in a washboard-like potential due to the periodic lattice distortions. (**c**) Fragments of the conductivity–temperature curves along the CDW direction in $K_{0.3}MoO_3$ for samples with different lengths. Arrows indicate the direction of temperature sweeps, showing a quantized conductivity hysteresis. Dashed lines indicate the missed reversible segments. The quantization is caused by a temperature-dependent change in the CDW wavevector and the slipping of an integer number of CDW periodicities bounded by the metallic contacts. (Adapted with permission from Ref. [4], APS (**a**); Ref. [5], APS (**b**); Ref. [6], Springer Nature Ltd (**c**))

Below a critical temperature, a CDW system undergoes a phase transition where electrons and phonons cooperatively condense into a spatially ordered ground state. It is marked by an energy gap at the Fermi level and the spontaneous formation of charge density modulation accompanying a periodic ionic distortion. CDWs attracted tremendous attention in the 1970s to 1980s because the CDW condensate significantly modifies the transport properties of the host material system (Fig. 2.1). For example, conductivity quantization and strong hysteresis have been observed under the application of a small electric field. Beyond a threshold field, the condensate can be set to move relative to the underlying lattice, a collective behavior known as CDW sliding. As a historical note, the sliding CDW was once considered a candidate for superconductivity by Herbert Fröhlich in 1954 [1], a few years before the BCS theory was developed in 1957 [2]. In this sliding state, a plethora of exotic phenomena have been reported, such as current oscillations, broadband noise, and (sub)harmonic mode-locking [3].

Strong interest in CDW materials also developed from the proximity of the CDW phase to other ground states in the phase diagram, with superconductivity, in particular, being an adjacent phase of much interest (Figs. 1.7a and 2.2). This observation comes as no surprise because strong electron–phonon coupling (EPC) can facilitate both phase transitions. CDW often competes with superconductivity, as both require the electronic density of states at E_F to form their respective condensates: Cooper pairs for superconductivity and electron–hole pairs for CDW

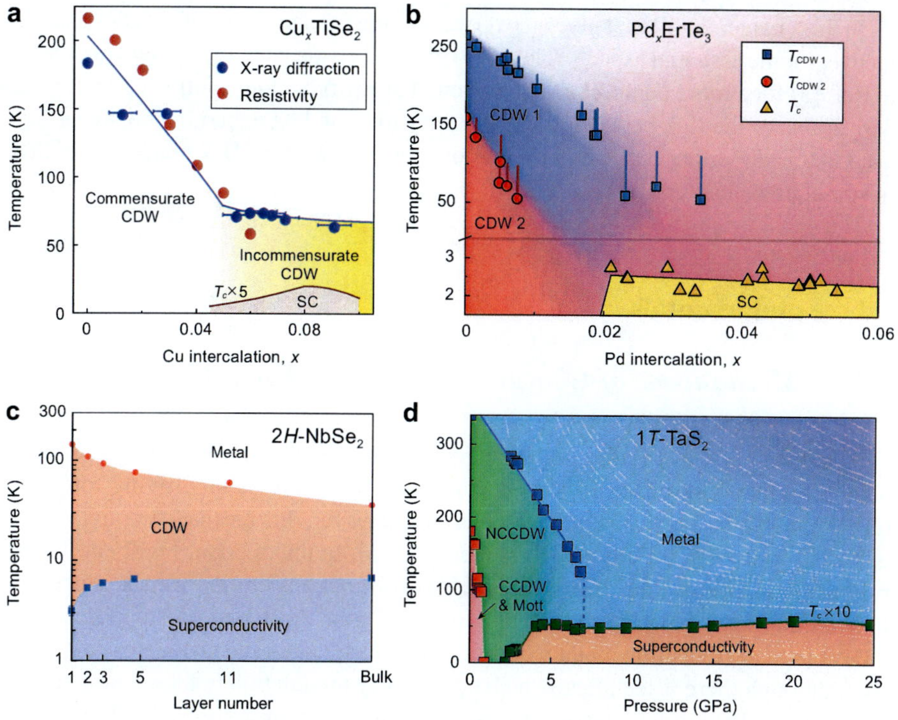

Fig. 2.2 Proximal phases near a charge density wave. Phase diagrams of several families of quasi-2D CDW materials: Cu_xTiSe_2 (**a**) Pd_xErTe_3 (**b**) $2H$-$NbSe_2$ (**c**) and $1T$-TaS_2 (**d**). SC: superconductivity; NCCDW: nearly commensurate CDW; T_c: superconducting transition temperature. (Adapted with permission from Ref. [9], APS (**a**); Ref. [10], APS (**b**); Ref. [11], Springer Nature Ltd (**c**); Ref. [12], Springer Nature Ltd (**d**))

[7]. There is an important distinction between these two types of condensations. The former results in a superfluid that has off-diagonal long-range order (ODLRO) [8], while the latter results in the formation of long-range order in coordinate space, such as a superlattice. This is known as the diagonal long-range order (DLRO), which is not accompanied by superfluidity.

With the advancement of ultrafast techniques, CDW materials have emerged as a rich platform on which a variety of questions can be explored in a nonequilibrium setting, for example, (i) How do a complex order parameter and its collective excitations evolve after photoexcitation? (ii) What is the relationship between electron and lattice dynamics, given their strong interaction in the CDW phase? (iii) As the density wave often lies in close proximity to other ground states, what is the interplay between these coupled or potentially competing orders? In Chaps. 4, 5, 6, 7, and 8, we will see that CDW materials not only serve to answer these fundamental questions about phase transitions out of equilibrium, they also exhibit intriguing photoinduced metastable states that may harbor potential device applications.

In this chapter, I will first give an overview of the fundamentals of a CDW system. It covers some essential aspects relevant to our discussion later and readers may refer to Ref. [3] for a more comprehensive review. The introduction is followed by formal derivations of the link between the CDW transition and its experimental signatures used in this dissertation. We end by introducing the specific CDW materials studied in subsequent chapters.

2.1 Fundamentals

2.1.1 *Mechanisms of Charge Density Wave Formation*

Peierls instability, often used interchangeably with Fermi surface nesting (FSN), is quoted by many introductory texts and research literature as the driving force for CDW transitions (Fig. 2.3a). In real materials, however, this mechanism rarely takes a leading role, even for quasi-1D systems [13, 14]. In this section, I will clarify the role of Fermi surface nesting and discuss two important driving forces: electron–phonon coupling and electron correlations. To limit the scope of our discussion, CDWs in many transition metal oxides, such as manganites and nickelates, are omitted. Although there is technically a charge density modulation in these compounds, it is usually known as charge ordering that often accompanies orbital and spin orderings. Nonetheless, as Daniel Khomskii pointed out, the distinction between these charge-ordering phenomena and CDWs discussed in this section can be quite blurry [15].

Lindhard Susceptibility, Peierls Instability, and Fermi Surface Nesting

We start by recalling the zero-frequency Lindhard susceptibility $\chi(\mathbf{q}) = \chi' + i\chi''$ that describes how charge density responds to a time-independent lattice potential,

$$\rho_{\text{ind}}(\mathbf{q}) = \chi(\mathbf{q})\,\phi(\mathbf{q}). \tag{2.1}$$

Here, we use the convention that χ is negative because a positive potential ϕ implies more electrons attracted; hence, the charge density becomes more negative. The total potential experienced by the electron, $\phi(\mathbf{q})$, consists of an externally applied potential and an induced lattice potential due to the change in charge density, $\phi(\mathbf{q}) = \phi_{\text{ext}}(\mathbf{q}) + \phi_{\text{ind}}(\mathbf{q})$. For small changes, we assume a linear coupling between charge density and lattice potential,

$$\phi_{\text{ind}}(\mathbf{q}) = -g\,\rho_{\text{ind}}(\mathbf{q}), \tag{2.2}$$

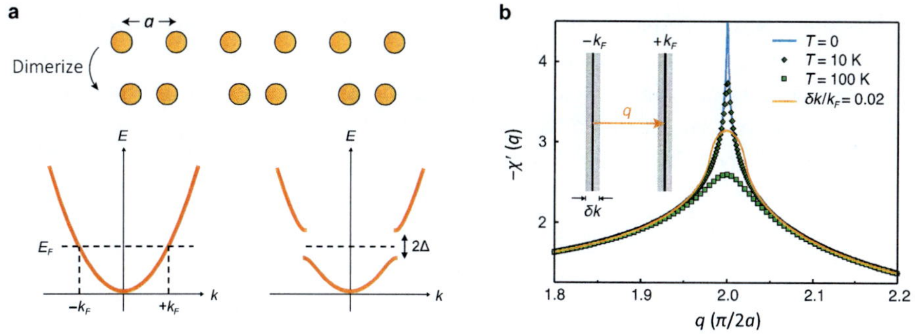

Fig. 2.3 Peierls instability and its fragility. (**a**) Peierls instability in a 1D atomic chain, resulting in a spontaneous dimerization and gap opening (2Δ) at $\pm k_F$ in the electronic dispersion. In real materials, however, this mechanism is rarely the main driving force of CDWs. (**b**) Real part of Lindhard susceptibility, $\chi'(q)$, for a 1D noninteracting electron gas at various temperatures. A logarithmic divergence happens at $q = 2k_F$ at zero temperature. For imperfect nesting (*inset*), such as in a quasi-1D system, the divergence is significantly reduced for even a small value of nesting uncertainty δk at $T = 0$ (orange curve). (Adapted with permission from Ref. [13], APS (**b**))

where g is a positive electron–phonon coupling term that is assumed to be independent of the scattering wavevector $\mathbf{q}$. Rearranging terms yields

$$\rho_{\text{ind}}(\mathbf{q}) = \frac{\chi(\mathbf{q})}{1 + g\chi(\mathbf{q})}\phi_{\text{ext}}(\mathbf{q}). \tag{2.3}$$

For an arbitrary $\mathbf{q}$, the product $g\chi(\mathbf{q})$ is a small negative number in the real part. However, for large susceptibility or strong electron–phonon coupling, we may reach the condition $1 + g\chi(\mathbf{q}) = 0$, where an arbitrarily small perturbation in the lattice potential $\phi_{\text{ext}}(\mathbf{q})$ will yield a large change in the charge density ρ_{ind}. In other words, the entire system including both electrons and lattice is unstable toward a density-wave formation at $\mathbf{q}$.

To see a concrete example of such an instability, let us recall the real and imaginary parts of the Lindhard susceptibility [13]:

$$\chi'(\mathbf{q}) = \sum_{\mathbf{k}} \frac{f(\epsilon_{\mathbf{k}}) - f(\epsilon_{\mathbf{k+q}})}{\epsilon_{\mathbf{k}} - \epsilon_{\mathbf{k+q}}}, \tag{2.4}$$

$$\lim_{\omega\to 0} \chi''(\mathbf{q}, \omega)/\omega = \sum_{\mathbf{k}} \delta(\epsilon_{\mathbf{k}} - \mu)\delta(\epsilon_{\mathbf{k+q}} - \mu), \tag{2.5}$$

where $f(\cdot)$ is the Fermi–Dirac distribution, μ is the chemical potential, and $\epsilon_{\mathbf{k}}$ is the electronic dispersion. Note that $\chi'(\mathbf{q})$ depends on the entire band structure, while $\chi''(\mathbf{q})$ only concerns the Fermi surface. Based on this definition, the notion of FSN by a wavevector $\mathbf{q}$ can be formalized as a peak in χ'' at this wavevector, which does not necessarily indicate a peak in χ' at the same $\mathbf{q}$.

Despite a lack of FSN-driven CDW transitions in real materials, it is instructive to follow Peierls' original reasoning of a 1D free-electron chain (Fig. 2.3a) and see what went wrong. In this case, the imaginary part of the susceptibility $\chi''(q)$ contains a δ-peak at $q = \pm 2k_F = \pm\pi/a$, where a is the spacing between neighboring atoms. This indicates a perfect FSN. The real part of the susceptibility $\chi'(q)$ is given by

$$\chi'(q) = \frac{1}{q} \ln \left| \frac{q - 2k_F}{q + 2k_F} \right|, \tag{2.6}$$

where we chose an overall normalization such that $\chi'(q \to 0) = -1/k_F$. At $q = \pm 2k_F$, χ' diverges logarithmically (Fig. 2.3b, blue curve), and this divergence was used by Peierls to argue for the spontaneous dimerization of the 1D chain. However, this conclusion is rather hasty for a few reasons [13].

First, this logarithmic divergence is very fragile. If we go above zero temperature where $f(\epsilon_k)$ is no longer a step function in Eq. (2.4), the divergence is readily removed (Fig. 2.3b). More importantly, real materials are only quasi-1D, such as several families of paradigmatic CDW materials: $A_{0.3}MoO_3$ (A=K, Rb, Tl), MX_3 (M = Nb, Ta, X = S, Se), $(MX_4)_nY$ (M = Ta, Nb, X = S, Se, Y = I, Br, Cl), and various organic linear chain compounds [3]. As each 1D chain has some weak coupling with its neighbors, there are residual dispersions in the direction perpendicular to the chains, resulting in imperfect nesting. We could estimate the effect of imperfect nesting by considering two Fermi sheets at $\pm k_F$, and each sheet has some spread δk due to the weak dispersion (Fig. 2.3b, inset). After averaging the susceptibility in Eq. (2.6) over δk,

$$\chi'_{\text{avg}}(q) = \frac{1}{\delta k} \int_{k_F - \frac{1}{2}\delta k}^{k_F + \frac{1}{2}\delta k} dk \, \frac{1}{q} \ln \left| \frac{q - 2k}{q + 2k} \right|, \tag{2.7}$$

we plot $\chi'_{\text{avg}}(q)$ in Fig. 2.3b (orange curve) for the case of $\delta k / k_F = 0.02$. For such a small deviation from the perfect nesting condition, the divergence in $\chi'(q)$ at $\pm 2k_F$ is removed.

Second, even if χ' diverges at some wavevector q, the $\chi'/(1 + g\chi')$ term in Eq. (2.3) does not diverge, so the coupled system involving both the electrons and the lattice may not experience an instability at q. Hence, divergence in $\chi'(q)$ alone without considering the lattice contribution is not a sufficient condition for a CDW transition in the original Peierls picture. It has been shown by density functional theory calculations that in an ideal 1D Na atomic chain, dimerization does not occur even under a perfect nesting condition [13]. If we allow the atoms to move in the other dimensions perpendicular to the chain, a zigzag pattern is formed instead.

Lastly, as a clarification to misnomers in the literature, the term FSN only concerns the divergence in the imaginary part of the susceptibility [Eq. (2.5)] and should not be confused with a divergence or peak in the real part. It happens that the divergence in χ' and χ'' coincides at $q = \pm 2k_F$ in a 1D chain. In general, however, their structure as a function of **q** can be vastly different. This is expected because χ' can have significant contributions away from the Fermi level that are neglected in χ''. For example, in the rare-earth tritelluride family, which will be introduced in Sect. 2.3, the best FSN wavevector suggested by $\chi''(\mathbf{q})$ is not observed experimentally. Instead, the actual CDW wavevector is found at the location of the peak value in $\chi'(\mathbf{q})$ [13].

Electron–Phonon Coupling

The preceding section highlights the importance of considering electron–phonon coupling (EPC) in addition to the Lindhard susceptibility. This interaction was approximated as a constant g in Eq. (2.3), but the general vertex that couples electronic states $\mathbf{k}_1$ and $\mathbf{k}_2$ by scattering a phonon of wavevector $\mathbf{q}$ is $g(\mathbf{k}_1, \mathbf{k}_2)$, where $\mathbf{q} = \mathbf{k}_2 - \mathbf{k}_1$. Under the assumption of $|g(\mathbf{k}_1, \mathbf{k}_2)|^2 = |g(\mathbf{q})|^2$, this electron–phonon scattering vertex is $\mathbf{q}$ dependent and can be highly selective for specific branches of phonons. This coupling also leads to a reduction of the phonon frequency from its original value $\omega_0(\mathbf{q})$ without considering EPC. The renormalized phonon frequency, $\omega(\mathbf{q})$, is given by [14]

$$\omega(\mathbf{q})^2 = \omega_0(\mathbf{q})^2 + 2\omega_0(\mathbf{q})|g(\mathbf{q})|^2\chi'(\omega, \mathbf{q}). \tag{2.8}$$

A particularly large $g(\mathbf{q})$ and $\chi'(\mathbf{q})$ at some wavevector hence results in a significant phonon softening, known as Kohn anomaly, which is also accompanied by phonon linewidth broadening. As the frequency dips below zero, an instability occurs and a CDW is formed.

The mechanism of momentum- and mode-selective EPC is demonstrated in, for example, the rare-earth tritelluride family. Figure 2.4a shows the *ab initio* calculation of phonon dispersion in $DyTe_3$, which exhibits significant softening at the CDW wavevector along the $(0\ 0\ L)$ direction in the two in-plane transverse phonon branches (red and blue). This theoretical dispersion is also verified experimentally by inelastic X-ray scattering [16], showing excellent agreement. It is worth noting that not all phonon branches at q_{CDW} display a Kohn anomaly, indicating the selective nature of the scattering vertex $g(\mathbf{q})$. This is to be contrasted with the mechanism purely driven by FSN or a large value in $\chi'(\mathbf{q})$, which does not discriminate between different phonon branches at the same scattering wavevector.

Electron Correlations

In systems with strong electron correlations, the single-particle picture described in the preceding section may no longer be sufficient to account for the formation of CDWs. This point is best exemplified by charge ordering in cuprate HTSCs [14, 18, 19] (Fig. 1.7a), where EPC is not strong enough to drive the CDW transition. The origin of incommensurate CDWs in cuprates, in contrast to the stripe order at $p = 1/8$ doping, remains unresolved and is believed to have

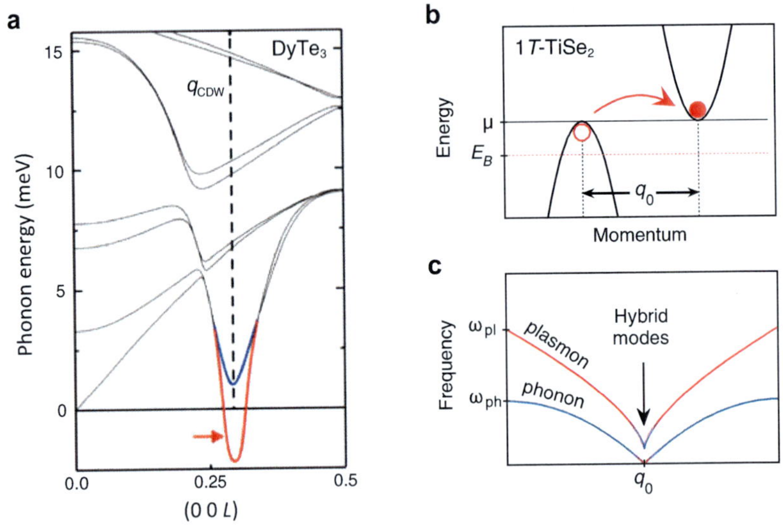

Fig. 2.4 Different mechanisms of CDW transitions. (**a**) Calculated dispersion for in-plane transverse phonons along the [001] direction for $DyTe_3$, using density functional perturbation theory (DFPT). The red and blue curves highlight the soft phonon modes that occur at the CDW wavevector (dashed line). (**b**) Schematic of normal-state bands of $1T$-$TiSe_2$, featuring the hole-like Se $4p$ band and the electron-like Ti $3d$ band. Excitons spontaneously condense when electrons and holes bind between the two bands near the chemical potential μ, where the exciton binding energy is E_B. (**c**) Schematic of collective excitations of an exciton condensate coupled to the lattice distortion at wavevector q_0. Both plasmon frequency (ω_{pl}) and phonon frequency (ω_{ph}) soften at q_0 near the transition temperature, revealing a CDW formation driven by exciton condensation. (Adapted with permission from Ref. [16], APS (**a**); Ref. [17], AAAS (**b**, **c**))

significant contributions from electron correlation and disorder effects. Nonetheless, there are indications of phonon participation that may drive the precursor of the CDW instability [19]. For example, phonon softening or linewidth broadening is observed across different families of cuprates, such as $YBa_2Cu_3O_{6+\delta}$ [20, 21], $La_{2-x}Ba_xCuO_4$ [22], and $Bi_2Sr_2CaCu_2O_{8+\delta}$ [23]. However, unlike a usual Kohn anomaly, the phonon frequency never reaches zero, which may be reconciled with the short-range nature of CDW correlations. Future theoretical and experimental works are still needed to clarify the role of EPC, if any, in the charge-ordering phenomena in cuprates.

A better-understood case of correlation-driven CDW transitions is found in a putative excitonic insulator (EI), $1T$-$TiSe_2$ [17, 24]. In the normal state, electrons and holes in two separate bands near the Fermi level form an exciton, whose large binding energy drives a spontaneous formation of an exciton condensate across the transition temperature (Fig. 2.4b). As the electron–hole pairs are connected by a finite wavevector q_0, this exciton condensation is accompanied by a broken translational symmetry by the same wavevector, leading to a CDW in the ground state. One signature of an exciton-driven CDW transition can be found in the associated collective excitation, which bears both electronic and lattice characters.

As a reference, the hallmark in an EPC-driven transition is the softening of a specific phonon mode at the CDW wavevector, whose energy reaches zero at T_c. In an excitonic insulator, the collective excitation is an electronic, plasmon-like mode, which displays a similar feature as a Kohn anomaly and softens near T_c (Fig. 2.4c). The softening of this plasmon-like mode was experimentally measured by momentum-resolved EELS, which sensitively detects bosonic excitations of valence electrons [17].

Another strong piece of evidence for this exciton-driven picture in $1T$-$TiSe_2$ is offered by time-resolved studies that clock how fast the CDW phase disappears after photoexcitation [25, 26]. As we discussed in Sect. 1.1.3, the initial response time in a photoinduced phase transition is bottlenecked by the characteristic timescale of the microscopic interaction, which in this case is proposed to have a leading contribution from exciton formation as opposed to lattice distortion (Fig. 1.4d). By tracking the fluence dependence of CDW melting time that is well below 100 fs, it was revealed that this nonequilibrium phase transition proceeds by screening the Coulomb attraction between electrons and holes due to the excited free carriers. The time to build up the screening effect is given by the inverse plasma frequency, $\omega_{\mathrm{pl}}^{-1}$. This timescales with $n^{-1/2}$, where n is the excited free carrier density, in excellent agreement with observations in tr-ARPES experiments [25].

2.1.2 Ginzburg–Landau Theory of Elementary Excitations

On a phenomenological level, many incommensurate CDW transitions can be described by the simplest Ginzburg–Landau theory of second-order phase transitions, where the translational symmetry of the lattice is spontaneously broken. The free energy of the ground state resembles a Mexican hat (Fig. 2.5a), and the order parameter is represented by a complex number $\psi = \Delta e^{i\phi}$ for $\Delta > 0$ and $\phi \in [-\pi, \pi)$. The amplitude of the CDW order parameter (Δ) is the ionic displacement relative to the normal state, or equivalently, the deviation of charge density from the normal state. We will discuss how to measure Δ in Sect. 2.2. The phase of the order parameter (ϕ) is the phase of the density wave relative to the original lattice. For an incommensurate CDW, in the absence of impurity pinning, ϕ can take any value and all states are degenerate. Despite its simplicity, Ginzburg–Landau theory has been particularly successful in describing the spatial and temporal evolutions of the complex order parameter. In Chaps. 5, 6, and 7, we will use various forms of the Ginzburg–Landau formalism to describe the CDW state after photoexcitation. When competing CDWs are present, the theory can also capture their interactions in space and time.

In this section, we use the Ginzburg–Landau theory to introduce the two elementary excitations of a CDW ground state: amplitudon and phason. They have tangible consequences in the observables of time-resolved experiments, as we will see in Chaps. 4 and 8. These excitations can be thought of as two orthogonal

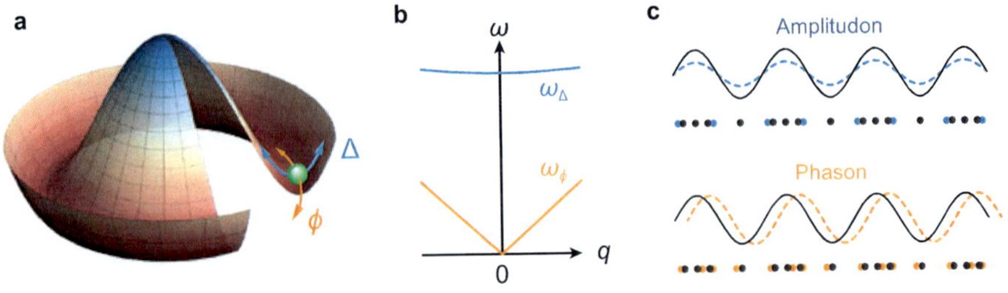

Fig. 2.5 Complex CDW order parameter and its elementary excitations. (**a**) Mexican hat potential of a complex CDW order parameter $\Delta e^{i\phi}$. Double-sided arrows represent the elementary excitations of Δ and ϕ. (**b**) Dispersion of amplitudon (ω_Δ) and phason (ω_ϕ) near $q = 0$ in the CDW ground state. (**c**) Schematic of amplitude and phase excitations in the $q \to 0$ limit. Dashed curves show the change in the electron density, while circles show the corresponding change in the atomic positions. It is worth noting that the atomic movement corresponding to the phason is *not* a uniform displacement that follows the electron density [27]. (Adapted with permission from Ref. [4], APS (**b**))

deviations from the energy minimum of the Mexican hat potential (Fig. 2.5a). As such, we can rewrite the CDW order parameter as

$$\psi(x,t) = [\Delta_0 + \delta(x,t)]\, e^{i[\phi_0 + \varphi(x,t)]}, \tag{2.9}$$

where Δ_0 and ϕ_0 are the equilibrium amplitude and phase, and $\delta(x,t)$ and $\varphi(x,t)$ are their respective fluctuations that vary with space and time. Here, x is along the direction of the CDW wavevector.

We begin by considering the Ginzburg–Landau free energy density

$$f\,[\psi(x,t)] = -\alpha|\psi|^2 + \frac{1}{2}\beta|\psi|^4 + \kappa|\psi_x|^2 + \zeta|\psi_t|^2, \tag{2.10}$$

where subscripts x and t denote partial derivatives. The $|\psi_x|^2$ term describes the energy cost of any spatial variation of the order parameter, while $|\psi_t|^2$ can be considered as the kinetic energy arising from its temporal fluctuations [3]. $\alpha, \beta, \kappa, \zeta$ are the phenomenological parameters that are all positive in the CDW ordered phase. In the ground state, we retrieve the standard relation $\Delta_0 = \alpha/\beta$. The Lagrangian density associated with $f[\psi(x,t)]$ is

$$\begin{aligned} \mathcal{L} &= \zeta|\psi_t|^2 - \left(-\alpha|\psi|^2 + \frac{1}{2}\beta|\psi|^4 + \kappa|\psi_x|^2\right) \\ &\approx \zeta\left(\delta_t^2 + \Delta_0^2\varphi_t^2\right) - \kappa\left(\delta_x^2 + \Delta_0^2\varphi_x^2\right) + 2\alpha\left(\frac{1}{4}\Delta_0^2 - \delta^2\right), \end{aligned} \tag{2.11}$$

where we used the definition of $\psi(x,t)$ in Eq. (2.9) for the second line and kept terms up to the second order in δ and φ, assuming small fluctuations and slow variations in space and time. The equations of motion for the amplitude and phase fluctuations are given by the Euler–Lagrange equations

$$\frac{\partial \mathcal{L}}{\partial \delta} - \frac{\partial}{\partial x}\frac{\partial \mathcal{L}}{\partial \delta_x} - \frac{\partial}{\partial t}\frac{\partial \mathcal{L}}{\partial \delta_t} = 0 \quad \text{and} \quad \frac{\partial \mathcal{L}}{\partial \varphi} - \frac{\partial}{\partial x}\frac{\partial \mathcal{L}}{\partial \varphi_x} - \frac{\partial}{\partial t}\frac{\partial \mathcal{L}}{\partial \varphi_t} = 0, \tag{2.12}$$

which are evaluated to

$$\kappa \delta_{xx} - \zeta \delta_{tt} - 2\alpha\delta = 0 \quad \text{and} \quad \kappa \varphi_{xx} - \zeta \varphi_{tt} = 0. \tag{2.13}$$

Seeking solutions of the form $\delta(x, t) = \delta_0 e^{i(\omega_\Delta t - qx)}$ and $\varphi(x, t) = \varphi_0 e^{i(\omega_\phi t - qx)}$, we obtain the dispersion relations in the long-wavelength limit,

$$\omega_\Delta = \left(\frac{2\alpha + \kappa q^2}{\zeta}\right)^{1/2}, \tag{2.14}$$

$$\omega_\phi = \left(\frac{\kappa}{\zeta}\right)^{1/2} q. \tag{2.15}$$

These dispersions are plotted in Fig. 2.5b, where the amplitudon (ω_Δ) and phason (ω_ϕ) branches resemble those of optical and acoustic phonons, respectively. Indeed, these excitations often show up as new phonon modes when a material enters its CDW state. Note that the distinction between an amplitudon and a phason is only meaningful far below T_c; close to T_c, the amplitudon softens and these two excitations are hybridized.

Figure 2.5c depicts a schematic of these CDW excitations in the $q \to 0$ limit. The amplitudon alters the amplitude of the density modulation, or equivalently, the deviation of atoms from their high-symmetry position. Phasons change the spatial location of the CDW condensate relative to the underlying lattice, and the ions respond to this shift of electron density accordingly. Both modes may be excited by an ultrafast laser pulse and readily detected in pump–probe experiments [28, 29]. In Chaps. 4 and 8, we will see two examples of amplitude modes examined by transient optical spectroscopy and ultrafast electron diffraction, respectively.

2.2 Signatures of the Phase Transition

As both the electron density and ionic positions undergo a periodic modulation across the CDW phase transition, multiple experimental probes that are sensitive to these two degrees of freedom bear a signature of the transition [3]. For example, the specific heat and electrical resistivity both show anomalies at T_c, the latter of which may display a metal–insulator transition if the Fermi surface completely vanishes due to the CDW. The change in the ionic positions leads to a modification of the nuclear magnetic resonance (NMR) spectrum, whose frequency linewidth near a resonance increases with the periodic lattice distortion. The structural change also gives rise to new lattice modes, such as the amplitudon discussed in Sect. 2.1.2, which are readily detectable by Raman spectroscopy. In the following, we introduce two important experimental signatures that are considered as smoking-gun evidence for the lattice and electronic change associated with the CDW transition. These are also the most used techniques in this dissertation.

2.2.1 *Superlattice Peak*

One hallmark of a CDW transition is the appearance of satellite peaks in a diffraction experiment using X-rays, neutrons, or electrons (Fig. 2.6). Theses satellite peaks arise from a periodic lattice distortion that changes the nuclear and electronic densities. In this section, we formalize the link between the diffraction intensity and the CDW order parameter. By considering the intensity change in both lattice Bragg peaks and CDW satellite peaks, one can also learn about the phase fluctuations in a CDW.

Following the treatment in Ref. [31], we start by expressing the modulation of lattice site positions as

$$\mathbf{r}_n \rightarrow \mathbf{r}_n + \mathbf{u}\cos(\mathbf{q}\cdot\mathbf{r}_n), \tag{2.16}$$

where $\mathbf{u}$ and $\mathbf{q}$ are the amplitude and the wavevector of the modulation; $\mathbf{r}_n$ are high-symmetry lattice positions. Here, we consider a lattice with a monoatomic basis, and we leave the general case of polyatomic crystals to Sect. 8.5.1. Assuming that the amplitude $|\mathbf{u}|$ is much smaller than the lattice spacing, we can Taylor expand the electron density

$$\begin{aligned}\rho(\mathbf{r}) &= \sum_{\mathbf{r}_n} \delta\left[\mathbf{r} - \mathbf{r}_n - \mathbf{u}\cos(\mathbf{q}\cdot\mathbf{r}_n)\right] \\ &\approx \rho_0(\mathbf{r}) - \sum_{\mathbf{r}_n} (\mathbf{u}\cdot\nabla_{\mathbf{r}})\delta(\mathbf{r}-\mathbf{r}_n)\cos(\mathbf{q}\cdot\mathbf{r}_n) \\ &\quad + \frac{1}{2}\sum_{\mathbf{r}_n} (\mathbf{u}\cdot\nabla_{\mathbf{r}})^2\delta(\mathbf{r}-\mathbf{r}_n)\cos^2(\mathbf{q}\cdot\mathbf{r}_n) + \cdots,\end{aligned} \tag{2.17}$$

where $\rho_0(\mathbf{r}) = \sum_{\mathbf{r}_n} \delta(\mathbf{r} - \mathbf{r}_n)$ corresponds to the density of an unmodulated lattice. After performing a Fourier transform, we obtain

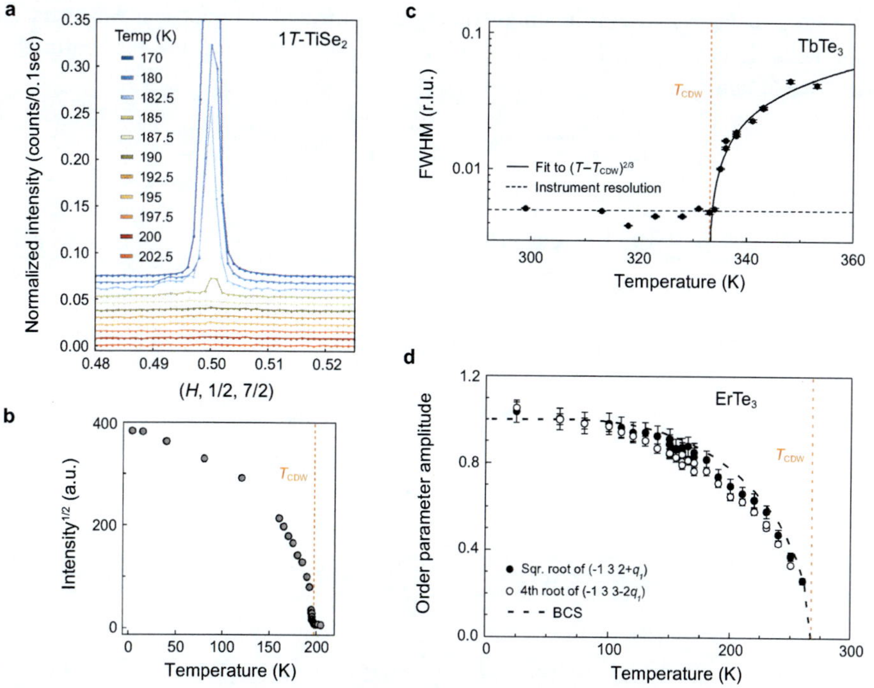

Fig. 2.6 Emergence of superlattice peaks across the CDW phase transition. (**a**) X-ray intensity along the H cut for the superlattice peak of $1T$-$TiSe_2$ across its T_{CDW} for the $2\times2\times2$ lattice distortion. It shows the growth of a resolution-limited peak at $(\frac{1}{2},\frac{1}{2},\frac{7}{2})$, indicating the emergence of long-range order (LRO). Data was taken with 12-keV hard X-rays. (**b**) Square root of the X-ray intensity of the superlattice peak at $(\frac{1}{2},\frac{1}{2},\frac{23}{2})$ in $1T$-$TiSe_2$, showing the evolution of the CDW order parameter amplitude. T_{CDW} is approximately 198 K. Data was taken with 38.5-keV X-rays. (**c**) Width of the superlattice peak $(1, 7, q)$ in the out-of-plane direction across the high-temperature CDW transition of $TbTe_3$, where the incommensurate CDW wavevector is $q = 0.296(4)c^*$. The solid curve is a fit to the inverse CDW correlation length, $\xi^{-1} \sim (T - T_{CDW})^{\gamma}$ with $T_{CDW} = 332.8(5)$ K and $\gamma = 2/3$, consistent with the X-Y model universality class. The peak width below T_{CDW} is resolution-limited, and the CDW correlation length is estimated to be at least 1.8 μm within the Te plane and 0.5 μm perpendicular to the plane. (**d**) Temperature evolution of the order parameter amplitude for the c-axis CDW in $ErTe_3$. The plot shows the square root of the integrated intensity of the first-order superlattice peak $(\bar{1}, 3, 2+q)$, and the fourth root of the integrated intensity of the second-order superlattice peak $(\bar{1}, 3, 3-2q)$ [see Eqs. (2.21) and (2.22)]. The integrated intensities were first normalized to the main Bragg peak $(\bar{1}, 3, 2)$ and then normalized to unity at $T = 0$. Both quantities exhibit a BCS-like temperature dependence, with $T_{CDW} = 267(3)$ K. Data in (**c**) (**d**) was taken with X-rays at 9–13 keV. (Adapted with permission from Ref. [9], APS (**a**); Ref. [30], APS (**c, d**))

$$\rho_{\mathbf{k}} = \rho_{0\mathbf{k}} - \int d^3\mathbf{r}\, e^{-i\mathbf{k}\cdot\mathbf{r}} \sum_{\mathbf{r}_n} (\mathbf{u}\cdot\nabla_{\mathbf{r}})\delta(\mathbf{r}-\mathbf{r}_n)\cos(\mathbf{q}\cdot\mathbf{r}_n)$$
$$+ \frac{1}{2}\int d^3\mathbf{r}\, e^{-i\mathbf{k}\cdot\mathbf{r}} \sum_{\mathbf{r}_n} (\mathbf{u}\cdot\nabla_{\mathbf{r}})^2\delta(\mathbf{r}-\mathbf{r}_n)\cos^2(\mathbf{q}\cdot\mathbf{r}_n) + \cdots, \qquad (2.18)$$

where $\rho_{0\mathbf{k}} = F_{\mathbf{k}} \sum_{\mathbf{G}} \delta_{\mathbf{k},\mathbf{G}}$ is a sum of sharp peaks located at reciprocal wavevectors $\mathbf{G}$ of the underlying crystal lattice. Here, $F_{\mathbf{k}}$ is the lattice form factor. Integrating by parts, we obtain

$$\rho_{\mathbf{k}} = \rho_{0\mathbf{k}} - i(\mathbf{u}\cdot\mathbf{k}) \sum_{\mathbf{r}_n} e^{-i\mathbf{k}\cdot\mathbf{r}_n} \cos(\mathbf{q}\cdot\mathbf{r}_n)$$

$$- \frac{(\mathbf{u}\cdot\mathbf{k})^2}{4} \sum_{\mathbf{r}_n} e^{-i\mathbf{k}\cdot\mathbf{r}_n} [1 + \cos(2\mathbf{q}\cdot\mathbf{r}_n)] + \cdots . \quad (2.19)$$

The terms in Eq. (2.19) can be combined as follows:

$$\rho_{\mathbf{k}} = \left[1 - \frac{(\mathbf{u}\cdot\mathbf{k})^2}{4}\right] \rho_{0\mathbf{k}} + \rho_{\mathbf{k}}^{\mathbf{q}} + \rho_{\mathbf{k}}^{2\mathbf{q}} + \cdots , \quad (2.20)$$

where

$$\rho_{\mathbf{k}}^{\mathbf{q}} = -i(\mathbf{u}\cdot\mathbf{k}) \sum_{\mathbf{r}_n} e^{-i\mathbf{k}\cdot\mathbf{r}_n} \cos(\mathbf{q}\cdot\mathbf{r}_n)$$

$$= -\frac{i}{2}(\mathbf{u}\cdot\mathbf{k}) F_{\mathbf{k}} \sum_{\mathbf{G}} (\delta_{\mathbf{k},\mathbf{G}+\mathbf{q}} + \delta_{\mathbf{k},\mathbf{G}-\mathbf{q}}), \quad (2.21)$$

$$\rho_{\mathbf{k}}^{2\mathbf{q}} = -\frac{(\mathbf{u}\cdot\mathbf{k})^2}{4} \sum_{\mathbf{r}_n} e^{-i\mathbf{k}\cdot\mathbf{r}_n} \cos(2\mathbf{q}\cdot\mathbf{r}_n)$$

$$= -\frac{(\mathbf{u}\cdot\mathbf{k})^2}{8} F_{\mathbf{k}} \sum_{\mathbf{G}} (\delta_{\mathbf{k},\mathbf{G}+2\mathbf{q}} + \delta_{\mathbf{k},\mathbf{G}-2\mathbf{q}}). \quad (2.22)$$

These terms describe the appearance of CDW peaks with wavevectors $n\mathbf{q}$, where $n = 1, 2, \ldots$ In particular, the electron density associated with the n-th-order CDW peak scales as $|\mathbf{u}|^n$. As diffraction intensity is proportional to $|\rho_{\mathbf{k}}|^2$, we conclude that the $2n$-th root of intensity of the n-th-order CDW peak scales as the atomic displacement and therefore proportional to the order parameter amplitude (Fig. 2.6b, d).

Besides the appearance of satellite peaks, the first term in Eq. (2.20) reveals that the CDW suppresses the density amplitude of lattice Bragg peaks by an amount

$$\delta\rho_{\mathbf{k}}^{\text{Bragg}} = -\frac{(\mathbf{u}\cdot\mathbf{k})^2}{4} \rho_{0\mathbf{k}}. \quad (2.23)$$

Note that Eq. (2.23) remains valid even when CDW correlations are only short-ranged, in the absence of a truly long-range order. To show this, we consider a more general expression for the electron density

$$\rho(\mathbf{r}) = \sum_{\mathbf{r}_n} \delta\left[\mathbf{r} - \mathbf{r}_n - \mathbf{u}\cos(\mathbf{q}\cdot\mathbf{r}_n + \phi(\mathbf{r}_n))\right]. \tag{2.24}$$

Here, $\phi(\mathbf{r}_n)$ is the phase of the order parameter, whose variation leads to the excitation of phasons. When $\phi(\mathbf{r}_n)$ varies as a function of $\mathbf{r}_n$, the CDW order weakens, or disappears completely, and becomes replaced by short-range correlations. To see the effect of CDW phasons on the Bragg peak, we generalize Eq. (2.19) and obtain

$$\begin{aligned}\rho_{\mathbf{k}} = \left[1 - \frac{(\mathbf{u}\cdot\mathbf{k})^2}{4}\right]\rho_{0\mathbf{k}} - i(\mathbf{u}\cdot\mathbf{k})\sum_{\mathbf{r}_n} e^{-i\mathbf{k}\cdot\mathbf{r}_n}\cos(\mathbf{q}\cdot\mathbf{r}_n + \phi(\mathbf{r}_n))\\ - \frac{(\mathbf{u}\cdot\mathbf{k})^2}{4}\sum_{\mathbf{r}_n} e^{-i\mathbf{k}\cdot\mathbf{r}_n}\cos(2\mathbf{q}\cdot\mathbf{r}_n + 2\phi(\mathbf{r}_n)) + \cdots.\end{aligned} \tag{2.25}$$

Thus, in the presence of a phase variation $\phi(\mathbf{r}_n)$, the intensity of the Bragg peaks remains unchanged, while the CDW peaks $\rho_{\mathbf{k}}^{\mathbf{q}}$ and $\rho_{\mathbf{k}}^{2\mathbf{q}}$ are modified to

$$\rho_{\mathbf{k}}^{\mathbf{q}} = -i(\mathbf{u}\cdot\mathbf{k})\sum_{\mathbf{r}_n} e^{-i\mathbf{k}\cdot\mathbf{r}_n}\cos(\mathbf{q}\cdot\mathbf{r}_n + \phi(\mathbf{r}_n)), \tag{2.26}$$

$$\rho_{\mathbf{k}}^{2\mathbf{q}} = -\frac{(\mathbf{u}\cdot\mathbf{k})^2}{4}\sum_{\mathbf{r}_n} e^{-i\mathbf{k}\cdot\mathbf{r}_n}\cos(2\mathbf{q}\cdot\mathbf{r}_n + 2\phi(\mathbf{r}_n)). \tag{2.27}$$

This distinction between the CDW and the Bragg peaks will be leveraged when we study the photoinduced dynamics of the complex order parameter in Chap. 4.

2.2.2 *Energy Gap*

When the lattice acquires a new periodicity through a CDW transition, the electronic structure is renormalized accordingly. For an incommensurate CDW, translational symmetry is broken and fundamental concepts such as Brillouin zones and crystal momentum are, strictly speaking, not well defined. Nevertheless, clear band dispersions can still be detected by photoemission measurements in the CDW state. In this section, I will formalize the change in the electronic structure across the transition. Using a simple model [32–34], we will see that the CDW transition is marked by the opening of an energy gap and the redistribution of spectral weight from the original dispersion to a CDW shadow band.

To illustrate the core idea, we consider a 1D parabolic band dispersion, ϵ_k, before the CDW transition (Fig. 2.7a, solid curve). A model Hamiltonian that describes this bare dispersion and electron–phonon coupling reads

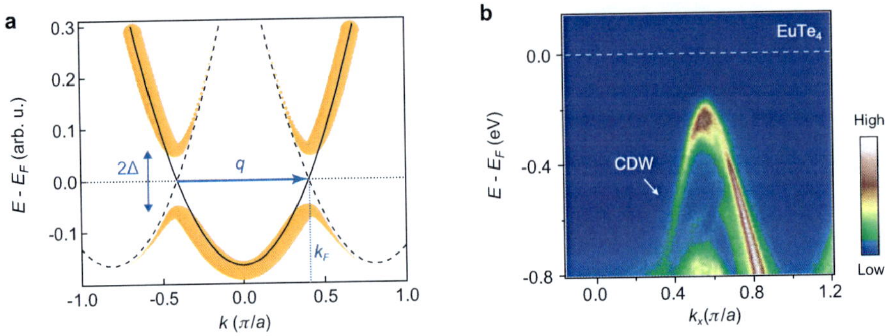

Fig. 2.7 Energy gap due to CDW transition. (**a**) Spectral weight redistribution and energy gap opening due to a CDW transition in a 1D metal. The calculation assumes a parabolic bare band in the metallic state (solid curve) and a CDW wavevector $q = 2k_F = 0.813\pi/a$. Dashed curves are CDW shadow bands displaced from the original band by $\pm q$. The area of the yellow markers corresponds to the spectral weight. (**b**) Photoemission intensity along the $\bar{\Gamma}$–$\bar{X}$ cut of $EuTe_4$ in its CDW state, featuring a large energy gap and the CDW shadow band (arrow). Data was taken at 55 K with 24 eV photon energy

$$H = \sum_k \epsilon_k c_k^\dagger c_k + \sum_{k,q} g_q c_{k+q}^\dagger c_k (b_q + b_{-q}^\dagger), \tag{2.28}$$

where g_q is the electron–phonon coupling vertex at wavevector q, and $c_k^\dagger(c_k)$ and $b_q^\dagger(b_q)$ are the creation (annihilation) operators for electrons and phonons. The CDW transition takes place when a permanent lattice distortion at some wavevector q occurs. Namely, there is a macroscopic, coherent occupation of phonons at this wavevector q, $\langle b_{\pm q}\rangle = \langle b_{\pm q}^\dagger\rangle \neq 0$. Taking a mean-field approach, this lattice distortion leads to a coupling between the electronic states $|k\rangle$ and $|k \pm q\rangle$ with an interaction strength

$$\Delta = 2g_q\langle b_q\rangle. \tag{2.29}$$

The new electronic eigenstates $|\psi_k\rangle$ can be written as a superposition of the original states. Keeping only the first harmonic in q, we have

$$|\psi_k\rangle = u_{k-q}|k-q\rangle + u_k|k\rangle + u_{k+q}|k+q\rangle, \tag{2.30}$$

which are the eigenstates of the matrix

$$M = \begin{pmatrix} \epsilon_{k-q} & \Delta & 0 \\ \Delta & \epsilon_k & \Delta \\ 0 & \Delta & \epsilon_{k+q} \end{pmatrix}. \tag{2.31}$$

The eigenvalues of this matrix yield the new electronic dispersion (Fig. 2.7a, orange curve), whose spectral weight (area of the circles) is given by $|u_k|^2$.

Compared to the original band without the CDW, the new dispersion opens a gap of size 2Δ. From Eq. (2.29), Δ scales as the atomic displacement in the periodic lattice distortion, so one could consider the size of the energy gap as an indicator of the CDW order parameter amplitude. This gap opens symmetrically around the E_F because in the 1D example considered here, the Fermi surface (just two points at $\pm k_F$) is perfectly nested by q. In general, the gap can open asymmetrically around E_F due to imperfect nesting, as we will see in Sect. 2.3.2; in some cases, the CDW gap may not cross E_F at all. Accompanying this gap opening is a spectral weight redistribution from the original $|k\rangle$ states to the CDW shadow bands $|k \pm q\rangle$, leading to an apparent back bending in the band structure. However, these shadow bands only gain spectral weights near the CDW gap and they quickly disappear as we move to higher momenta. An example band dispersion is shown in Fig. 2.7b for a layered compound EuTe_4 in its incommensurate CDW phase. The CDW shadow band is clearly resolved, and an energy gap of more than 0.3 eV is observed.

2.3 Rare-Earth Tritellurides $R\text{Te}_3$

The rare-earth tritelluride family is a series of layered compounds, all of them having a CDW ground state. As shown in Fig. 2.8b, for light rare-earth element (La to Gd), the CDW state occurs below T_{c1} and is characterized by a unidirectional stripe pattern. For heavy elements starting from Tb, an additional CDW appears below a lower transition temperature, T_{c2}, and the two perpendicular CDWs coexist and form a checkerboard pattern. The $4f$ electrons from the rare-earth element are localized and stay far away from the Fermi level. Different rare-earth ions, with their shrinking ionic radii from La to Tm, mainly serve to tune the chemical pressure across the series without affecting the low-energy physics of the density waves [30]. In the following sections, I will highlight important structural and electronic properties of the $R\text{Te}_3$ series and offer some insights into the trends across the rare-earth elements. A full review of their equilibrium properties can be found in Ref. [35].

2.3.1 Crystal Structure, Orthorhombicity, and CDW Direction

All $R\text{Te}_3$ share a similar structure, and the primary difference lies in the lattice parameters that track the trend of ionic radii of the rare-earth element (Fig. 2.9b). The normal-state crystal structure belongs to the orthorhombic space group *Cmcm* (No. 63) [39]. In today's convention, the in-plane axes are labeled by a and c, while b is the out-of-plane axis (Fig. 2.8a). The unit cell contains two building blocks: R-Te slabs sandwiched by Te sheets. The former acts as electron donors (R^{3+} vs. Te^{2-}),

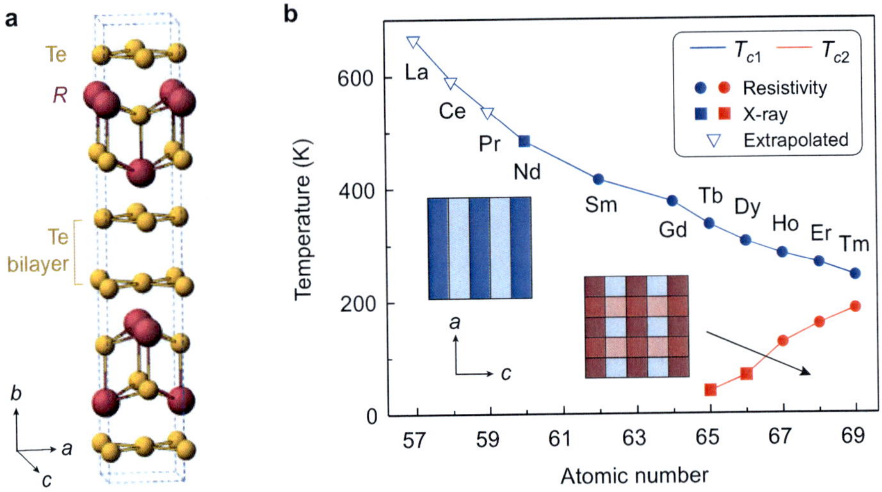

Fig. 2.8 **Crystal structure and CDW transitions in** ***R*****Te**$_3$**.** (**a**) Schematic of the RTe$_3$ crystal structure, where dashed lines indicate the unit cell corresponding to the orthorhombic space group *Cmcm*. The structure is quasi-2D, consisting of nearly square-shaped Te bilayers sandwiched by R-Te slabs. (**b**) Summary of the two CDW transition temperatures, T_{c1} and T_{c2}, across the rare-earth series. Insets are schematics of the unidirectional CDW below T_{c1} and the bidirectional CDW below T_{c2}, respectively. Transition temperatures are quoted from Refs. [16, 30, 36–38]

while the latter gives rise to electronic states near E_F and is the origin of the CDW distortion. The structure is quasi-2D, with weak van der Waals bonds between the Te bilayers. The weak inter-layer coupling allows one to exfoliate RTe$_3$ into atomically thin flakes, with a cleavage plane between the two Te sheets.

The unit cell is almost tetragonal, with a slight difference between the a and c lattice parameters (Fig. 2.9b). Their distinction arises from a glide plane at $b/2$: if we translate the top half of the unit cell by $c/2$ and reflect it across this plane, it overlaps completely with the bottom half. The same symmetry operation does *not* work along the a-axis. This glide plane results in a slightly larger c-axis lattice constant compared to the a-axis for most RTe$_3$ measured (Fig. 2.9b). This difference between the two in-plane axes dictates that the first CDW transition across T_{c1} always happens along the c-axis, with a wavevector $q_c \approx 2/7c^*$. Here, $c^* \equiv 2\pi/c$, and we adopt the convention of writing 2/7 instead of 5/7. The formation of the c-axis CDW further breaks the equivalence between the in-plane axes and enhances the orthorhombicity, as revealed by X-ray diffraction around T_{c1} (Fig. 2.9a). This observation implies that the direction of the CDW in the nearly square-shaped Te sheets sensitively depends on the degree of orthorhombicity. For heavier rare-earth elements where $T_{c2} > 0$, the wavevector of the second CDW is slightly different, $q_a \approx 1/3a^*$, and we will comment on this difference in Sect. 2.3.3.

Note that the distinction in the glide plane symmetry between the a- and c-axis, hence the orthorhombicity, is caused by the R-Te slab as opposed to the Te sheets. As the CDW is primarily driven by the Te sheets, in principle, if we

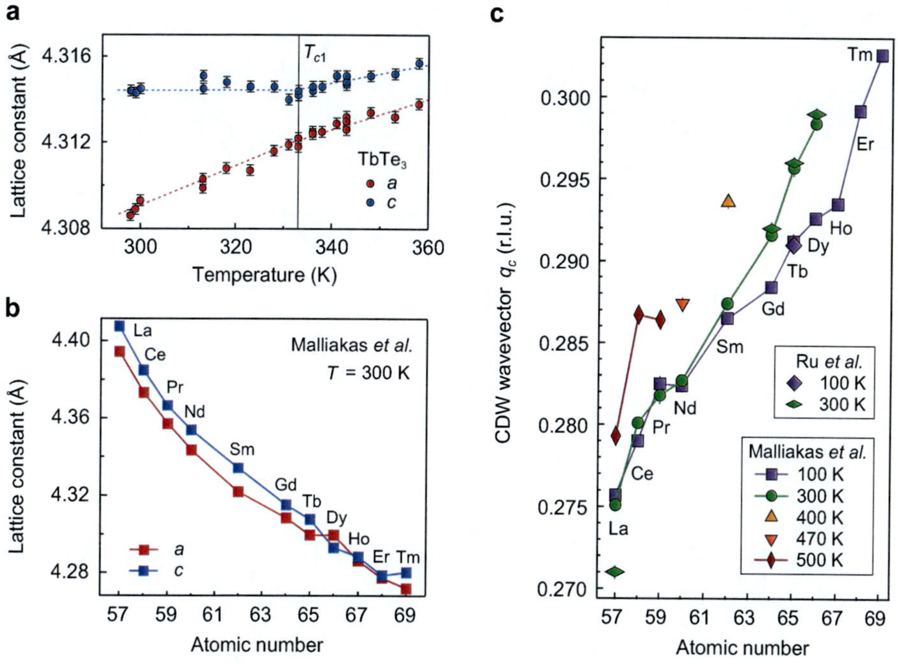

Fig. 2.9 Temperature and element dependence of in-plane lattice constants and *c*-axis CDW wavevector across RTe_3. (**a**) Temperature evolution of in-plane lattice parameters a and c across T_{c1} of $TbTe_3$. Dashed lines are guides to the eye. Orthorhombicity is enhanced due to the CDW formation along the c-axis. (**b**) Lattice constants, a and c, for different RTe_3, taken from X-ray measurements at 300 K by Malliakas et al. [37]. Error bars are smaller than the symbol size. (**c**) CDW wavevector q_c at various temperatures, taken from X-ray measurements by Ru et al. [35] and Malliakas et al. [37]. (Adapted with permission from Ref. [30], APS (**a**))

strain the lattice such that a is larger than c, the CDW direction may be switched even if the glide plane symmetry remains unchanged. This CDW switching has recently been demonstrated by elastocaloric and elastoresistivity measurements in $ErTe_3$ and $TmTe_3$ under uniaxial stress [40]. The switching behavior adds further proof for the intimate relation between the CDW ordering direction and the lattice orthorhombicity.

2.3.2 *Electronic Structure*

For all RTe_3 in the series, the electronic states near the Fermi level arise from the $5p_{\tilde{x}}$ and $5p_{\tilde{z}}$ orbitals of the Te bilayers (Fig. 2.10a). Here, we use $\tilde{x}$ and $\tilde{z}$ to denote the direction between neighboring Te atoms in the nearly square-shaped grid, which are 45° rotated from the definition of the unit cell. As each R^{3+}-Te^{2-} slab donates

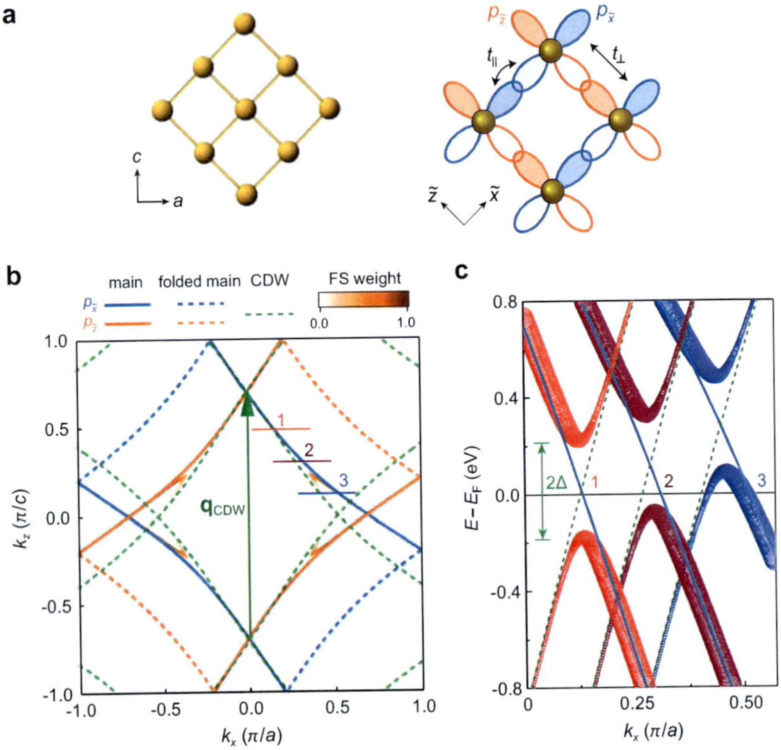

Fig. 2.10 Electronic structure of $R\text{Te}_3$. (**a**) Nearly square-shaped Te sheets (*left*), whose $5p$ orbitals give rise to the electronic states near E_F. The simplest tight-binding (TB) model considers two nearest-neighbor hoppings in the σ bond ($t_{\parallel}$) and π bond ($t_{\perp}$). Note that the unit vectors along the a- and c-axis are 45° rotated from the $\tilde{x}/\tilde{z}$ direction of the $5p$ orbitals. (**b**) Typical Fermi surface of $R\text{Te}_3$ calculated from a tight-binding model, using $t_{\parallel} = -1.9$ eV and $t_{\perp} = 0.35$ eV. The Fermi level is set to $E_F = -2t_{\parallel}\sin(\pi/8)$, which approximates the band filling of 5/8, assuming $t_{\perp}/t_{\parallel} \to 0$. If $t_{\perp}$ is taken into account, this filling gives $E_F = 1.44$ eV, about 1% smaller than $-2t_{\parallel}\sin(\pi/8) = 1.45$ eV used. Main bands from $p_{\tilde{x}}$ and $p_{\tilde{z}}$ orbitals are shown as solid curves, and their folded bands due to inter-layer coupling are shown as dashed curves in the respective color. CDW bands (green dashes) are obtained by translating the main bands through the wavevector $\mathbf{q}_{\text{CDW}}$. In the CDW phase, some residual Fermi surface remains around $k_z = 0$ due to imperfect nesting. (**c**) Band dispersions along cuts 1–3 in (**b**) calculated from the interacting TB model with $\Delta = 0.2$ eV. Main and CDW bands are shown as solid and dashed lines before the CDW gap opening, using the same color code as in (**b**). Gapped dispersion is shown by circular markers, whose size corresponds to the spectral weight. (Adapted from Ref. [34] (**b**, **c**))

one electron to two Te sheets and the $5p_y$ orbital of Te is fully filled [35], the $5p_{\tilde{x}}$ and $5p_{\tilde{z}}$ orbitals have 1.25 electrons each; equivalently, the band filling is 5/8.

The band structure can be well approximated by a tight-binding (TB) model that considers two types of nearest-neighbor hoppings (Fig. 2.10a). The hopping integrals are denoted by $t_\parallel$ for the σ bond and $t_\perp$ for the π bond, where $|t_\parallel| \gg |t_\perp|$. Next nearest-neighbor hoppings may be added to model the mixing between the $p_{\tilde{x}}$ and $p_{\tilde{z}}$ orbitals, which we ignore in this introduction. Under this tight-binding approximation, there are two sets of bands, one from $p_{\tilde{x}}$ and the other from $p_{\tilde{z}}$ [33]

$$E_{p_{\tilde{x}}} = -2t_\parallel \cos\left[\frac{\pi}{2}(k_x + k_z)\right] - 2t_\perp \cos\left[\frac{\pi}{2}(k_x - k_z)\right] - E_F, \tag{2.32}$$

$$E_{p_{\tilde{z}}} = -2t_\parallel \cos\left[\frac{\pi}{2}(k_x - k_z)\right] - 2t_\perp \cos\left[\frac{\pi}{2}(k_x + k_z)\right] - E_F, \tag{2.33}$$

where energy is expressed with respect to E_F, and k_x (k_z) are written in units of π/a (π/c). The Fermi surface arising from these two bands is shown as solid curves in Fig. 2.10b with $t_\parallel = -1.9\,\text{eV}$ and $t_\perp = 0.35\,\text{eV}$, determined from photoemission data [33]. They form a large hole-like pocket around the Brillouin zone (BZ) center and small electron-like pockets at $(k_x, k_z) = (\pm 1, 0)$ and $(0, \pm 1)$. As the unit cell is 45° oriented with respect to the Te–Te bond because of the glide plane symmetry, these main bands are folded at the BZ boundary, shown as dashed curves with corresponding colors as the main bands. In the first BZ, these folded bands are very weak in photoemission intensity as their strength depends on the inter-layer coupling, which is intrinsically small for this quasi-2D crystal [41].

This normal-state electronic structure is almost identical along both k_x and k_z directions, except for the tiny orthorhombicity from the crystal structure. Therefore, one would expect that any instability would also be symmetric between the a- and c-axis. This is indeed the case seen in the electronic susceptibility based on *ab initio* calculation of the band structure [13], where the real part $\chi'(\mathbf{q})$ peaks at $(0, \pm q_{\text{CDW}})$ and $(\pm q_{\text{CDW}}, 0)$ in the a–c plane, where q_{CDW} is the experimentally observed CDW wavevector along the c-axis. This wavevector is indicated on the Fermi surface in Fig. 2.10b. A similar symmetry between the a- and c-axis is observed in the phonon instability at the same wavevector: an example of the unstable phonon dispersion along the c-axis is shown in Fig. 2.4a, and an almost identical dispersion is found along the a-axis [16]. In Chaps. 6 and 7, we will explore the consequence of these two almost equivalent instabilities, which result in a light-induced CDW order out of equilibrium.

It is important to note that the CDW wavevector found in χ' and in the phonon dispersion is parallel to k_x or k_z, as opposed to being rotated by 45°, a direction that better nests the normal-state Fermi surface between the main and folded main bands (Fig. 2.10b). Therefore, as we discussed in Sect. 2.1.1, Fermi surface nesting is not the main driving mechanism for the CDW formation in $R\text{Te}_3$. Instead, states away from the Fermi surface, which contribute to peaks in the real part of the susceptibility χ' but not the imaginary part χ'', and wavevector-dependent electron–phonon coupling are thought to play the leading roles.

In reality, the small orthorhombicity breaks the *a*/*c* symmetry so the CDW forms along the *c*-axis first. The resulting band structure can be calculated by the same interacting model presented in Sect. 2.2.2. In short, we could translate the main bands by the CDW wavevector to form a set of CDW bands, shown as dashed green curves in Fig. 2.10b. As the CDW bands cross the main bands, a hybridization gap of 2Δ opens, a value proportional to the electron–phonon coupling strength. Some spectral weight of the main band is redistributed to the CDW bands, especially near their crossings; example dispersions are shown in Fig. 2.10c. Due to imperfect nesting by the CDW wavevector, the gap opening is not symmetric around E_F. This also results in remnant metallic pockets around $k_z = 0$, and the system remains conducting in the CDW phase. Due to the formation of the CDW gap, the instability along the *a*-axis is weakened by the lack of density of state at E_F. For heavy rare-earth element, a second gap eventually opens at T_{c2} when an orthogonal CDW sets in [42, 43], whose band dispersion will be discussed in Sect. 6.2.1.

2.3.3 CDW Trends Across Rare-Earth Elements

The $R\text{Te}_3$ family displays several trends as one moves from La to Tm: (i) T_{c1} decreases; (ii) T_{c2} first becomes nonzero in Tb and increases afterward (Fig. 2.8b); (iii) the unidirectional CDW wavevector, q_c, increases (Fig. 2.9c); and (iv) the second CDW wavevector, q_a, decreases [44]. In the following, we will explain these trends by considering the decreasing ionic radii of the rare-earth element from La to Tm.

Transition Temperatures

We first discuss the trend in T_{c1}. As atomic radii decrease, the in-plane lattice constants show the same decreasing trend (Fig. 2.9b). Therefore, one would expect the hopping integral $|t_\parallel|$ to be larger due to better overlaps between adjacent Te *p* orbitals. The transverse hopping $|t_\perp|$ would similarly be larger, but its change can be neglected as $|t_\perp/t_\parallel| \ll 1$. Based on the TB band structure, Vorobyev and coworkers computed the electronic susceptibility $\chi'(\mathbf{q}, T)$ using Eq. (2.4) [45]. At each temperature, one could identify the wavevector $\mathbf{q}_{\text{CDW}}$ where $|\chi'|$ has a maximum; this is the CDW wavevector [13]. From Eq. (2.3), the CDW transition occurs at a temperature T_{c1} when $1 + g\chi'(\mathbf{q}_{\text{CDW}}, T_{c1}) = 0$. Recall that $\chi' < 0$ and $|\chi'|$ decreases with increasing temperature (Fig. 2.3b); $g > 0$ is the electron–phonon coupling strength, which is assumed to be constant for this analysis. As $|t_\parallel|$ increases from La to Tm, it is found that $|\chi'(\mathbf{q}_{\text{CDW}})|$ decreases; hence, T_{c1} decreases across the series accordingly.

The decreasing trend of T_{c1} is indicative of a weaker CDW instability as one moves from La to Tm. As expected from mean-field behavior, the associated CDW gap size Δ_1 is found to have the same trend as T_{c1} from both optical and photoemission studies [33, 38]. The trend of T_{c1} also explains the opposite behavior in T_{c2}: as the gap due to the first CDW becomes smaller, more density of states near E_F is available for the formation of the second CDW along the a-axis; hence, the second order strengthens from Tb to Tm.

CDW Wavevectors

We start by examining the wavevector of the first CDW, q_c. From the same susceptibility calculation using the TB model, one could track how the peak position in $\chi'(\mathbf{q})$ shifts as a function of $t_\parallel$. It was found that q_c increases with $t_\parallel$, in good agreement with experimentally determined q_c and $t_\parallel$ [45]. Another perspective comes from the conservation of occupied electrons before and after the CDW gap opening. This is an important consideration because the gap does not open symmetrically around E_F and part of the Fermi surface remains ungapped. This conservation law may be approximated by the condition that the $k_x = 0$ crossing between the main band and the CDW band takes place at an energy $-\Delta_1$ relative to Fermi level [33]. Using the TB dispersions in Eqs. (2.32)–(2.33), this condition translates to

$$q_c = 1 - \frac{2}{\pi} \arccos\left[\frac{\Delta_1 - E_F}{2\left(t_\parallel + t_\perp\right)}\right]. \tag{2.34}$$

Therefore, from La to Tm, Δ_1 decreases, and hence q_c increases. In this case, the dominant contribution comes from changes in Δ_1 as T_{c1} is reduced by almost a factor of 3 from La to Tm; variations in $t_\parallel$ and $t_\perp$ only serve as small corrections.

We now turn to q_a, which has an opposite trend compared to q_c [44]. By the same argument of an anti-correlation between q_c and T_{c1} shown in Eq. (2.34), one would expect a similar anti-correlation between q_a and T_{c2} if subtle details of Fermi surface reconstruction below T_{c1} are neglected. In addition, it is observed that $q_a \approx 1/3a^*$, which is slightly larger than $q_c \approx 2/7c^*$ [30]. This slight difference between q_a and q_c in the same $R\mathrm{Te}_3$ compound is indicative of the Fermi surface anisotropy introduced by the opening of the c-axis CDW gap. Specifically, when the second CDW forms along the a-axis, the anisotropy results in $q_a > q_c$. As we move from heavier to lighter rare-earth element (Tm to La), one would expect the anisotropy to grow due to an increasing c-axis CDW gap. Correspondingly, one would expect q_a to increase from Tm to La as well.

2.4 Tantalum Disulfide $1T$-TaS_2

$1T$-TaS_2 belongs to a group of quasi-2D transition metal dichalcogenides (TMDs) that display one or more CDW transitions. They had been extensively studied by the 1970s [46], but the past decade has witnessed a resurgence of interest in $1T$-TaS_2

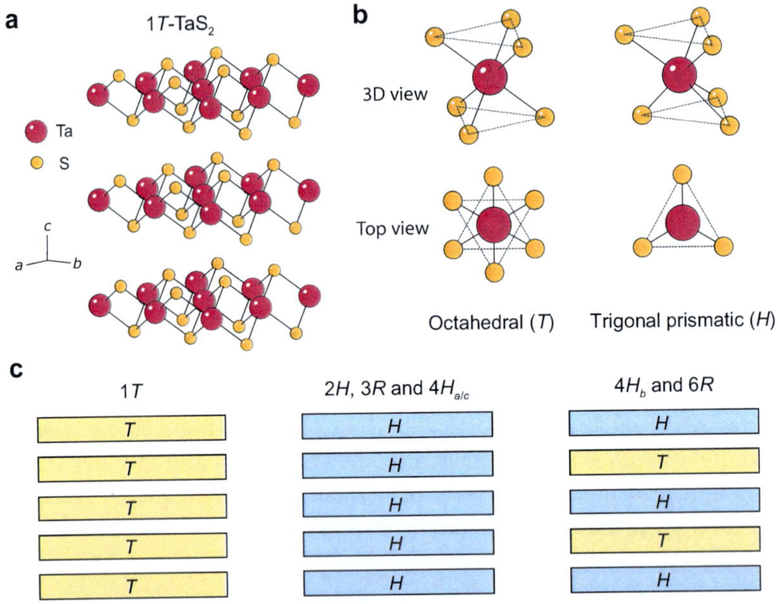

Fig. 2.11 Lattice structure of $1T$-TaS_2 and associated polytypes. (**a**) Crystal structure of $1T$-TaS_2. Neighboring layers are bonded by weak van der Waals forces. (**b**) Schematic of two types of coordinations between Ta and S atoms. (**c**) Several examples of polytypes of TaS_2 and $TaSe_2$, showing pure octahedral coordination, pure trigonal prismatic coordination, or a mixture of both [46]. The $\sqrt{13} \times \sqrt{13}$ CDW superlattice originates from the T layer [61–63]

because it possesses a complex phase diagram as one tunes temperature, pressure [12], doping [47, 48], layer number [49, 50], or if one applies electric or sub-picosecond optical pulses [51–56]. In addition to CDWs, $1T$-TaS_2 hosts a Mott insulating ground state [57], superconductivity [12] (Fig. 2.2d), a hidden metallic state [51] (Fig. 1.6a), or putative orbital textures [58], a quantum spin liquid state [59], and Wigner crystallization [60].

In this section, I will first introduce the CDWs in $1T$-TaS_2 and highlight their planar chiral property. As discussed in Chap. 8, an ultrafast laser pulse can reversibly change the sense of chirality. I will conclude by some comments about its fascinating ground state.

2.4.1 Multiple CDW Transitions

In the normal state, $1T$-TaS_2 has a trigonal structure with space group $P\bar{3}m1$ (No. 164) [64]. A schematic of the crystal structure is drawn in Fig. 2.11a, showing layers bonded by weak van der Waals forces. Like many other TMDs, there are two types of coordinations between the Ta and S atoms: octahedral (T) and

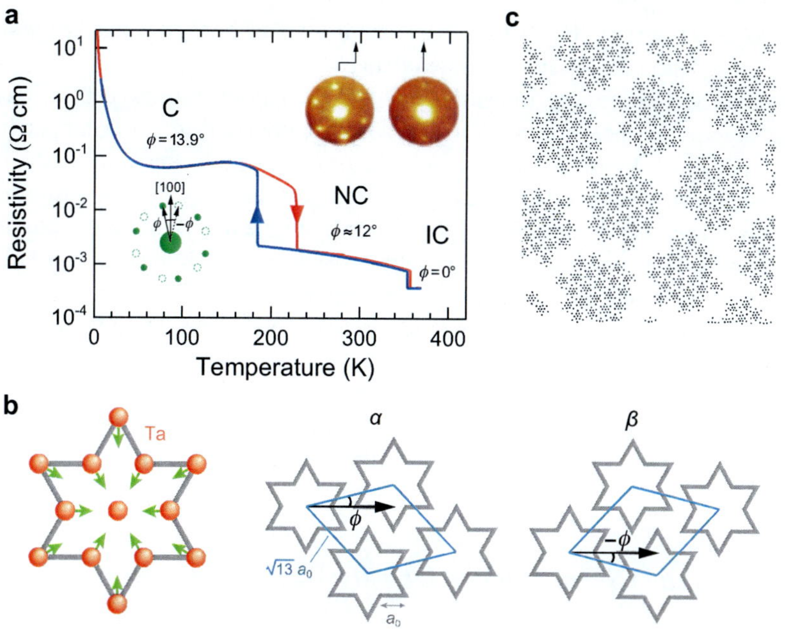

Fig. 2.12 Structural phases of $1T$-TaS$_2$ and its two mirror-symmetric commensurate states. (**a**) Temperature-dependent resistivity measured during cool-down (blue) and warm-up (red), showing hysteretic discontinuities as $1T$-TaS$_2$ transitions through incommensurate (IC), nearly commensurate (NC), and commensurate (C) CDW phases. The triclinic phase during warm-up (223 to 283 K) is omitted. *Lower inset*, schematic diffraction pattern in CDW phases, with a central Bragg peak surrounded by six first-order superlattice peaks. Filled (dashed) circles represent α (β) orientation in the C or NC phase. Upper inset, static diffraction patterns of the (2 0 0) Bragg and satellite peaks taken at 295 and 370 K. Only the α domain is present throughout the sample. (**b**) Schematics of in-plane atomic arrangements in commensurate regions for α or β state that breaks the in-plane mirror symmetry. Orange spheres denote Ta atoms, which form clusters of regular hexagrams. Blue diamonds represent unit cells of the $\sqrt{13} \times \sqrt{13}$ superlattice. (**c**) Calculated positions of Ta atoms in a particular layer in the NC phase, based on X-ray refinement data. Only atoms belonging to complete hexagrams are plotted, while regions with a domain-wall-like structure are shown as blank. Each cluster has a diameter of approximately 5 hexagrams. (Adapted with permission from Ref. [64] (**c**))

trigonal prismatic (H). The $1T$ polytype has every layer following an octahedral coordination (Fig. 2.11b, c). The $2H$ form is the stable polytype at room temperature [46], so the $1T$ polytype is usually prepared by a fast quench from a very high temperature. We will consider the effect of this fast quench when we study photoinduced phenomena in Chap. 8.

Figure 2.12a illustrates the series of CDW transitions in resistivity and diffraction measurements. An incommensurate CDW (IC) first occurs below $T_{\mathrm{IC}} = 550\,\mathrm{K}$. As $1T$-TaS$_2$ is further cooled, a weakly first-order transition to a nearly commensurate CDW (NC) occurs at $T_{\mathrm{IC\text{-}NC}} = 354\,\mathrm{K}$. In this transition, the previously incommensurate superlattice forms small patches of commensurate hexagrams with a

$\sqrt{13} \times \sqrt{13}$ periodicity (Fig. 2.12b). An example of such hexagram patches is shown in Fig. 2.12c based on X-ray refinement data [64].

In reciprocal space, the transition into the NC phase is marked by a rotation (ϕ) of superlattice peaks away from the [100] direction (Fig. 2.12a, upper inset). Either counterclockwise (α) or clockwise (β) rotation can occur (Fig. 2.12a, lower inset), depending on the relative hexagram arrangements (Fig. 2.12b). Thus, the NC phase spontaneously breaks the in-plane mirror symmetry, yielding the possibility of two equivalent domains. Cooling below $T_{\text{NC-C}} = 184$ K results in a strongly first-order transition to a Mott insulating commensurate CDW (C) phase, where hexagram clusters inherit their orientation from the NC phase. Despite the apparent equivalence between α and β configurations, only one orientation is typically observed for the entire sample of macroscopic size [55]. This observation may be attributed to an extrinsic lattice strain field that lifts the degeneracy between the two orientations and we will comment on it further in the next section.

2.4.2 *Planar Chiral Charge Density Wave*

The α and β orientations in the NC and C phase of $1T$-TaS_2 are formally known as a pair of planar chiral structures [65], which cannot be mapped into one another through translation and proper rotation in 2D (Fig. 2.13). They should be distinguished from a 3D chiral structure, also referred to as enantiomerism, which is present in other CDWs [66].

Despite the degeneracy between α and β domains in $1T$-TaS_2, it is intriguing to note that in an equilibrium IC-to-NC phase transition, only one orientation is typically observed [46, 55, 67, 68]. Only under special circumstances, such as in de-intercalated potassium-doped $1T$-TaS_2 [69], two domains were spotted, probably due to crystallographic defects. In a closely related compound, $1T$-$TaSe_2$, α/β boundaries are less rare [46, 67, 70]. The structural properties of $1T$-$TaSe_2$ and $1T$-TaS_2 are almost identical, except that the former lacks the NC phase and its

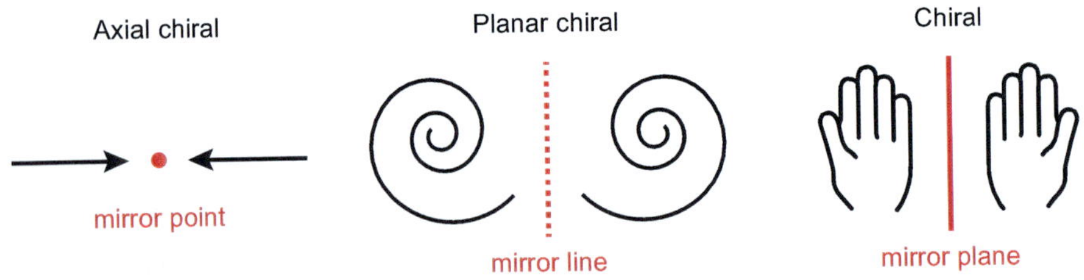

Fig. 2.13 Chiral objects in different dimensions. An arrow, a spiral, and a hand are chiral in 1D, 2D, and 3D, respectively. An object is chiral in N-dimension if translation and proper rotation within the dimension cannot map the object to its mirror image. Note that if an object is chiral in N-dimension, it is no longer chiral in M-dimension for $M > N$ [65]

transition temperature from the IC to the C phase, $T_{IC\text{-}C}$, is much higher at 473 K [46]. A systematic study is required to compare the average size of mirror domains in $1T$-$TaSe_2$ and $1T$-TaS_2, as it is known that α/β domain walls may form due to lattice impurities [71].

By contrast, coexisting α/β domains are routinely observed in other polytypes such as $4H_b$- and $6R$-TaS_2 or $TaSe_2$ [46, 61, 72]. In these polytypes, octahedrally coordinated (T) layers are separated by trigonal prismatic (H) layers (Fig. 2.11c). Since the $\sqrt{13} \times \sqrt{13}$ superlattice originates in the T layers [61–63], it is suggested that the large separation between adjacent T layers allows the stacking of both α and β domains with minimal energy penalty, giving rise to inter-layer α/β domain walls [46, 61]. Without the presence of H layers, the T–T coupling is stronger in the $1T$ polytype. This may explain the rare coexistence of both planar chiral domains in $1T$-TaS_2 due to possibly the high energy cost.

2.4.3 Multiple Facets of the Ground State

The commensurate CDW ground state in $1T$-TaS_2 is accompanied by a Mott insulating phase, setting it apart from many other TMDs that also possess CDWs. In this section, I will briefly explain the reason behind the Mott gap formation and remark on its consequences.

We can employ the same interacting TB model outlined in Sect. 2.2.2 to approximate the ground state band structure. Figure 2.14a shows an example, where the Ta $5d$ orbital dominates the electronic state near E_F. Starting from the $5d$ band in the normal state [panel (i)], we can first translate it by the CDW wavevectors corresponding to the $\sqrt{13} \times \sqrt{13}$ superstructure [panel (ii)]. The resulting bands are complicated, though we learned from Sect. 2.2.2 that spectral weights of the CDW bands concentrate around their crossings with the original band, where CDW gaps form. In the TB calculation shown in Fig. 2.14a, hopping integrals were also adjusted to reflect the changing bond lengths due to the formation of hexagrams; spin–orbit coupling (SOC) was further added to split the Ta $5d$ orbitals [57, 74]. The resulting band in the commensurate CDW phase is shown in panel (iii), where a flat band (width $\approx$ 80 meV) with a half-filling appears at the Fermi level. This narrow band is susceptible to a Mott–Hubbard transition, leading to lower and upper Hubbard bands in the ground state.

As discussed in Sect. 1.2.2, this Mott gap is rather fragile and is closely tied to the formation of the CDW hexagrams, leading to a variety of metallic states when it is subjected to perturbations such as an optical or electrical pulse. An example of such a metastable state is shown in Fig. 2.14b. A voltage pulse of 100 ms in duration and $\geq$2.0 V in amplitude from an STM tip results in a metallic textured phase at 4.3 K, which features meandering domain walls in an originally pristine C phase. Zooming into one such boundary (Fig. 2.14c), one could see that the relative phase of the hexagrams changes abruptly across the domain wall, even though the planar chirality remains intact. A careful spectroscopy scan from this pulse-induced

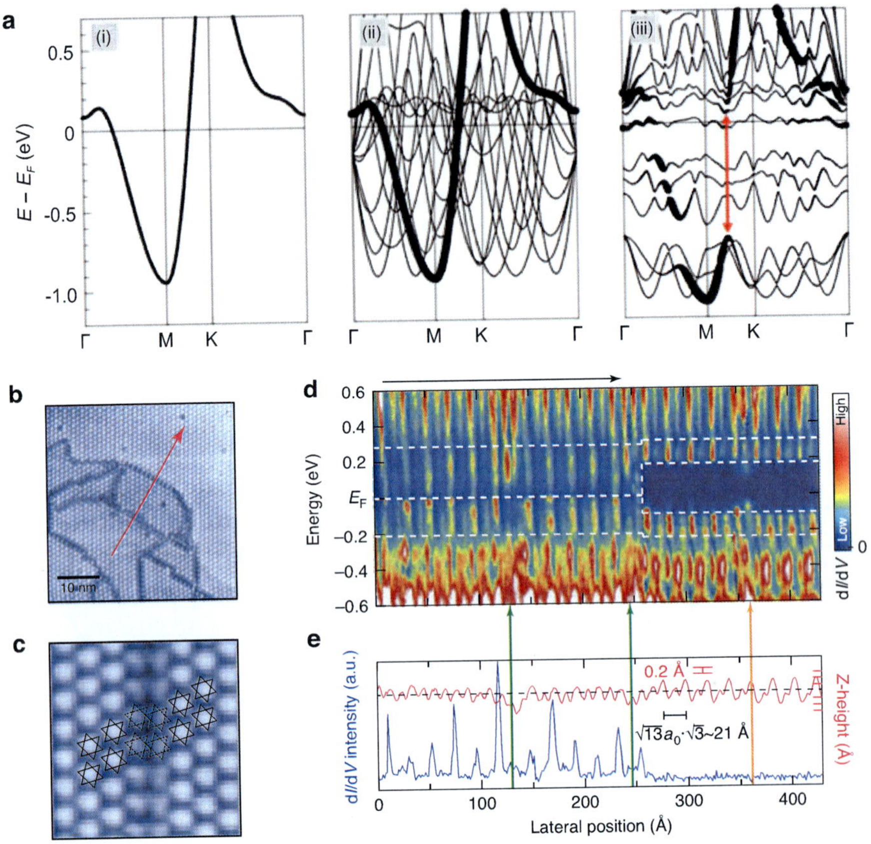

Fig. 2.14 Mott insulating ground state of 1*T*-TaS$_2$. (**a**) Electronic structure of 1*T*-TaS$_2$ in the commensurate CDW state via a tight-binding (TB) calculation [73, 74]. (i) Original Ta 5*d* band without the $\sqrt{13} \times \sqrt{13}$ superlattice; (ii) reconstructed bands after translating the original band (thick black) by the CDW wavevectors; (iii) reconstructed bands after gap opening, adopting the interacting model in Sect. 2.2.2, including the effects of bond strengthening within the hexagram and Ta spin–orbit coupling (SOC). A flat band near E_F appears. Thickness corresponds to spectral weight. Red arrow marks a pronounced gap in the spectral weight distribution. (**b**) STM map of a commensurate CDW region (*upper right*) and a neighboring patch (*lower left*) created by a voltage pulse from the tip. (**c**) Zoomed-in view of a domain wall, showing the phase shift of hexagrams from the neighboring domains. (**d**) Differential conductance dI/dV map along the arrow in (**b**). Dashed lines highlight the broadening of the Hubbard states near E_F. (**e**), STM topographic line profile (magenta) and the dI/dV intensity at E_F (blue) as a function of lateral position. Two domain walls and a point defect are marked by green and orange arrows, respectively. (Adapted with permission from Ref. [57], IOP Publishing (**a**); Ref. [52] (**b–e**))

texture to the pristine C phase indicates significant broadening of the upper and lower Hubbard bands in the textured region (Fig. 2.14d, e), leading to a finite density of states at E_F. This local collapse of the Mott gap may explain various types of pulse-induced hidden metallicity in 1*T*-TaS$_2$.

Even without using an intense pulse, one could expect that the fragile Mott gap in $1T$-TaS_2 may be weakened whenever the commensurate CDW order is disrupted, for example, at a domain wall [75]. It has been found by STM that the Mott gap indeed disappears at the domain wall of the superlattice. In addition, a zigzag pattern of electrons emerges along the boundary, forming a putative 1D Wigner crystal [60]. This electronic crystallization is further evidenced by the characteristic telegraph noise pattern, which may indicate charge or spin fluctuations induced by the tunneling electrons. If verified, $1T$-TaS_2 may provide an attractive platform to study Wigner crystallization at more experimentally accessible temperatures—liquid ^{4}He temperature as opposed to milli-Kelvin conditions typically required for semiconductors [60].

Another dimension of the ground state concerns the possibility of $1T$-TaS_2 being a candidate quantum spin liquid (QSL) in its Mott insulating phase [59]. When the inter-layer coupling is small, the localized single electron per hexagram resides in the lower Hubbard band, forming a triangular lattice of spin-$1/2$. However, there is no evidence of any long-range spin ordering down to the tens of mK regime [76, 77]. This paramagnetic ground state with an odd number of electrons per unit cell is a tantalizing sign for a QSL, though more experimental work is needed to provide evidence for long-range entanglement and fractionalized spin excitations [78, 79].

References

1. Fröhlich, H.: On the theory of superconductivity: the one-dimensional case. Proc. R. Soc. Lond. A **223**, 296–305 (1954)
2. Bardeen, J., Cooper, L.N., Schrieffer, J.R.: Theory of superconductivity. Phys. Rev. **108**, 1175–1204 (1957)
3. Grüner, G.: Density Waves in Solids. Addison-Wesley, Boston (1994)
4. Grüner, G.: The dynamics of charge-density waves. Rev. Mod. Phys. **60**, 1129–1181 (1988)
5. Thorne, R.E., Lyons, W.G., Lyding, J.W., Tucker, J.R., Bardeen, J.: Charge-density-wave transport in quasi-one-dimensional conductors. I. Current oscillations. Phys. Rev. B **35**, 6348–6359 (1987)
6. Zybtsev, S.G., Pokrovskii, V.Y., Zaitsev-Zotov, S.V.: 'Quantized' states of the charge-density wave in microcrystals of $K_{0.3}MoO_3$. Nat. Commun. **1**, 85 (2010)
7. Kohn, W., Sherrington, D.: Two kinds of bosons and Bose condensates. Rev. Mod. Phys. **42**, 1–11 (1970)
8. Yang, C.N.: Concept of off-diagonal long-range order and the quantum phases of liquid He and of superconductors. Rev. Mod. Phys. **34**, 694–704 (1962)
9. Kogar, A., de la Pena, G.A., Lee, S., Fang, Y., Sun, S.X.-L., et al.: Observation of a charge density wave incommensuration near the superconducting dome in Cu_xTiSe_2. Phys. Rev. Lett. **118**, 027002 (2017)
10. Straquadine, J.A.W., Weber, F., Rosenkranz, S., Said, A.H., Fisher, I.R.: Suppression of charge density wave order by disorder in Pd-intercalated $ErTe_3$. Phys. Rev. B **99**, 235138 (2019)
11. Xi, X., Zhao, L., Wang, Z., Berger, H., Forró, L., et al.: Strongly enhanced charge-density-wave order in monolayer $NbSe_2$. Nat. Nanotechnol. **10**, 765–769 (2015)
12. Sipos, B., Kusmartseva, A.F., Akrap, A., Berger, H., Forró, L., et al.: From Mott state to superconductivity in $1T$-TaS_2. Nat. Mater. **7**, 960–965 (2008)

13. Johannes, M.D., Mazin, I.I.: Fermi surface nesting and the origin of charge density waves in metals. Phys. Rev. B **77**, 165135 (2008)
14. Zhu, X., Cao, Y., Zhang, J., Plummer, E.W., Guo, J.: Classification of charge density waves based on their nature. Proc. Natl. Acad. Sci. U. S. A. **112**, 2367–2371 (2015)
15. Khomskii, D.I.: Charge ordering in transition metal compounds. In: Transition Metal Compounds. Cambridge University Press, Cambridge (2014)
16. Maschek, M., Zocco, D.A., Rosenkranz, S., Heid, R., Said, A.H., et al.: Competing soft phonon modes at the charge-density-wave transitions in $DyTe_3$. Phys. Rev. B **98**, 094304 (2018)
17. Kogar, A., Rak, M.S., Vig, S., Husain, A.A., Flicker, F., et al.: Signatures of exciton condensation in a transition metal dichalcogenide. Science **358**, 1314–1317 (2017)
18. Comin, R., Damascelli, A.: Resonant X-ray scattering studies of charge order in cuprates. Annu. Rev. Condens. Matter Phys. **7**, 369–405 (2016)
19. Frano, A., Blanco-Canosa, S., Keimer, B., Birgeneau, R.J.: Charge ordering in superconducting copper oxides. J. Phys.: Condens. Matter **32**, 374005 (2020)
20. Blackburn, E., Chang, J., Said, A.H., Leu, B.M., Liang, R., et al.: Inelastic x-ray study of phonon broadening and charge-density wave formation in ortho-II-ordered $YBa_2Cu_3O_{6.54}$. Phys. Rev. B **88**, 054506 (2013)
21. Le Tacon, M., Bosak, A., Souliou, S.M., Dellea, G., Loew, T., et al.: Inelastic X-ray scattering in $YBa_2Cu_3O_{6.6}$ reveals giant phonon anomalies and elastic central peak due to charge-density-wave formation. Nat. Phys. **10**, 52–58 (2014)
22. Miao, H., Ishikawa, D., Heid, R., Le Tacon, M., Fabbris, G., et al.: Incommensurate phonon anomaly and the nature of charge density waves in cuprates. Phys. Rev. X **8**, 011008 (2018)
23. He, Y., Wu, S., Song, Y., Lee, W.-S., Said, A.H., et al.: Persistent low-energy phonon broadening near the charge-order q vector in the bilayer cuprate $Bi_2Sr_2CaCu_2O_{8+\delta}$. Phys. Rev. B **98**, 035102 (2018)
24. Jérome, D., Rice, T.M., Kohn, W.: Excitonic insulator. Phys. Rev. **158**, 462–475 (1967)
25. Rohwer, T., Hellmann, S., Wiesenmayer, M., Sohrt, C., Stange, A., et al.: Collapse of long-range charge order tracked by time-resolved photoemission at high momenta. Nature **471**, 490–493 (2011)
26. Hellmann, S., Rohwer, T., Kalläne, M., Hanff, K., Sohrt, C., et al.: Time-domain classification of charge-density-wave insulators. Nat. Commun. **3**, 1069 (2012)
27. Thorne, R.E.: Charge-density-wave conductors. Phys. Today **49**, 42–47 (1996)
28. Schmitt, F., Kirchmann, P.S., Bovensiepen, U., Moore, R.G., Rettig, L., et al.: Transient electronic structure and melting of a charge density wave in $TbTe_3$. Science **321**, 1649–1652 (2008)
29. Torchinsky, D.H., Mahmood, F., Bollinger, A.T., Božović, I., Gedik, N.: Fluctuating charge-density waves in a cuprate superconductor. Nat. Mater. **12**, 387–391 (2013)
30. Ru, N., Condron, C.L., Margulis, G.Y., Shin, K.Y., Laverock, J., et al.: Effect of chemical pressure on the charge density wave transition in rare-earth tritellurides $R\mathrm{Te}_3$. Phys. Rev. B **77**, 035114 (2008)
31. Dolgirev, P.E., Rozhkov, A.V., Zong, A., Kogar, A., Gedik, N., et al.: Amplitude dynamics of charge density wave in $LaTe_3$: theoretical description of pump-probe experiments. Phys. Rev. B **101**, 054203 (2020)
32. Voit, J., Perfetti, L., Zwick, F., Berger, H., Margaritondo, G., et al.: Electronic structure of solids with competing periodic potentials. Science **290**, 501–503 (2000)
33. Brouet, V., Yang, W.L., Zhou, X.J., Hussain, Z., Moore, R.G., et al.: Angle-resolved photoemission study of the evolution of band structure and charge density wave properties in $R\mathrm{Te}_3$ (R = Y, La, Ce, Sm, Gd, Tb, and Dy). Phys. Rev. B **77**, 235104 (2008)
34. Rettig, L., Cortés, R., Chu, J.-H., Fisher, I.R., Schmitt, F., et al.: Persistent order due to transiently enhanced nesting in an electronically excited charge density wave. Nat. Commun. **7**, 10459 (2016)
35. Ru, N.: Charge Density Wave Formation in Rare-Earth Tellurides. Ph.D. thesis, Stanford University, Stanford (2008)

36. Banerjee, A., Feng, Y., Silevitch, D.M., Wang, J., Lang, J.C., et al.: Charge transfer and multiple density waves in the rare earth tellurides. Phys. Rev. B **87**, 155131 (2013)
37. Malliakas, C.D., Kanatzidis, M.G.: Divergence in the behavior of the charge density wave in $RETe_3$ (RE = Rare-Earth element) with temperature and RE element. J. Am. Chem. Soc. **128**, 12612–12613 (2006)
38. Hu, B.F., Cheng, B., Yuan, R.H., Dong, T., Wang, N.L.: Coexistence and competition of multiple charge-density-wave orders in rare-earth tritellurides. Phys. Rev. B **90**, 085105 (2014)
39. Norling, B.K., Steinfink, H.: The crystal structure of neodymium tritelluride. Inorg. Chem. **5**, 1488–1491 (1966)
40. Straquadine, J.A.W., Ikeda, M.S., Fisher, I.R.: Evidence for realignment of the charge density wave state in $ErTe_3$ and $TmTe_3$ under uniaxial stress via elastocaloric and elastoresistivity measurements. arXiv 2005.10461 (2020)
41. Brouet, V., Yang, W.L., Zhou, X.J., Hussain, Z., Ru, N. et al.: Fermi surface reconstruction in the CDW state of $CeTe_3$ observed by photoemission. Phys. Rev. Lett. **93**, 126405 (2004)
42. Moore, R.G., Brouet, V., He, R., Lu, D.H., Ru, N., et al.: Fermi surface evolution across multiple charge density wave transitions in $ErTe_3$. Phys. Rev. B **81**, 073102 (2010)
43. Lee, C., Rohwer, T., Sie, E.J., Zong, A., Baldini, E., et al.: High resolution time- and angle-resolved photoemission spectroscopy with 11 eV laser pulses. Rev. Sci. Instrum. **91**, 043102 (2020)
44. Kogar, A., Zong, A., Dolgirev, P.E., Shen, X., Straquadine, J., et al.: Light-induced charge density wave in $LaTe_3$. Nat. Phys. **16**, 159–163 (2020)
45. Vorobyev, P.A., Grigoriev, P.D., Kesharpu, K.K., Khovaylo, V.V.: The evolution of electron dispersion in the series of rare-earth tritelluride compounds obtained from their charge-density-wave properties and susceptibility calculations. Materials **12**, 2264 (2019)
46. Wilson, J.A., Di Salvo, F.J., Mahajan, S.: Charge-density waves and superlattices in the metallic layered transition metal dichalcogenides. Adv. Phys. **24**, 117–201 (1975)
47. Ang, R., Tanaka, Y., Ieki, E., Nakayama, K., Sato, T., et al.: Real-space coexistence of the melted Mott state and superconductivity in Fe-substituted $1T$-TaS_2. Phys. Rev. Lett. **109**, 176403 (2012)
48. Li, L.J., Lu, W.J., Zhu, X.D., Ling, L.S., Qu, Z., et al.: Fe-doping-induced superconductivity in the charge-density-wave system $1T$-TaS_2. EPL **97**, 67005 (2012)
49. Yoshida, M., Suzuki, R., Zhang, Y., Nakano, M., Iwasa, Y.: Memristive phase switching in two-dimensional $1T$-TaS_2 crystals. Sci. Adv. **1**, e1500606 (2015)
50. Tsen, A.W., Hovden, R., Wang, D., Kim, Y.D., Okamoto, J., et al.: Structure and control of charge density waves in two-dimensional $1T$-TaS_2. Proc. Natl. Acad. Sci. U. S. A. **112**, 15054–15059 (2015)
51. Stojchevska, L., Vaskivskyi, I., Mertelj, T., Kusar, P., Svetin, D., et al.: Ultrafast switching to a stable hidden quantum state in an electronic crystal. Science **344**, 177–180 (2014)
52. Cho, D., Cheon, S., Kim, K.-S., Lee, S.-H., Cho, Y.-H., et al.: Nanoscale manipulation of the Mott insulating state coupled to charge order in $1T$-TaS_2. Nat. Commun. **7**, 10453 (2016)
53. Ma, L., Ye, C., Yu, Y., Lu, X.F., Niu, X., et al.: A metallic mosaic phase and the origin of Mott-insulating state in $1T$-TaS_2. Nat. Commun. **7**, 10956 (2016)
54. Vaskivskyi, I., Mihailovic, I.A., Brazovskii, S., Gospodaric, J., Mertelj, T., et al.: Fast electronic resistance switching involving hidden charge density wave states. Nat. Commun. **7**, 11442 (2016)
55. Zong, A., Shen, X., Kogar, A., Ye, L., Marks, C., et al.: Ultrafast manipulation of mirror domain walls in a charge density wave. Sci. Adv. **4**, eaau5501 (2018)
56. Gerasimenko, Y.A., Vaskivskyi, I., Litskevich, M., Ravnik, J., Vodeb, J., et al.: Quantum jamming transition to a correlated electron glass in $1T$-TaS_2. Nat. Mater. **18**, 1078–1083 (2019)
57. Rossnagel, K.: On the origin of charge-density waves in select layered transition-metal dichalcogenides. J. Phys. Condens. Matter **23**, 213001 (2011)
58. Ritschel, T., Trinckauf, J., Koepernik, K., Bü, Zimmermann, M.V., et al.: Orbital textures and charge density waves in transition metal dichalcogenides. Nat. Phys. **11**, 328–331 (2015)

59. Law, K.T., Lee, P.A.: $1T$-TaS_2 as a quantum spin liquid. Proc. Natl. Acad. Sci. U. S. A. **114**, 6996–7000 (2017)
60. Aishwarya, A., Howard, S., Padhi, B., Wang, L., Cheong, S.-W., et al.: Visualizing 1D zigzag Wigner crystallization at domain walls in the Mott insulator TaS_2. arXiv 1906.11983 (2019)
61. Lüdecke, J., van Smaalen, S., Spijkerman, A., de Boer, J.L., Wiegers, G.A.: Commensurately modulated structure of $4H_b$-$TaSe_2$ determined by x-ray crystal-structure refinement. Phys. Rev. B **59**, 6063–6071 (1999)
62. Ekvall, I., Kim, J.J., Olin, H.: Atomic and electronic structures of the two different layers in $4H_b$-TaS_2 at 4.2 K. Phys. Rev. B **55**, 6758–6761 (1997)
63. Kim, J.-J., Olin, H.: Atomic- and electronic-structure study on the layers of $4H_b$-TaS_2 prepared by a layer-by-layer etching technique. Phys. Rev. B **52**, R14388–R14391 (1995)
64. Spijkerman, A., de Boer, J.L., Meetsma, A., Wiegers, G.A., van Smaalen, S.: X-ray crystal-structure refinement of the nearly commensurate phase of $1T$-TaS_2 in $(3+2)$-dimensional superspace. Phys. Rev. B **56**, 13757–13767 (1997)
65. Arnaut, L.R.: Chirality in multi-dimensional space with application to electromagnetic characterisation of multi-dimensional chiral and semi-chiral media. J. Electromagn. Waves Appl. **11**, 1459–1482 (1997)
66. Xu, S.-Y., Ma, Q., Gao, Y., Kogar, A., Zong, A., et al.: Spontaneous gyrotropic electronic order in a transition-metal dichalcogenide. Nature **578**, 545–549 (2020)
67. Bovet, M., Popović, D., Clerc, F., Koitzsch, C., Probst, U., et al.: Pseudogapped Fermi surfaces of $1T$-TaS_2 and $1T$-$TaSe_2$: a charge density wave effect. Phys. Rev. B **69**, 125117 (2004)
68. Shiba, H., Nakanishi, K.: Phenomenological Landau theory of charge density wave phase transitions in layered compounds. In: Motizuki, K. (ed.) Structural Phase Transitions in Layered Transition Metal Compounds, pp. 175–266. D. Reidel, Dordrecht (1986)
69. Williams, P.M.: Phase transitions and charge density waves in the layered transition metal dichalcogenides. In: Lévy, F. (ed.) Crystallography and Crystal Chemistry of Materials with Layered Structures, vol. 37, pp. 51–92. Springer, Dordrecht (1976)
70. Brouwer, R., Jellinek, F.: The low-temperature superstructures of $1T$-$TaSe_2$ and $2H$-$TaSe_2$. Physica B+C **99**, 51–55 (1980)
71. Wu, X.L., Lieber, C.M.: Direct characterization of charge-density-wave defects in titanium-doped $TaSe_2$ by scanning tunneling microscopy. Phys. Rev. B **41**, 1239–1242 (1990)
72. Fung, K.K., Steeds, J.W., Eades, J.A.: Application of convergent beam electron diffraction to study the stacking of layers in transition-metal dichalcogenides. Physica B+C **99**, 47–50 (1980)
73. Smith, N.V., Kevan, S.D., DiSalvo, F.J.: Band structures of the layer compounds $1T$-TaS_2 and $2H$-$TaSe_2$ in the presence of commensurate charge-density waves. J. Phys. C: Solid State Phys. **18**, 3175–3189 (1985)
74. Rossnagel, K., Smith, N.V.: Spin-orbit coupling in the band structure of reconstructed $1T$-TaS_2. Phys. Rev. B **73**, 073106 (2006)
75. Ma, E.Y., Cui, Y.-T., Ueda, K., Tang, S., Chen, K., et al.: Mobile metallic domain walls in an all-in-all-out magnetic insulator. Science **350**, 538–541 (2015)
76. Klanjšek, M., Zorko, A., Žitko, R., Mravlje, J., Jagličić, Z., et al.: A high-temperature quantum spin liquid with polaron spins. Nat. Phys. **13**, 1130–1134 (2017)
77. Ribak, A., Silber, I., Baines, C., Chashka, K., Salman, Z., et al.: Gapless excitations in the ground state of $1T$-TaS_2. Phys. Rev. B **96**, 195131 (2017)
78. Alexandradinata, A., Armitage, N.P., Baydin, A., Bi, W., Cao, Y., et al.: The future of the correlated electron problem. arXiv 2010.00584 (2020)
79. Wen, J., Yu, S.-L., Li, S., Yu, W., Li, J.-X.: Experimental identification of quantum spin liquids. npj Quantum Mater. **4**, 12 (2019)

Chapter 3
Ultrafast Electron Diffraction

Abstract Integral to the exploration of nonequilibrium phenomena in solid state systems is the measurement of lattice motion after photoexcitation by a femtosecond laser pulse. For the past two decades, ultrafast electron diffraction (UED) has played a critical role in the field of photoinduced structural dynamics, including the study of non-adiabatic phase transitions, metastable states, and the electron–phonon interaction in correlated systems. This chapter introduces the fundamentals of the UED technique—one of the principal tools employed in this dissertation—accompanied by an overview of recent advances in instrumentation that significantly improve the temporal and momentum resolutions. The design, construction, and characterization of a table-top keV UED setup are discussed in detail. The quality of UED data in the transmission geometry critically depends on fabrication of single-crystalline thin films with a large lateral dimension, and three state-of-the-art fabrication techniques are presented. The chapter is concluded by the data analysis protocol used throughout the dissertation.

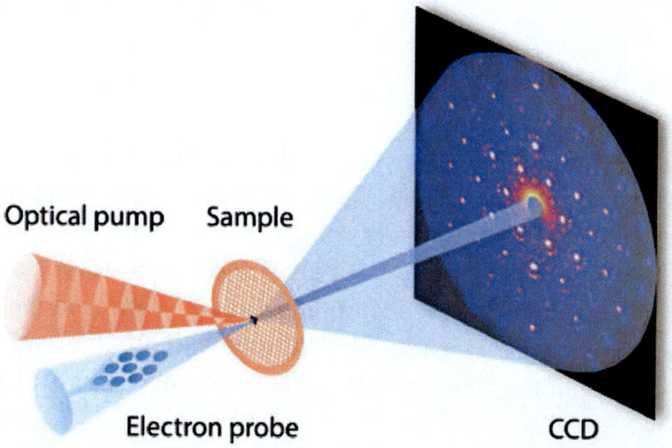

A schematic of an ultrafast electron diffraction setup in the transmission geometry

A. Zong, *Emergent States in Photoinduced Charge-Density-Wave Transitions*, Springer Theses, https://doi.org/10.1007/978-3-030-81751-0_3

3.1 Overview

Our ability to examine the nanoscopic world was greatly advanced by the discovery of electron diffraction in 1927 [1], 15 years after the demonstration of X-ray diffraction by Max von Laue [2]. Nowadays, the addition of a time axis for both diffraction techniques has enabled scientists to study dynamics of atoms on femtosecond timescales. Both electrons and X-rays have short wavelengths, which make it possible to discern atomic positions with sub-angstrom precision. However, there are some important differences, which offer ultrafast electron diffraction (UED) unique advantages over time-resolved X-ray diffraction (tr-XRD). In the following introduction of UED, I will highlight its key characteristics by comparing it to the closely related technique of tr-XRD. I will also review frontiers of UED instrumentation that have propelled the rapid progress in this field for the past five years.

3.1.1 *Scattering Cross Section*

Compared to X-rays, electrons have a much larger scattering cross section (10^5 to 10^6) [3]. Therefore, to obtain similar scattering statistics, it is more feasible to construct table-top diffraction setups with a low-flux electron beam as opposed to facility-based X-rays sources based on a free-electron laser (FEL). We can understand the difference in scattering rates through a back-of-the-envelope estimate of the cross section ratio. For X-rays, adopting the Thomson scattering model, the scattering happens when electrons in the solid oscillate in response to the incoming X-ray field and emit dipole radiation as a result. Mathematically, the Hamiltonian for a single non-relativistic electron interacting with an electromagnetic field is

$$H = \frac{1}{2m}\left(\mathbf{p} - \frac{e}{c}\mathbf{A}\right)^2, \tag{3.1}$$

where e is the electronic charge, m is the electron mass, c is the speed of light, and $\mathbf{A}$ is the vector potential for the X-ray. We can expand $\mathbf{A}$ in terms of photon creation and annihilation operators $\mathbf{A}(\mathbf{r}) \propto a_1\hat{a}^\dagger + a_2\hat{a}$ for some coefficients a_1 and a_2. Expanding the $(\cdot)^2$ term in Eq. (3.1) gives three terms associated with $\mathbf{A}$: the linear terms in $\mathbf{A}$ represent photon absorption and emission, and the quadratic term represents scattering, which is associated with the annihilation and the subsequent creation of an X-ray photon. The scattering matrix element $f_{\text{X-ray}}$ therefore contains the prefactor

$$f_{\text{X-ray}} = \frac{1}{m}\left(\frac{e}{c}\right)^2. \tag{3.2}$$

For electron diffraction, the incoming electron interacts with electrons inside a solid via Coulomb repulsion. Assuming the incoming electron is a plane wave, under the first Born approximation, the scattered electron by a charge e at the origin has a wavefunction

$$\psi_{\text{scattered}} = \frac{e^{i\mathbf{k}\cdot\mathbf{r}}}{r}\frac{2m}{\hbar^2}\int d^3\mathbf{r}'\, V(\mathbf{r}')e^{i\mathbf{q}\cdot\mathbf{r}'}, \tag{3.3}$$

where $\mathbf{q}$ is the scattering wavevector. Hence, the scattering matrix element contains a factor of

$$f_{\text{electron}} = \frac{m}{\hbar^2}V_{\mathbf{q}} = \frac{m}{\hbar^2}\frac{e^2}{q^2}, \tag{3.4}$$

and the ratio of scattering rates between X-ray and electron diffraction is

$$\left|\frac{f_{\text{X-ray}}}{f_{\text{electron}}}\right|^2 = (\alpha a_B q)^4 \approx \alpha^4, \tag{3.5}$$

where $\alpha \equiv e^2/(\hbar c) \approx 1/137$ is the fine structure constant and $a_B = \hbar^2/(me^2)$ is the Bohr radius. The approximation in Eq. (3.5) assumes $q \sim a_B^{-1}$ for a typical scattering wavevector. In contrast to the simple result in Eq. (3.5), actual scattering rates depend on beam energy, scattering geometry, atomic species, and many other factors. Nonetheless, this exercise reveals that the distinct cross sections arise from the fundamental difference in the scattering mechanisms: photon–electron interaction for X-ray diffraction, as opposed to electron–charge interaction for electron diffraction.

The disadvantage of a large scattering cross section for UED is the high probability of multiple scattering events when electrons traverse the sample thickness in a transmission diffraction geometry. This often results in a large diffuse background and the appearance of symmetry-forbidden Bragg peaks. These multiple scatterings make quantitative analysis of diffraction intensity challenging because the kinematic approximation, in which the intensity is simply given by the square of the structure factor, no longer applies. Two strategies have been adopted to remedy this problem: (i) reducing sample thickness, and (ii) increasing electron beam energy. For the first point, I will explain in Sect. 3.3 three methods used to prepare single crystals under 50 nm in thickness with more than 100 μm in the lateral dimensions. In addition, UED studies have also been carried out on atomically thin samples [4, 5]. As for the second point, several UED beamlines of up to a few MeV electron kinetic energy have been established [6–11]. They demonstrate a significant reduction of multiple scatterings compared to keV sources, allowing the application of the kinematic scattering theory. The suppression of multiple scatterings not only enables a quantitative evaluation of diffraction peak intensity [5, 6, 9, 12] but also allows higher order analysis of diffuse scattering dynamics [13–15]. In Chap. 7, I will also show an example of how diffuse peaks reveal critical information about phase competition in an out-of-equilibrium context.

3.1.2 Momentum Range and Resolution

The de Broglie wavelength of a relativistic electron is given by

$$\lambda = \frac{hc}{\sqrt{(eV)^2 + 2eVm_0c^2}} = \frac{h}{\gamma\beta m_0 c}, \tag{3.6}$$

where m_0 is the electron rest mass, h is the Planck constant, V is the accelerating voltage, γ is the Lorentz factor, and β is the normalized speed. For a 10-keV electron beam, $\lambda = 0.12$ Å. On the other hand, for an X-ray beam with the same energy, its wavelength is 1.2 Å, about ten times longer. For the same momentum transfer, the scattering angle θ is directly proportional to λ, so the consequence of a much shorter wavelength for electrons is the capability of capturing peaks over multiple orders given the same detector area. This property is advantageous as we can monitor multiple peaks at the same time in a typical UED experiment. Peaks at different orders offer distinct information about the material, such as the evolution of valence electrons compared to that of the nuclei [16–19]. We will utilize this large momentum span in the study of both Bragg and satellite peaks in Chaps. 4 and 8, as well as in the investigation of two spatially orthogonal CDWs in Chaps. 6 and 7.

The short de Broglie wavelength of high-energy electrons, however, results in a much worse momentum resolution compared to X-rays. Let s denote the momentum transfer, and then for small angle scatterings,

$$\Delta s = \frac{2\pi}{\lambda}\frac{\Delta r}{L}, \tag{3.7}$$

where L is the sample-to-detector distance, and Δr is the width of the diffraction spot on the detector. Given a fixed distance and point-spread function of the detector, a small value of λ undermines the ability to accurately determine the momentum transfer and the associated structural change. Nonetheless, the detection of multiple peaks allows us to perform statistical averaging, a procedure we will use to accurately determine the CDW wavevector in Chap. 6.

An alternative solution to counteracting the limited momentum resolution is to utilize low-energy electron diffraction (LEED) at energy $\lesssim$100 V ($\lambda \gtrsim 1.2$ Å), which offers additional advantages of studying surfaces and nanostructures [20]. However, as we will discuss in the next section, the low-energy electrons are extremely susceptible to pulse broadening during their propagation from the emission point to the sample. To solve this problem, miniaturized ultrafast LEED guns have been developed with a nanotip for photoemission, where the sample to electron gun distance is compressed to below 300 μm [21] (Fig. 3.1), at least

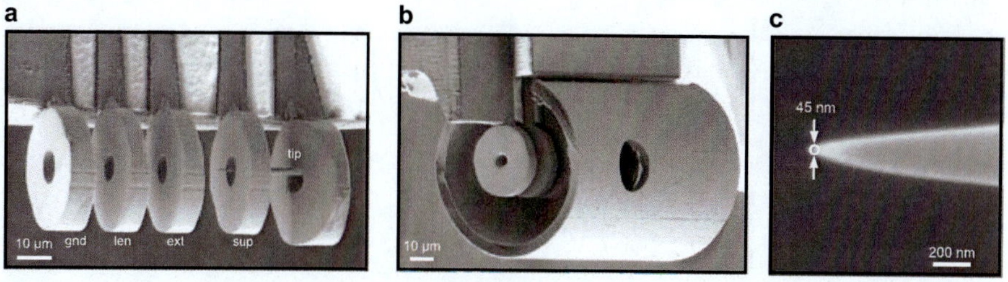

Fig. 3.1 Miniaturized UED with nanotip emission. The example shown here is developed for ultrafast low-energy electron diffraction (LEED) in the backscattering geometry. (**a**) Scanning electron micrographs of the miniaturized electron gun, showing contacted chromium electrodes connecting to the tip (−50 V), suppressor (−90 V), extractor (0 V), lens (90 V), and ground (0 V), where typical voltages are indicated in parentheses. (**b**) Micro-gun with electrostatic shielding attached. The side entrance allows optical coupling for photoemission at the tip. (**c**) Tungsten tip prepared by focused ion beam (FIB) etching. (Adapted from Ref. [21])

two orders of magnitude shorter than conventional UED designs. Implemented in a backscattering as opposed to transmission geometry, the nanotip-based ultrafast LEED also lifts the restriction of sample thickness, hence opening the opportunity to study a wider class of materials that cannot be fabricated into thin flakes.

3.1.3 Temporal Resolution

The most important difference between electrons and X-rays is that the former are charged, leading to significant pulse broadening, both spatially and temporally, due to the space charge effect. In addition to space charge broadening, the temporal resolution of the electron source is also affected by an initial energy spread of the photoelectrons and pulse-to-pulse energy jitters. In the single-electron limit, the temporal resolution is ultimately limited by the dispersion of a massive electron in vacuum and its finite bandwidth, where different energy portions of the electron wavepacket travel with different speeds [22]. Various strategies have been developed to tackle these issues and to compress the electron pulse. In the following, I will highlight a few examples.

The initial energy spread of photoelectrons arises from the mismatch between the photon energy and the photocathode work function. To get some insights into the effect of this energy spread, let us consider two photoelectrons with initial kinetic energies eV_1 and eV_2 right after their ejection from the cathode, and they are accelerated through a dc field of V over a distance d to the anode. Taking into account electron relativistic effects, while ignoring electron-electron Coulomb repulsion, we can determine the time it takes for the i-th electron to be accelerated through the dc field over a distance d:

$$t_{i=1,2} = d\left(\frac{m_0}{eV}\right)^{\frac{1}{2}}\left\{\left[\left(1+\frac{V_i}{V}\right)\left(2+\frac{e(V+V_i)}{m_0c^2}\right)\right]^{\frac{1}{2}}-\left[\frac{V_i}{V}\left(2+\frac{eV_i}{m_0c^2}\right)\right]^{\frac{1}{2}}\right\} \tag{3.8}$$

$$\approx d\left(\frac{m_0}{eV}\right)^{\frac{1}{2}}\left\{\left[\left(1+\frac{V_i}{V}\right)\left(2+\frac{eV}{m_0c^2}\right)\right]^{\frac{1}{2}}-\left[\frac{2V_i}{V}\right]^{\frac{1}{2}}\right\}, \tag{3.9}$$

where the second step assumes that the initial kinetic energy of the electron is negligible compared to its rest mass m_0c^2. Assuming an acceleration over $d = 6$ mm to a voltage of $V = 30$ kV, and assuming the initial energies of the two electrons are $eV_1 = 0$ and $eV_2 = 100$ meV, the second electron will arrive at the anode 213 fs earlier than the first one. Since the second electron has a slightly higher speed at the anode, it will also travel to the sample faster; however, the time difference for this segment of the trajectory is small for $V_i \ll V$ and for a typical distance between anode and sample less than 1 m. This example illustrates how the initial energy spread adversely impacts the temporal resolution, which may be mitigated by a better match between the cathode work function and photon energy via, for example, a wavelength-tunable photon source [23].

To overcome the space charge effect, besides reducing the number of electrons per pulse, there are a few other options. First, we may reduce the electron gun-to-sample distance using a compact geometry, so the time of flight and hence pulse broadening are reduced [24]. This strategy was validated through an N-body simulation and mean-field calculations that estimate the broadening of keV femtosecond electron pulses as they propagate in free space [25]. For example, starting from a 120-fs pulse with 10k electrons, 1.5 mrad initial beam divergence, and 75 μm initial transverse beam radius, after 0.4 ns of propagation (~4 cm for a 30-keV beam), the pulse duration is expanded to 400 fs; after 0.8 ns of propagation (~8 cm for a 30-keV beam), the pulse is further broadened to 1.7 ps. Hence, in this compact geometry scheme, the anode-to-sample distance needs to be restricted to a few centimeters. We have seen an extreme example of this implementation by a micron-sized electron gun with a nano-sized photoemission tip [21] (Fig. 3.1). In addition, our table-top keV setup introduced in Sect. 3.2 also adopts this design principle.

Second, we may increase the electron beam energy to MeV via the acceleration by an RF photoinjector, so that relativistic effects offset the Coulomb repulsion [9]. We can obtain some intuition of this strategy by considering the force between two electrons, both moving along the $+z$ axis with velocity v, but one electron lags behind by a distance l and has a transverse offset of distance a. In the lab frame, the paths of the two electrons are $z_1 = vt, x_1 = 0$ and $z_2 = vt - l, x_2 = a$. As the electric and magnetic fields of a moving charge are compressed along the direction of propagation, the repulsive force experienced by the trailing electron is [26]

$$F_z = -\frac{1}{4\pi\epsilon_0}\frac{e^2 l}{\gamma^2\left(l^2 + a^2/\gamma^2\right)^{3/2}}, \tag{3.10}$$

$$F_x = \frac{1}{4\pi\epsilon_0}\frac{e^2 a}{\gamma^4\left(l^2 + a^2/\gamma^2\right)^{3/2}}, \tag{3.11}$$

where $\gamma = 1/\sqrt{1-(v/c)^2}$ is the Lorentz factor and ϵ_0 is the permittivity of free space. We see that both longitudinal and transverse repulsions quickly diminish as γ increases, leading to a more compact electron pulse in both time and space.

The third strategy to correct space charge broadening is to actively compress the electron pulse after its generation, using an RF buncher cavity [28, 29], THz pulses [27, 30], or magnetic steering [11, 31]. For example, as an electron bunch enters an RF cavity, by adjusting the phase of the RF field relative to the electron arrival time, the bunch head is decelerated while the tail is accelerated so that the total pulse width is reduced when the electrons reach the sample position. While sub-10-fs electron pulses have been demonstrated using RF compression [29], this method suffers from inaccurate timing synchronization between the high power RF field and the electrons, leading to timing jitter between pulses and may worsen the temporal resolution to tens or hundreds of femtoseconds. Therefore, the THz compression scheme offers a competitive alternative, where the high electric field in a single-cycle THz pulse could similarly accelerate or decelerate electrons at different positions of the electron bunch (Fig. 3.2a–c). As the THz pulse is intrinsically synchronized with the laser pulse used for photoemission, the timing uncertainty present in the RF compression is eliminated. Another exciting development in jitter-free compression solutions leverages the achromatic bending magnets from accelerator physics [11, 31] (Fig. 3.2d–g). In this case, the electron beam path is curved so that electrons with higher energy will travel a longer distance, hence taking a longer time to traverse the same path. This bending has two effects. First, within a single pulse, the linear chirp introduced by the space charge effect is reversed as the faster electron at the bunch head acquires a longer flight time compared to the slower electron at the tail. Second, for pulses with different average kinetic energies, which may originate from instabilities during the electron acceleration stage, the bend leads to a similar arrival time of these pulses due to their different trajectories. Even with several million electrons per pulse, Kim and coworkers demonstrated that this scheme was able to achieve a 25-fs pulse width with sub-10-fs arrival time jitter for a 3 MeV UED setup [11].

3.2 Development of the keV UED Setup

In this section, I will describe the design and construction of the keV UED developed in the Gedik group. This UED setup has a long history, starting with the reflectron-based design for electron pulse compression [32]. Eventually a compact

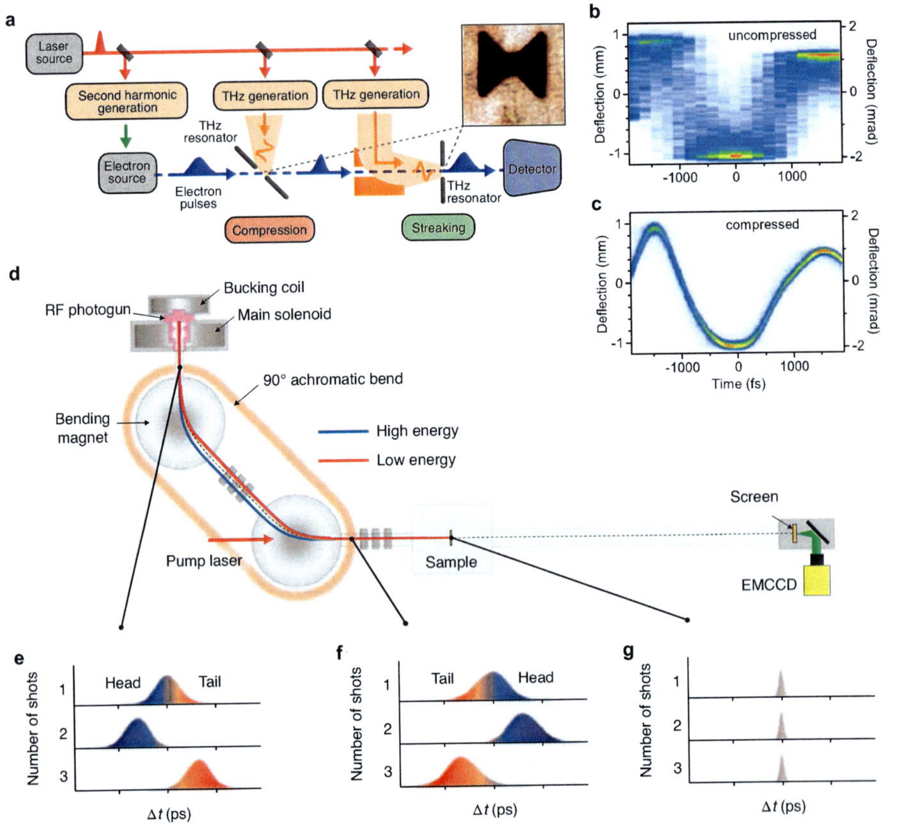

Fig. 3.2 **Electron pulse compression schemes for UED.** (**a**) Schematic of pulse compression and streaking by THz pulses. The 1-ps output of a Yb:YAG system is frequency doubled and used to generate photoelectrons from a thin-film gold or silver cathode. The laser also generates single-cycle THz pulses with energy up to 40 nJ using a butterfly resonant structure, which is made with 30-μm-thick aluminum foil (*inset*). The first resonator provides a time-dependent longitudinal field for pulse compression. The second resonator provides a time-dependent transverse deflection for streaking. (**b**, **c**) Electron deflection as a function of time delay between the electron pulse and the streaking THz pulse. The electron pulse width is 930 fs (uncompressed) and 75 fs (compressed) at delay times of 200 and 10 fs, respectively. (**d**) A double-bend achromatic pulse compression and jitter-suppression scheme for MeV UED. Electrons of different kinetic energies have different path lengths (blue and red). EMCCD, electron-multiplying charge-coupled device. (**e**–**g**) Schematics of the flight time for electron pulses with different energy distributions at various locations in (**d**). Blue indicates higher energy than red. Δt is the relative flight time between electrons. (Adapted with permission from Ref. [27], AAAS (**a**–**c**); Ref. [11], Springer Nature Ltd (**d**–**g**))

electron gun geometry was adopted with 26 keV electron kinetic energy in the transmission diffraction geometry. A schematic of the setup is shown in Fig. 3.3. Below, I will explain individual components and their performances.

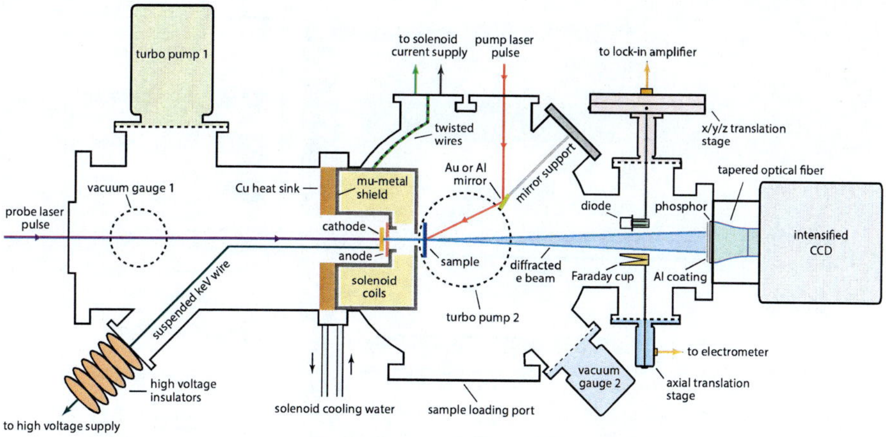

Fig. 3.3 Schematic of the keV UED setup. Solid black curves mark the boundary between the high vacuum and the atmosphere. The schematic shows the top view except for the solenoid cooling pipes, which are located at the bottom of the flange. Vacuum gauge 1 and turbomolecular pump 2 are located at the top and bottom of the marked positions, respectively. The electron gun and optical paths are detailed in Figs. 3.4 and 3.5

3.2.1 *Chamber Design and Specifications*

Vacuum Chamber

There are two cylindrical vacuum chambers: the electron gun resides in a small vacuum chamber attached to an 8-inch ConFlat (CF) flange of a larger diffraction chamber. Each chamber is individually evacuated by a turbomolecular pump (T-Station 75, Edwards), which has a N_2 pumping speed of 61 l/s and is attached via a 4.5-inch OD CF flange. Airflow between the two chambers is restricted to the 400-μm-diameter pinhole anode in the electron gun, and the typical vacuum level is $<3.9 \times 10^{-9}$ torr in the electron gun and $\lesssim 10^{-8}$ torr in the diffraction chamber. These vacuum levels reduce the arcing probability in the electron gun and ensure minimal condensation on the sample for low-temperature measurements.

Air pressures are monitored by two combination gauges—Pirani and cold cathode (MPG400, INFICON)—that are capable of continuous measurements from 3.8×10^{-9} torr to the atmosphere. For the gauge in the diffraction chamber (gauge 2 in Fig. 3.3), its permanent magnet is found to deflect the electron beam; the magnet of gauge 1 has a negligible effect on the electron trajectory. Hence, gauge 2 is typically disabled during data acquisition. The reading of gauge 1 is sampled every 0.25 s. A sudden jump of the reading indicates an arcing event, which is used to trigger a software protection routine that automatically ramps down the dc voltage for electron acceleration.

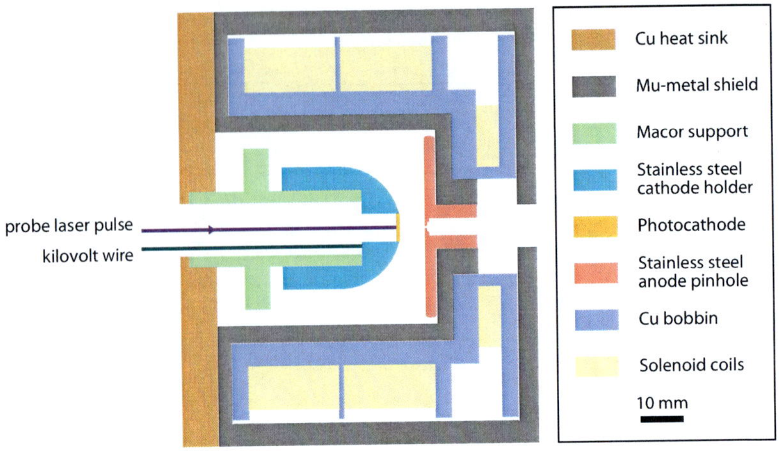

Fig. 3.4 Drawing of the compact keV electron gun. Photoemitted electrons are accelerated under a dc field of 26 kV for 5.7 mm toward a 400-μm-diameter anode pinhole and are subsequently focused by a solenoid. The total distance between the anode and the sample is approximately 30 mm

Electron Gun

The central piece of the instrument is a custom-machined electron gun detailed in Fig. 3.4. The photocathode consists of a gold film deposited on a 6.35-mm-diameter, 0.48-mm-thick circular sapphire substrate. Through electron beam evaporation, a 1-nm-thick adhesion layer of chromium is first deposited at 0.5 Å/s rate followed by 20-nm-thick gold deposited at 1.5 Å/s. The effective work function of the gold film under this condition is about 4.3 eV, determined by examining the quantum efficiency of photoemission as a function of probe photon energy [23, 33]. A previous iteration of the electron gun uses 25-nm-thick silver film as the cathode (without the chromium adhesion layer), which has an order-of-magnitude better quantum efficiency but poorer durability than gold. The photocathode is press fitted into a stainless steel photocathode holder, and the electrical connection is provided by trace amount of conducting vacuum-compatible grease (641-EV thermal grease, Aremco Products, Inc.). The holder is supported at the end of a thick Macor ceramic tube that provides high voltage electrical isolation. A high voltage line at −26 kV is directly connected to the photocathode holder, while the rest of the chamber is grounded. The high voltage supply is EPGS-4005A from Kimball Physics Inc., which is in turn connected to an electrical isolator (ISO/T-34 Kleen Line isolator conditioner, Electronic Specialists, Inc.).

After photoemission, photoelectrons are accelerated over a distance of 5.7 mm toward the anode (extraction field ~4.6 MV/m), which is a 400-μm-diameter pinhole made of stainless steel. The pinhole does not clip the electron beam and serves to assist beam alignment along the axial center of the electron gun. Though the extraction field is far below the theoretical vacuum breakdown value of ~3 GV/m [34], sharp edges need to be reduced to avoid excessive local build-up of high electric fields, which may lead to arcing events. Therefore, the photocathode holder, the mu-metal shielding, and the anode have been carefully polished to achieve a highly reflective surface finish. Under normal operation at 26 kV, arcing events are rare and are averaged to less than once per month.

Electron beam focusing is achieved through a magnetic lens [35]. To minimize the distance from the anode to the sample, the solenoid is placed in vacuum and it surrounds the photocathode. This is accomplished by wrapping the solenoid windings, 735 turns in total with 20 AWG wires, around an oxygen-free high thermal conductivity (OFHC) copper bobbin that encloses the photocathode. In order to confine the magnetic field to a short segment after the anode, the solenoid wiring is enclosed within a mu-metal housing that consists of an inner and an outer cylindrical shell. Under typical operations, a dc current of 1.5 A (PSP-603, Instek) is supplied to the solenoid, producing an axial magnetic field of ~0.01 T at the exit of the solenoid, measured without the mu-metal casing. The electron beam is focused onto a phosphor screen in the detector plane, which is 29 cm away from the gun exit. The current-induced thermal energy from the solenoid windings is dissipated through an OFHC copper heat sink, which is cooled by a ~6 l/s water flow held at 8 °C. To ensure good thermal conductivity for the heat management, thermal grease (641-EV, Aremco Products, Inc.) is applied to the solenoid wires as well as between the surface contacts of the copper bobbin, the mu-metal casing, and the copper heat sink.

Sample Manipulator and Cryostat

A commercial cryogenic manipulator stage (VACGEN), flange-mounted to the top of the diffraction chamber, serves the major roles of supporting and positioning a vertically hanging sample mount. It features an ultra-high vacuum (UHV) welded bellows, allowing three translational degrees of freedom: x, y, and z. A fourth degree of freedom is provided by a vacuum rotational feedthrough that can be driven over 180°, which allows, in principle, both transmission and reflection diffraction geometries. The cryostat (ST-400, Janis), compatible with both liquid N_2 and ^4He, is mounted on a 2.75-inch CF flange that is a part of a welded bellows assembly. The sample holder made of OFHC copper is directly screwed into the cold head, and it can hold up to 18 standard 3-mm-diameter TEM grids. Two diode temperature sensors (DT-670, LakeShore) are attached, one at the cold head, and the other at the far end of the sample holder. The highest attainable temperature is ~100 °C, achieved through a local heater at the cold head. The lowest attainable temperature with liquid ^4He cooling is 4.4 K at the cold head, corresponding to 18.2 K at the far end; their difference is within 3 K when the cold head temperature is above 30 K. As samples are screw-secured to the holder in between these two sensors, the sample temperature is expected to fall within their bounds.

Imaging System

The electron beam current can be characterized by a custom-machined Faraday cup mounted on a single-axis vacuum translation stage. The Faraday cup is made of OFHC copper, with an acceptance radius of 1.9 mm. It is electrically connected to an external electrometer (Model 6514, Keithley) via a BNC vacuum feedthrough. In addition, an ultra-high speed photodiode optimized for keV electron detection (AXUVHS5, Opto Diode) is mounted on an xyz-translation stage and is connected via an SMA feedthrough, allowing for both electron beam characterization and real-time pump–probe data acquisition of a specific Bragg peak, to be detailed in Sect. 3.2.2. Though the diode is intended for electron detection, its sensitivity to scattered NIR photons typically used in the pump pulse may overwhelm the electron signal. Therefore, the diode is housed in an aluminum casing, which has a 400-μm-diameter pinhole in front of the diode active area and several vent holes on the side. To block the NIR photons while allowing electron passage, the pinhole is covered by a freestanding 200-nm-thick aluminum foil (Lebow, Co.). The inner side of the casing is also carbon painted (DAG-T-502, Ted Pella) to reduce diffuse scatterings of NIR photons that enter through the vent holes.

The full electron diffraction image is captured on a 41-mm-diameter, 15-μm-thick phosphor screen (P46, Beam Imaging Solutions), which is directly coated on an optical fiber bundle. The phosphor is over-coated with 400-nm aluminum to avoid charging and to block undesirable NIR pump photons. The P46 phosphor has a central emission line around 530 nm and a fast fluorescence decay (0.5 μs for decay to 10% and 2 μs to 1%), ensuring that fluorescence events are independent from pulse to pulse even when we operate at several hundred kHz repetition rate. A fiber taper bundle (Incom, Inc.) is used to couple fluorescence light to the CCD camera. Compared to lens optics, the fiber coupling is advantageous as it captures photons in the entire 2π solid angle and experiences minimal aberration. Each fiber in the bundle is approximately 5 μm at the coupling to the CCD sensor, which is more than three times smaller than the size of an individual CCD pixel. The fiber taper also demarcates the vacuum/air boundary through a clamped Viton o-ring around its waist. To minimize interfacial reflection and image distortion, gaps between neighboring optical fibers are filled with optical index-matching liquid ($n_D^{25\,^\circ\mathrm{C}} = 1.8000(5)$, Cargille Labs).

The camera is an intensified CCD (PI-MAX II, Princeton Instruments), which consists of an image intensifier (V9501D-75-G137, Hamamatsu Photonics) and a CCD array with a 20-μm pixel size. The incoming light from the optical fiber is first incident on a photocathode, from which photoelectrons are accelerated through 218 V to a microchannel plate (MCP), where the electron number is multiplied and the gain can be adjusted by increasing or decreasing the voltage at the MCP output. As electrons exit the MCP, they are further accelerated by 5 kV to a P43 phosphor screen, and the resulting fluorescence is captured by the CCD array. The image intensifier is optimized for spectral response from 400 to 700 nm via the selection of specific photocathode materials. This range covers the entire emission profile of the P46 phosphor, which is used to image the electron diffraction patterns (not to be confused with the P43 phosphor *after* the MCP). However, this spectral response also excludes any pump pulse wavelength within the 400–700 nm range as any residual pump scattering will be amplified and will overwhelm the electron signal.

3.2.2 Optical System

Pump and probe pulses are generated by a commercial Yb:KGW regenerative amplified laser system (PHAROS SP-10-600-P, Light Conversion). The fundamental output is 1038 nm (1.19 eV), with a pulse duration of 190 fs and a maximum power of 10 W. The repetition rate is variable from 1 Hz up to 600 kHz, granting considerable flexibility of different acquisition modes—high repetition rate for diode detection or small pump fluence and low repetition rate for high pump fluence when laser-induced steady-state heating is significant. Four different optical paths have been constructed for different experimental capabilities (Fig. 3.5), which are detailed below.

Figure 3.5a is the most commonly used configuration. The fundamental output is split at BS1, where 30% enters the harmonic generation unit (HIRO, Light Conversion) and passes through two frequency-doubling stages, leading to the 260-nm (4.78-eV) 4th harmonic output. The 4th harmonic intensity can be adjusted by a half-wave plate (WP3). After passing through the expanding telescope (L3, L4), the beam is focused onto the photocathode by a plano-convex lens (L8). The remaining 70% of the fundamental output constitutes the pump branch, whose intensity is independently adjusted by a motorized half-wave plate (WP2) and a thin-film polarizer (BS2). The optical delay stage (150 mm travel range, ILS150CC, Newport) is placed in the pump branch instead of the probe branch because electron generation is extremely sensitive to pointing stability of the beam. The pump beam is finally focused by L7, and a fixed in situ gold-coated mirror (Fig. 3.3) redirects the beam to the sample. Spatial overlap between the pump photon and the probe electron is achieved by aligning a 50-μm-diameter copper pinhole at the sample position. The electron count on the CCD camera helps to overlap the pinhole and the electron beam, while a telescopic lens through a chamber viewport helps to overlap the pinhole and the pump beam.

Figure 3.5b describes the configuration using an alternative pump wavelength at 260 nm, the 4th harmonic of the fundamental. Note that the 2nd and the 3rd harmonics are not viable as they fall within or close to the spectral range of the CCD image intensifier, as discussed in the preceding section. This configuration may be used to study the effect of carrier excitation of large band-gap materials ($\geq$1.2 eV). The light intensity of the pump and probe can also be individually adjusted through two half-wave plates (WP3 and WP4). The in situ mirror is switched to an aluminum coating in this configuration.

Figure 3.5c shows the layout of a probe arm with variable wavelengths, generated by a combination of optical parametric amplification (OPA) and second harmonic generation (SHG) using a commercial unit (ORPHEUS/LYRA, Light Conversion).

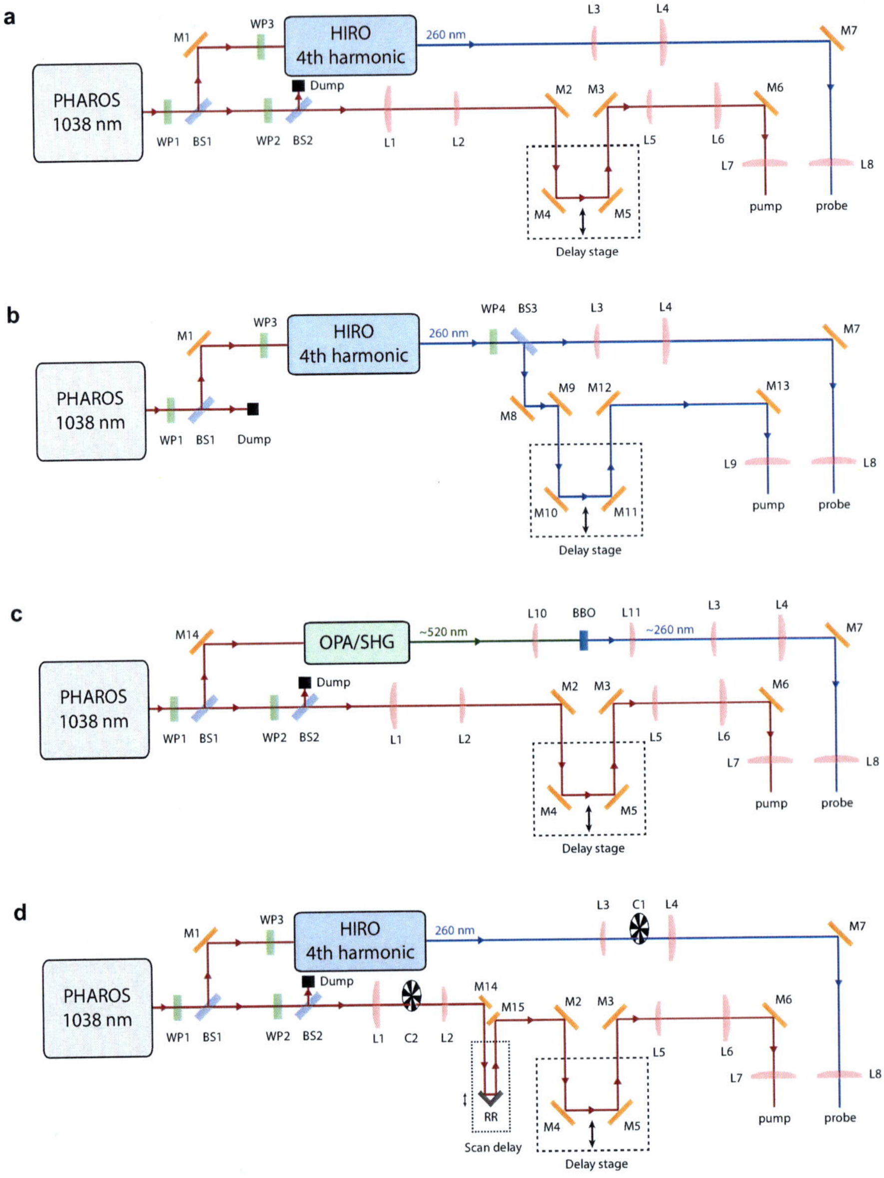

Fig. 3.5 Schematic of different optical configurations. (**a**) Fundamental (1038 nm) pump and 4th harmonic (260 nm) probe. (**b**) 4th harmonic pump and probe. (**c**) Fundamental pump and variable-wavelength probe. (**d**) Diode detection with fast scan delay. *M* dielectric mirror, *L* plano-convex lens, *WP* half-wave plate, *BS* thin-film polarizer, *RR* retroreflector, *C* mechanical chopper, *BBO* β-BaB_2O_4. Some steering mirrors are omitted

The variable output around 520 nm is frequency doubled through a nonlinear crystal, and the resulting UV pulse is focused to the photocathode. As discussed in Sect. 3.1.3, matching the photon energy with the cathode work function minimizes the initial energy spread of the photoelectron, hence improving the temporal resolution. This configuration therefore permits the shortest electron pulse duration when space charge is not the limiting factor [33]. Furthermore, the nonlinear crystal may be removed for electron generation via two-photon photoemission (2PPE) close to the work function. 2PPE has the additional advantage of minimizing emission duration and effective source size at the photocathode, leading to improvements in beam coherence, duration, and stability [23].

While the setups in Fig. 3.5a–c are used when electrons are imaged by the CCD camera, Fig. 3.5d describes the operation using a fast diode detection scheme, allowing the use of lock-in amplifiers (SR830, Stanford Research Systems) to enhance the signal-to-noise ratio. It also accommodates a rapid-scanning optical delay line (scanDelay 150, silver-coated retroreflector, Angewandte Physik & Elektronik GmbH), ideal for real-time data acquisition and for improved sensitivity [36–38]. For lock-in detection, a mechanical chopper, C2 (2971 Hz) in Fig. 3.5d, is inserted into the pump branch, providing the reference frequency for a lock-in amplifier (LIA2) to resolve changes in the electron count with and without the pump pulse. As the diode is also sensitive to scattered pump light in addition to the diffraction signal, another mechanical chopper, C1 (9991 Hz), is inserted into the probe path and is connected to another lock-in amplifier (LIA1). In the frequency domain, the double lock-in scheme separates pump-induced changes in the diffraction signal from pump scattering changes. To reduce high-frequency noise, two low-pass filters with cut-off frequencies of 40 and 17 kHz, respectively, are placed at the input of LIA1 and LIA2. The rapid scanning delay line operates at 10 Hz (peak-to-peak delay up to 138 ps), providing the third reference frequency. The demodulated diode signal from the two LIAs is triggered by the scan delay reading on an oscilloscope (DPO7054, Tektronix) for real-time data acquisition and averaging.

3.2.3 Calibration and Performance

Electron Beam

Figure 3.6a shows an example of a focused electron beam at 26 keV captured by the CCD detector, with horizontal and vertical cuts shown in Fig. 3.6b, c. The electron beam has a Lorentzian profile with FWHM of $\sim$150 μm at the P46 phosphor screen location. This width corresponds to 0.045 Å^{-1} in the momentum space for a typical sample-to-detector distance (283 mm) at this beam energy. The relative ratio between widths along the two directions can be adjusted by the current in the solenoid lens.

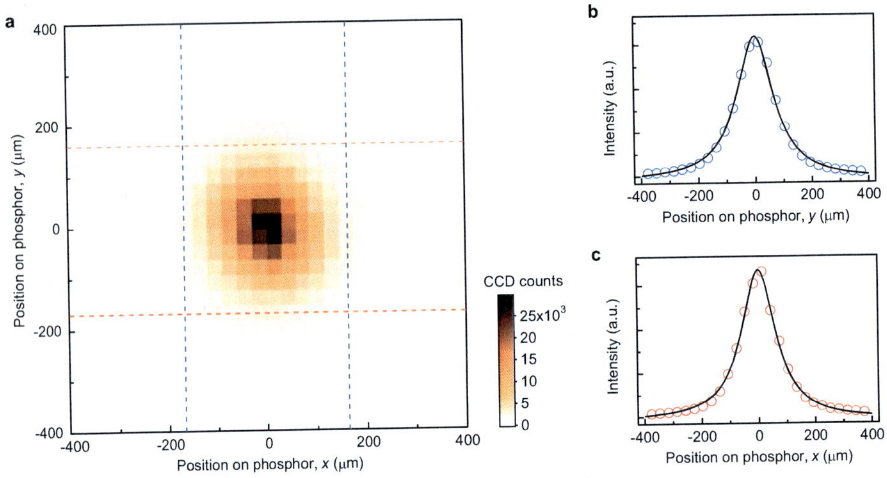

Fig. 3.6 Electron beam profile at the detector. (**a**) CCD image of a focused 26 keV beam at the phosphor screen, with approximately 870 electrons per pulse at a repetition rate of 100 kHz. (**b**, **c**) Integrated intensities within the vertical and horizontal boundaries indicated in (**a**) fitted with a Lorentzian profile. The FWHMs are 156 μm along the y-axis and 144 μm along the x-axis

The quantum efficiency of photoemission at the cathode differs between the type of metallic films used and the input photon energy and power (Fig. 3.7a). Empirically, for 260-nm incident photons, a 25-nm silver film has one order of magnitude higher efficiency than a 20-nm gold film. However, this leads to a fast degradation of the silver cathode and necessitates more frequent replacements. The number of electrons generated increases linearly with the number of photons per pulse if the latter is less than 10^{10}, a condition used in most experiments. Nonlinear dependence at higher photon flux may indicate potential damages to the cathode. Increasing the UV wavelength from 260 nm (4.78 eV) to 285 nm (4.35 eV) leads to a dramatic decrease in the quantum efficiency as the photon energy approaches the estimated work function of the gold film, about 4.3 eV [23, 33]. In most data shown in this dissertation, unless otherwise specified, the gold cathode with 260-nm incident photons was used. It should be noted that the photoemission efficiency sensitively depends on the location on the cathode due to inhomogeneity of the metallic coating and degradation. This demands a stable pointing of the UV beam as well as minimal mechanical vibration between the vacuum chambers and the optical table (see Sect. 3.4.1).

When the electron beam is magnetically focused at the detector, its footprint at the sample location is in the range of several hundred microns (Fig. 3.7b). Due to the space charge effect, increasing the electron number per pulse results in significant transverse broadening. Increasing the beam energy helps reduce the beam size, which may be attributed to the reduced Coulomb repulsion in the transverse direction, as predicted by Eq. (3.11).

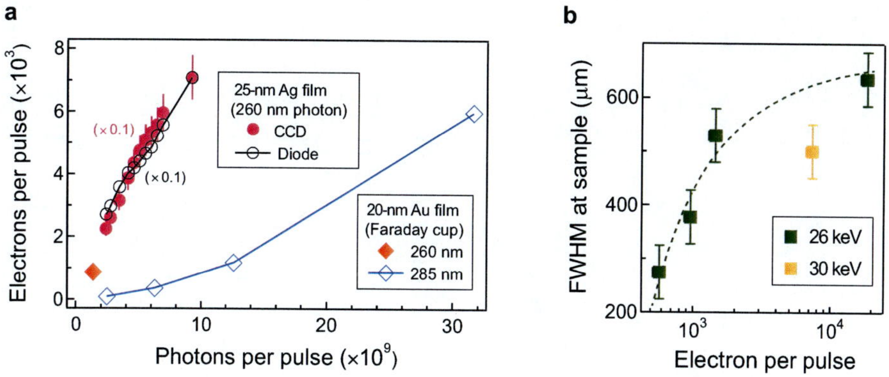

Fig. 3.7 Photocathode efficiency and electron beam size at the sample location. (**a**) The number of electrons generated by Ag and Au photocathodes as a function of incident UV photons per pulse. The plot for the Ag film is vertically scaled by 0.1. Electrons generated from the Ag film were accelerated by 30 kV, while those from the Au film were accelerated by 26 kV. Different detection methods are labeled: (i) intensity counts from a CCD, (ii) diode reading from a lock-in amplifier, which references the 250 kHz laser repetition rate, and (iii) Faraday cup current measured by an electrometer. The CCD count was calibrated by the Faraday cup measurement and the diode reading was normalized to the CCD value at a photon count of 9.2×10^9. For the Au film, UV photons of two different wavelengths were used (260 and 285 nm). (**b**) Width of the electron beam generated by a Au photocathode, measured at the sample position by scanning a 50-μm-diameter pinhole. FWHM is taken as the geometric mean of the two transverse directions. Electron counts per pulse are estimated from CCD intensities. Dashed curve is a guide to the eye. For each electron count, the magnetic lens was adjusted to achieve the optimal focus of the beam at the CCD camera

Temporal Resolution

The overall instrument temporal resolution of the UED setup, τ, can be approximated as [8]

$$\tau = \sqrt{\tau_e^2 + \tau_{\text{photon}}^2 + \tau_{\text{TOA}}^2 + \tau_{\text{VM}}^2}, \tag{3.12}$$

where τ_e and τ_{photon} are the pulse durations of the probe electrons and pump photons, τ_{TOA} is the time-of-arrival jitter between the pump and probe pulses, and τ_{VM} is the velocity mismatch between the two pulses as they meet at the sample. In the following, we evaluate the individual contributions.

In our keV setup, $\tau_{\text{photon}} = 190$ fs. The time-of-arrival jitter is mainly caused by two factors. First, the output of the dc high voltage supply may fluctuate, leading to slightly different accelerations between electron pulses and hence their arrival times. We can estimate this fluctuation using Eq. (3.8), assuming an acceleration voltage of $V = 26$ kV. For a 5.7-mm electron acceleration path and a 30-mm anode-to-sample distance, a ±0.02% uncertainty in the acceleration voltage (EPGS-4005A, Kimball Physics Inc.) leads to $\tau_{\text{TOA},1} = 84$ fs. Second, as the electron and the photon beams approach the sample from opposite sides, any mechanical vibration of the sample

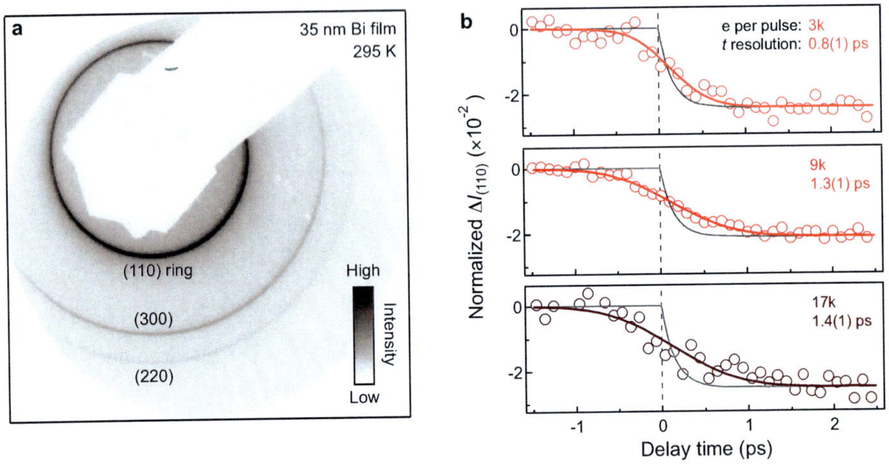

Fig. 3.8 Instrumental temporal resolution under the operation with 26 keV electrons. (**a**) Static diffraction pattern at room temperature of a 35-nm Bi film, which was deposited on a 10-nm-thick silicon nitride window via electron beam evaporation. The three most prominent Debye–Scherrer rings are labeled. (**b**) Photoinduced change in the (110) ring intensity, normalized to its value before photoexcitation. Data was taken at 295 K, with a repetition rate of 1 kHz and incident pump fluence of 0.54 mJ/cm^2. Following Ref. [8], the instrumental temporal resolution is determined by fitting the raw data (circles) with an exponential decaying function (time constant 150 fs, gray curve) convolved by a Gaussian kernel. The temporal resolution is taken as the fitted FWHM of the Gaussian function. The number of electrons per pulse is estimated from the CCD intensity counts

along the direction of the electron beam may result in the time-of-arrival jitter. Based on sample images from a telescopic lens mounted outside the chamber, the vibration is less than 10 μm. With a 26-keV beam, this vibration contributes at most $\tau_{\mathrm{TOA},2} = 140$ fs. Next, the temporal broadening due to the velocity mismatch occurs because the photon pulse (traveling at c) and the 26-keV electron pulse (traveling at 0.3c) traverse the sample thickness with different times. However, as typical sample thickness (~50 nm) is orders of magnitude smaller than the mechanical vibration of the sample, we conclude that $\tau_{\mathrm{VM}} \ll \tau_{\mathrm{TOA},2}$ and hence neglect the τ_{VM} contribution. Therefore, the total contribution from factors other than the electron pulse duration is $(190^2 + 84^2 + 140^2)^{1/2} = 250$ fs.

In this keV setup, the major contribution to the temporal resolution comes from the electron pulse width, τ_e. Using 260-nm photons and a gold photocathode with 4.3 eV work function [23, 33], the initial kinetic energy spread of photoelectrons is 470 meV. From Eq. (3.8), for a 26-keV beam, this initial energy spread translates to 510 fs temporal broadening at the sample position. In addition, with 1k to 10k electrons per pulse used in typical experiments, space charge broadening is significant, which may further broaden the pulse to more than 1 ps [25]. To estimate the overall temporal resolution of the setup, we measure the photoinduced reduction of diffraction intensity in a 35-nm bismuth film, which is known to have a fast initial decay with a time constant of 150 ± 50 fs [8]. Figure 3.8b shows three intensity evolutions of the (110) Debye–Scherrer ring of bismuth, with varying number of electrons per pulse. The intensity traces are fitted with a single-exponential decay (gray curves) convolved with a Gaussian kernel, whose FWHM yields the temporal resolution. The resolution significantly improves if we reduce electrons per pulse from 17k to 3k, indicating the dominant contribution of the space charge effect, which is consistent with the trend of transverse electron profiles in Fig. 3.7b. When there are fewer than 1k electrons per pulse, we expect a dominant role from the initial kinetic energy spread of the photoelectrons. In this case, variable-wavelength photons from the setup in Fig. 3.5c may be leveraged to further push the overall temporal resolution.

Sensitivity

To assess the sensitivity of the UED setup in detecting photoinduced changes, we show in Fig. 3.9 three time traces of single-crystalline samples, measured with optical setups in Fig. 3.5a, b using the CCD camera. The data acquisition took any time between 1 to 30 hours, depending on the signal strength of the sample. Each trace demonstrates one particular aspect of a time-dependent diffraction peak: intensity, width, and momentum position. The noise level of these traces before photoexcitation indicates the smallest relative change that may be detected: 8×10^{-4} for intensity, 3×10^{-3} for width, and 6×10^{-5} for position relative to the undiffracted beam. These high sensitives enable one to detect very small photoinduced signals, such as the initial Debye–Waller decay in Ta_2NiSe_5 under a small pump fluence (Fig. 3.9a), and the coherent acoustic phonon in $FePS_3$ that modulates the peak width and position (Fig. 3.9b, c).

A unique capability of this keV UED setup is the diode detection mechanism introduced in Sect. 3.2.1 (Fig. 3.3). Coupled with a rapid-scanning delay line, it enables efficient data acquisition of diffraction intensity with extremely high sensitivity in a lock-in detection scheme. Figure 3.10a shows an example time trace of the zeroth order Bragg peak of $SrTiO_3$. With only 20 min of data acquisition (10,000 averages), it can detect relative intensity changes down to the 5×10^{-5} level. The diode can also be used with a stepped linear translation stage. For this mode of operation, Fig. 3.10b shows a representative intensity trace of the (200) Bragg peak in a gold crystal, revealing a coherent acoustic phonon at 67 GHz. With 12 hours of data acquisition, this stepped mode also achieves a noise level of 5×10^{-5}, at least one order of magnitude better than the CCD with comparable acquisition time. Despite the high sensitivity, the diode may only monitor the intensity change of Bragg peaks while neglecting the width and position information. In addition, if the electron beam has a long-time position drift, the diode stage is currently not motorized to track the moving beam. Nonetheless, for short-time acquisition using a high repetition rate, the diode greatly outperforms the CCD for detecting tiny changes under a small pump fluence.

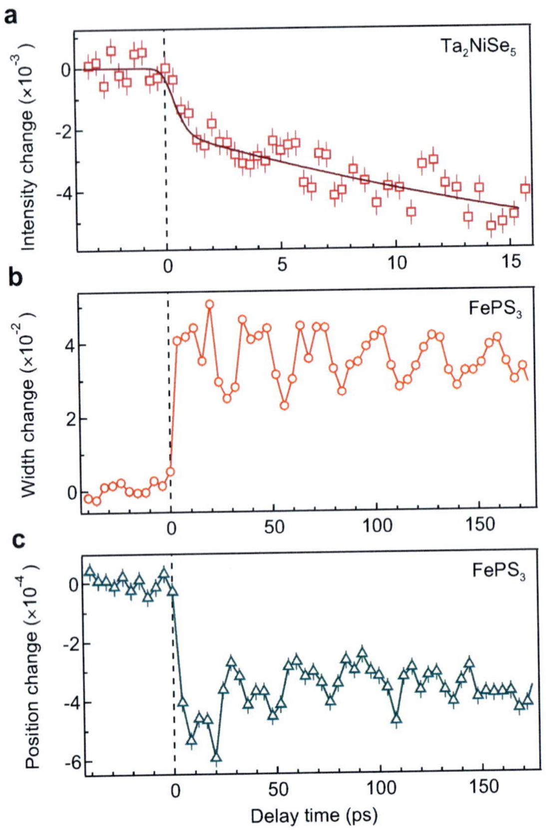

Fig. 3.9 Sensitivity of changes in diffraction peak with CCD data acquisition. (**a**) Evolution of averaged integrated intensity of (200) and ($\bar{2}00$) Bragg peaks in Ta_2NiSe_5. Solid curve is a fit to a double-exponential decay. Data was acquired for 26 hours at 295 K with 1038 nm pump wavelength, 160 μJ/cm^2 incident fluence, and 100 kHz repetition rate. (**b, c**) Evolutions of the width and position of the ($\bar{3}\bar{3}1$) Bragg peak in $FePS_3$, showing photoinduced coherent acoustic phonon oscillations. Peak width is taken as the geometric mean of the FWHMs along the two detector axes. Peak position is the distance from ($\bar{3}\bar{3}1$) to the undiffracted beam. Data was acquired for 1.3 hours at 79 K with 260 nm pump wavelength, 1 mJ/cm^2 incident fluence, and 1 kHz repetition rate. For all panels, a 26-keV electron beam generated from a gold-film photocathode was used. The change in all quantities is normalized to their averaged values before photoexcitation. Error bars are one standard deviation of values before photoexcitation, which indicate the smallest detectable change in the diffraction peak profile: 0.08% for intensity (**a**) 0.3% for width (**b**) and 0.006% for position (**c**)

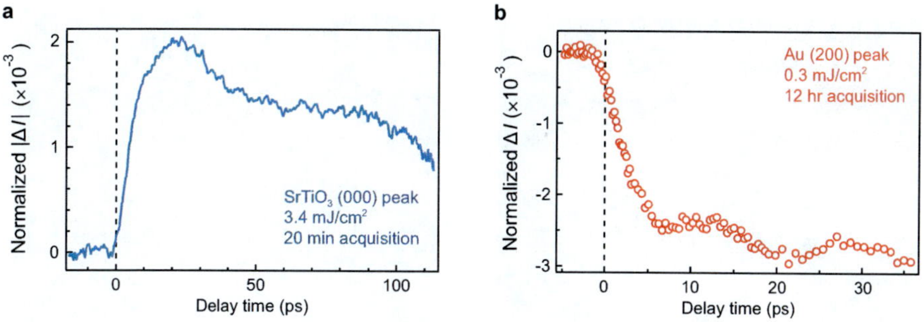

Fig. 3.10 Sensitivity of peak intensity change with diode data acquisition. (**a**) Absolute intensity evolution of the undiffracted electron beam through an 80-nm-thick single-crystalline $SrTiO_3$ film, acquired with a 30-keV electron beam using a rapid scanning delay line operating at 10 Hz. (**b**) Intensity evolution of the (200) Bragg peak of a Au single crystal (02904-AB, SPI Supplies), acquired with a 26-keV electron beam using a stepped linear translation stage. A 67-GHz coherent acoustic phonon is excited. For both panels, experiments were conducted at room temperature, with 250 kHz repetition rate, 1038 nm pump wavelength (with incident fluence labeled), and 9991 Hz and 2971 Hz optical chopper frequencies in the probe and pump arms, respectively (Fig. 3.5d). Changes in intensities are normalized to values before photoexcitation. The sensitivity to intensity change, as determined by the noise level before photoexcitation, is approximately 5×10^{-5} for both panels

3.3 Sample Fabrication

The electron beam spot at the sample location is an important parameter of any UED setup. In the transmission geometry, while a smaller beam spot is preferred to save the effort of preparing electron-transparent samples of large lateral size, focusing electrons to a tiny region exacerbates the problem of space charging, hence undermining the temporal resolution [39]. To overcome this problem, one way is to reduce the number of electrons per pulse while increasing the repetition rate to maintain the average beam current and diffraction statistics. Another way is to increase the electron kinetic energy up to several MeV, where space charge forces are significantly reduced for relativistic particles [Eqs. (3.10)–(3.11)]. However, in either case, a tight focusing of the electron beam at the sample intrinsically results in an increased divergence and a large beam spot in the far field, leading to a poor momentum resolution. Therefore, it remains a significant challenge to optimize beam spot, temporal resolution, and momentum resolution at the same time.

An alternative solution to the dilemma lies in obtaining ultrathin single-crystalline samples of a lateral size on the order of hundreds of microns, which corresponds to most common beam spots in UED without special focusing. It is an equally challenging task, considering the sample thickness is only tens of nanometers, a length scale set by (i) the requirement of electron transparency and the application of kinematic approximation and (ii) the pump laser penetration depth, where one needs to match the photoexcited and the probed volumes. Many methods have been developed, such as ultramicrotomy, reactive ion etching or ion

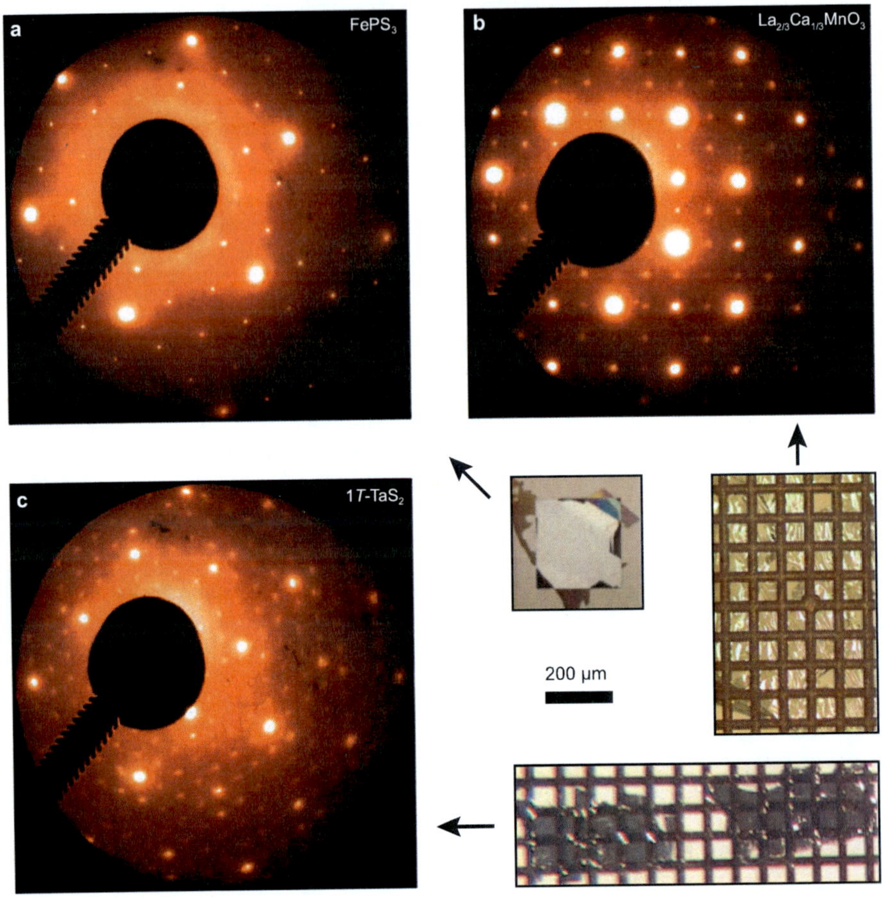

Fig. 3.11 Static diffraction patterns at 26 keV for samples prepared by different methods. (**a**) Mechanically exfoliated $FePS_3$ transferred to a 10-nm-thick Si_3N_4 window. (**b**) 20-nm film of $La_{2/3}Ca_{1/3}MnO_3$ grown by pulsed laser deposition (PLD) on a sacrificial water-soluble layer [40] and transferred to a Cu TEM grid. (**c**) $1T$-TaS_2 cut by ultramicrotome and transferred to a Cu TEM grid. The diffraction for $1T$-TaS_2 was taken at 78 K in its commensurate CDW phase, while others were taken at room temperature. Corresponding sample images were recorded by an optical microscope

milling for bulk crystals [41], and wet transfer [4, 42, 43] or substrate etching [44–46] for thin films. In this section, I will discuss three different fabrication methods that covered a wide range of material classes we studied in the past few years. All of them can yield high-quality single crystals larger than 100 μm in lateral size and smaller than 50 nm in thickness, giving rise to sharp diffraction spots in our keV setup with a high signal-to-background ratio (Fig. 3.11).

3.3.1 Exfoliation and Viscoelastic Stamping

This method was developed in collaboration with the Jarillo-Herrero group at MIT, leveraging recent advances in viscoelastic stamping for deterministic stacking of 2D van der Waals heterostructures [47–49]. In contrast to other protocols that involve etching or ion milling, this method does not introduce dopants, structural deformation, or residual chemicals, hence working very well with air-sensitive samples and producing clean, flat single crystalline films. The effectiveness of this method relies on a carefully selected trio: a thin film of amorphous Si_3N_4 as a substrate (SiMPore Inc.), Gel-Film® (Gel-Pak, Delphon) as a transfer medium, and polydimethylsiloxane (PDMS) gel as a cushion layer. In this section, I will first describe the steps of depositing UED samples with desired dimensions onto a standard TEM window made of Si_3N_4. I will then comment on reasons for the procedures and the choice of materials.

Figure 3.12 summarizes the main steps. First, thin flakes of single crystals are tape-exfoliated to a transparent Gel-Film and are identified under an optical microscope (Fig. 3.12b). Given the large number of flakes, typically a few candidates of desired lateral dimensions and thickness can be found [50], where thickness is readily estimated through the optical contrast between the flake and the Gel-Film [51]. Second, a transparent transfer arm is prepared (Fig. 3.12c), which consists of a flat piece of PDMS gel ($5 \times 5 \times 1\,\text{mm}^3$, ratio 1:20) deposited on one end of a glass slide. The Gel-Film with the desired sample is stacked onto the PDMS gel (Fig. 3.12d), which serves as an elastic cushion when the flake comes into physical contact with the target Si_3N_4 window. The last step is the transfer (Fig. 3.12e–g), where the transfer arm is gradually lowered and then lifted up. As both the sample and the Si_3N_4 have clean, flat surfaces, the contact area and adhesion forces between the two can be very large [48]. If the flake adheres to Si_3N_4 more strongly than to the Gel-Film, the sample is deposited onto the Si_3N_4 window.

Several factors are critical in the transfer step. Setup-wise, we use an air-floated motorized manipulator (Fig. 3.12a) to achieve submicron stability and precision. In addition, a long working distance objective (>1 cm) is necessary for observing the movement of the transfer arm. The optical transparency of the glass slide, the PDMS gel, and the Gel-Film further allows one to monitor the location of the sample relative to the Si_3N_4 window: the real-time microscopic image makes it possible to deposit the flake on a target location with submicron accuracy. The tilt angle of the transfer arm (Fig. 3.12a) may also be adjusted in the transfer process. A large tilt angle reduces the undesired contact between the Si_3N_4 window and the Gel-Film but may induce excessive strain on the Si_3N_4. During a typical transfer, an angle of $\sim 2°$ is used to balance the two competing effects.

The selection and treatment of the Gel-Film and the Si_3N_4 membrane are also essential to ensuring that the flake has more affinity to the latter than the former upon physical contact. In this aspect, the commercially available Gel-Film that comes with different retention levels is ideal, where a lower retention level is used for weak flake-Si_3N_4 bonding. In a typical experiment, a PF film with retention level

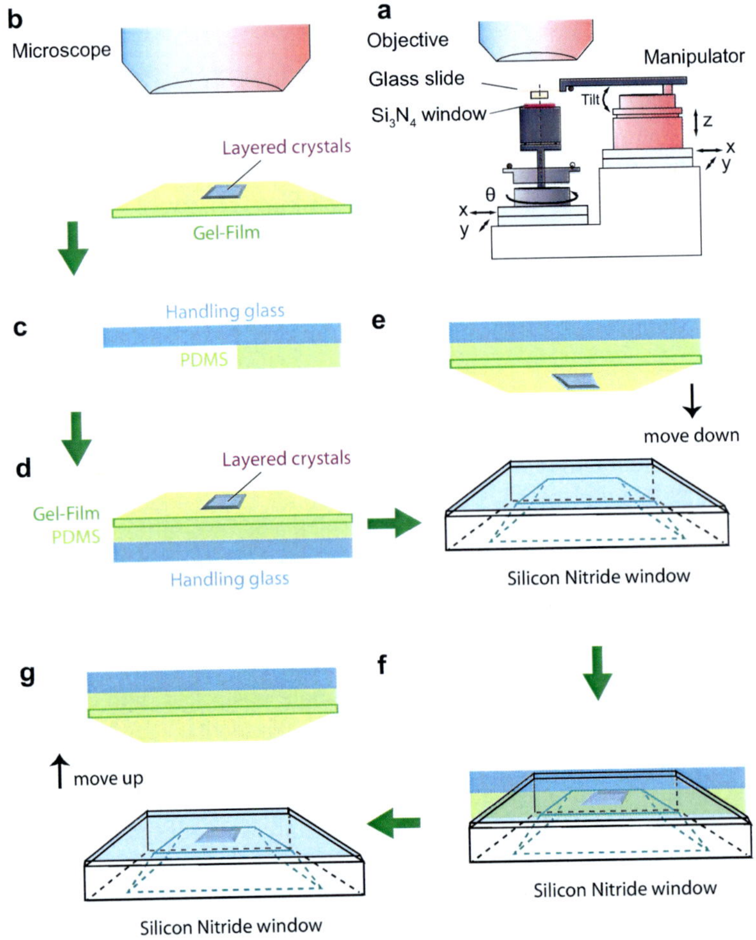

Fig. 3.12 Sample preparation steps using viscoelastic stamping. (**a**) Schematic of the transfer setup, including the microscope objective, a transfer arm mounted on a four-axis manipulator, and a TEM Si_3N_4 substrate secured on a rotatable stage. (**b–g**) Steps of preparing UED samples, including exfoliation onto a Gel-Film® (**b**) preparation of the transfer arm with a selected sample flake (**c**, **d**), and sample transfer (**e–g**)

X0 is used. On the Si_3N_4 side, we can adjust its sample affinity by changing its hydrophilicity: a more hydrophilic surface bonds more strongly with most inorganic samples as stray polar molecules are often found to adhere to dangling bonds on the sample surface [48]. For this purpose, Si_3N_4 windows are exposed with UV illumination in an ozone environment for a variable period of time (typically minutes). This treatment removes organic impurities and creates dangling oxygen bonds on the Si_3N_4 surface, thus making it mildly more hydrophilic.

The above considerations allow one to adjust the relative sample retention levels between the Gel-Film and the Si_3N_4 membrane. Hence, it is feasible to directly transfer sample flakes without additional chemical reagents. Compared to methods that require the removal of polymers through wet chemistry processes, our protocol is much more efficient and free from contaminants [48, 49]. In addition, this dry transfer protocol is well suited for preparing samples that are air-sensitive or reactive: both Gel-Film and Si_3N_4 are chemically inert, and the entire protocol can be executed in a glovebox filled with inert gas.

Lastly, I will comment on the choice of Si_3N_4 as the substrate. It is chemically inert and mechanically robust and can survive extreme temperatures well over 1000 °C that may be transiently induced by an intense pump laser pulse in the UED experiment. For low-stress Si_3N_4, it is routine to form suspended films with a thickness-to-lateral size ratio of 2×10^{-5} down to ≤ 10 nm in thickness, which serves as an ideal flat substrate that causes minimal background scattering for electrons [52]. Unlike (semi)conducting TEM substrates or support grids, the insulating Si_3N_4 with a large gap size (from 2.43 to 4.74 eV) makes it transparent to most optical to infrared photons [53]. This electromagnetic transparency largely reduces carrier excitation in the Si_3N_4 by the typical femtosecond lasers in the optical to infrared regime and makes data interpretation simpler as only the sample carriers are photoexcited.

Figure 3.11a shows the static diffraction of a $250 \times 250\,\mu m^2$ film of $FePS_3$ prepared using this method. Due to the structural support of the flat Si_3N_4, the $FePS_3$ film is much more flat compared to freestanding films deposited on TEM grids (Fig. 3.11b, c). This leads to very sharp Bragg peaks, whose width is only limited by that of the raw electron beam. In addition to $FePS_3$, the viscoelastic stamping method has been successfully applied to materials such as transition metal dichalcogenides (TMDs) and rare-earth tritellurides (RTe$_3$). The latter class of materials is air-sensitive and all operations were carried out in a glovebox. An immediate extension of this method is to fabricate heterostructures with electrical contacts, which work well with the insulating Si_3N_4 substrate. This enables the concurrent measurements of photoinduced structural dynamics and in situ electrical transport, offering insights into both lattice and electronic properties after photoexcitation. It also adds electrical current or gating voltage as a tuning knob, enabling a greater control of materials in an out-of-equilibrium regime.

3.3.2 *Sectioning with an Ultramicrotome*

Compared to exfoliation, ultramicrotomy is a much more efficient method to produce thin samples with a large lateral area. No chemicals are involved in the process except for water, making it a relatively clean method without sample contamination [41], provided that the sample is not sensitive to water or air. Figure 3.13a shows a schematic of the setup, which is placed on a vibration-free platform to achieve thickness precision of tens of nanometers. Samples are first embedded into a resin. After curing, the resin near the sample is carefully trimmed away with a scalpel, exposing the cleavage plane. The sample is mounted

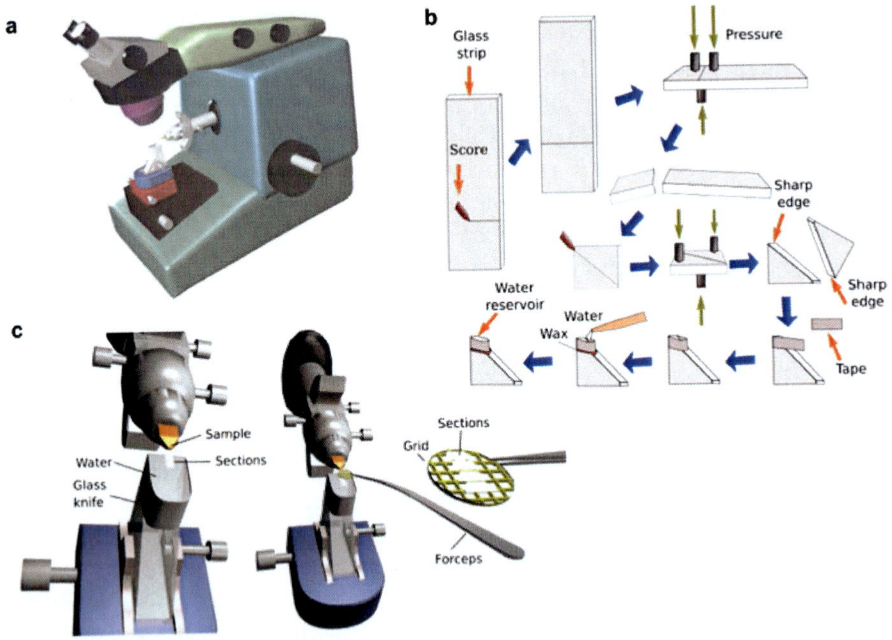

Fig. 3.13 **Sample sectioning with an ultramicrotome.** (**a**) Overview of an ultramicrotome setup. (**b**) Steps to produce a sharp glass knife and to attach a water reservoir near the knife edge. (**c**) Sectioning and sample collection onto a TEM grid. (Adapted from Ref. [54])

on the moving arm of the ultramicrotome such that its motion is parallel to the cleavage plane. A sharp glass blade (Fig. 3.13b) or a diamond knife is fixed at the bottom, where a reservoir of water is attached, wetting the knife edge (Fig. 3.13c). Sectioning proceeds with the oscillatory cutting motion of the moving arm, while the relative distance between the sample and the knife edge is adjusted for each cut. With a state-of-the-art ultramicrotome instrument, the distance is only changed by tens of nanometers per step, producing sample sections of the corresponding thickness. Each sectioned piece slides into the water reservoir and floats flat due to the surface tension (Fig. 3.13c), which are then scooped onto a TEM grid and dried in air.

If the sample cleaves well, multiple pieces can be produced within minutes. The most time-consuming steps are resin trimming and sample mounting. These procedures need to be carefully executed so that minimal resin is attached to the sectioned sample and the sample is correctly oriented with respect to the direction of cutting. An optical image of ultramicrotomed $1T$-TaS_2 is shown in Fig. 3.11c with its corresponding static electron diffraction pattern. Typically, ultramicrotomy works well with samples that cleave easily, including layered materials such as 2D magnets MPX_3 ($M = $ Mn, Fe, Ni; $X = $ S, Se), or needle-like materials such as Ta_2NiSe_5. Compared to exfoliated samples, ultramicrotomed pieces have a rugged surface due to the uneven cleavage when the knife edge traverses the sample. Additional curvature is introduced when the sample dries on the TEM grid. Therefore, diffraction spots appear broader as signals are averaged over different orientations of the sample.

3.3.3 Thin Film Growth and Lift-Off

The previous two methods work well with quasi-2D or quasi-1D crystals, but many important material systems have an intrinsic 3D structure, for example, perovskites, spinels, and garnets. While it is difficult to obtain a thin film of these compounds via mechanical exfoliation or crystal sectioning, various thin film growth techniques have been developed to prepare wafer-sized single-crystalline membranes with pulsed laser deposition (PLD), molecular beam epitaxy (MBE), and sputtering deposition serving as just a few examples of such methods. The task is therefore to lift off these membranes from the substrate and to transfer them to a TEM-compatible grid or window for UED measurements. This is a challenging task as epitaxial films are usually strongly bonded to the substrate. Nevertheless, recent advances have demonstrated two successful approaches.

First, one can grow the desired film on a sacrificial layer, which may be chemically etched away [40, 56–59]. In particular, the Hwang group at Stanford University discovered that the hygroscopic oxide, $Sr_3Al_2O_6$, is a suitable substrate for growing various oxide films via PLD. It is readily dissolved in deionized water, yielding freestanding oxide membranes, down to the monolayer limit [59] and without chemical contamination (Fig. 3.14a, b). With this water etching, films such as $\{Sr,Ba\}TiO_3$ and $La_{1-x}\{Sr,Ca\}_xMnO_3$ have been successfully transferred to TEM grids or Si_3N_4 windows. An example is shown in Fig. 3.11b for a 20-nm-thick $La_{2/3}Ca_{1/3}MnO_3$ film, whose diffraction pattern is slightly broadened due to wrinkling of the freestanding membrane over the 85-μm holes in the underlying grids.

A second method does not involve any chemical lift-off. Instead, two monolayers of graphene are first transferred to the substrate before the desired oxide films are grown on top (Fig. 3.14c–f). As the atomic potential can penetrate through the bilayer graphene, the target oxide films can still be epitaxially grown with the seeding of the substrate [55]. However, the bonding between the oxide and the substrate is sufficiently weakened by the graphene layers. Hence, the oxide films can be mechanically exfoliated and then transferred to other substrates. For films prepared by sputtering, a process that damages the graphene, another intermediate layer, $SrRuO_3$, was also found to aid the physical lift-off of complex oxides (Fig. 3.14f).

The rapid development in thin film synthesis and lift-off technologies may greatly expand the type of materials studied by UED beyond layered materials. Combined with expertise in device fabrication from the 2D-material community, UED may stand to provide new insights into photoinduced dynamics of a wide range of artificially assembled structures.

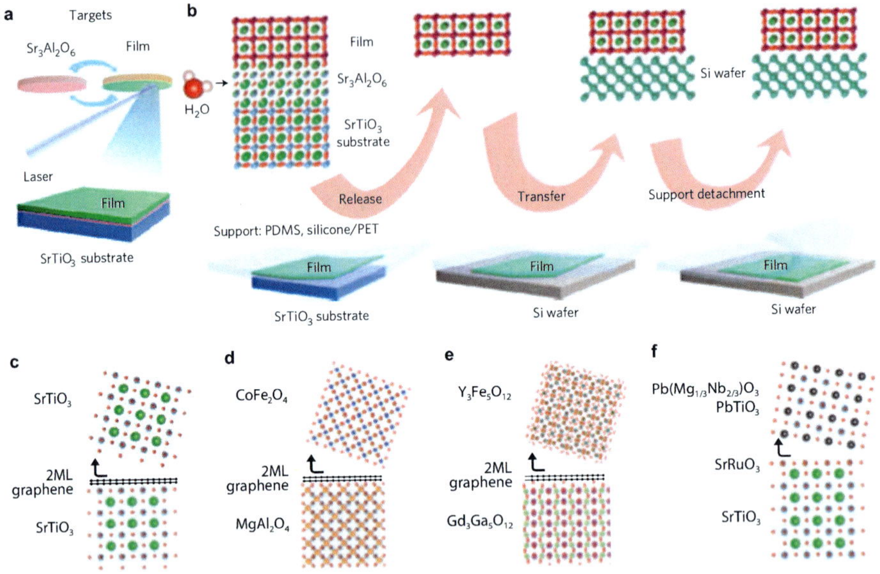

Fig. 3.14 Methods for lifting off complex oxide thin films. (**a**) Pulsed laser deposition (PLD) of sacrificial water-soluble $Sr_3Al_2O_6$ followed by an oxide thin film on a $SrTiO_3$ substrate. (**b**) Schematic steps for releasing and transferring an oxide membrane to a silicon wafer. PDMS: polydimethylsiloxane; silicone/PET: silicone-coated polyethylene terephthalate. (**c**–**f**) Schematics of mechanical lift-off of complex-oxide membranes from graphene-coated or $SrRuO_3$-coated substrates. Monolayers (MLs) of graphene were epitaxially grown via silicon sublimation from SiC (0001) and transferred to the respective substrates. Various oxide films were grown by PLD, molecular beam epitaxy (MBE), or sputtering deposition. (Adapted with permission from Ref. [40], Springer Nature Ltd (**a**, **b**); Ref. [55], Springer Nature Ltd (**c**–**f**))

3.4 UED Data Analysis

In a typical UED measurement with CCD acquisition, diffraction images are taken at each pump–probe delay step, a process hereafter referred to as a *scan*. The delay steps need not be sequential; randomized steps are sometimes adopted to eliminate spurious time-resolved signals due to long-term drift. To accumulate statistics, the same scan is repeated multiple times, generating approximately 10–30 GB of data for a 12-hour period, using the typical setting of 16-bit or 32-bit images with approximately 1 million pixels, where each image is taken with a camera exposure of a few seconds. Extracting scientific insight from this vast amount of data therefore requires efficient and scalable data processing schemes; an example is provided by the open-source software developed in the Siwick group at McGill University [60]. In this section, I will explain analysis methodologies developed for data presented in this dissertation. All software programs were written in Igor Pro (WaveMetrics Inc.) but are readily generalizable to other scientific programming languages.

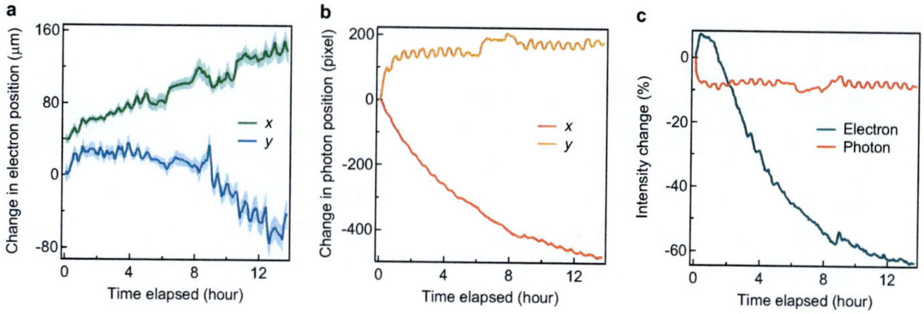

Fig. 3.15 Long-term stability of the 26-keV electron beam. (**a**) x and y positions of the electron beam on the phosphor screen (30 μm corresponds to 1 CCD pixel), monitored over 13.7 hours. The x position change is vertically offset by 40 μm for clarity. (**b**) x and y positions of the 260-nm photon beam monitored during the same period. Oscillatory features correspond to cycles of the air conditioner in the lab. Total pointing change is estimated to be $\lesssim$0.3 mrad. (**c**) Percentage change in the electron and the 260-nm photon beam intensity. Temperature and relative humidity changes during this period are within 0.5 °C and 1%, respectively. In all panels, shaded areas represent fluctuations, taken as one standard deviation of respective quantities

3.4.1 Preprocessing of Diffraction Images

For data acquisition that lasts for several hours, it is important to correct for the long-term drift of the electron beam. Figure 3.15 shows the position and intensity stabilities of the 26-keV electron beam and the 260-nm UV beam over 13.7 hours. As discussed in Sect. 3.2.3, the quantum efficiency of photoemission is extremely sensitive to the emission spot on the cathode, so any small drift of the photon beam position (Fig. 3.15b, orange curve) can result in a significant change in the electron beam intensity (Fig. 3.15c, green curve). Hence, intensity normalization protocols are necessary so that the unstable beam intensity does not mask any photoinduced signals. In addition, any slightly altered photoemission condition could result in a few-pixel drift of the electron position on the detector (Fig. 3.15a), so image alignment is also needed. Another consequence of long-term data acquisition is the unavoidable detection of errant high-energy particles, predominantly muons [61]. They appear as sharp, bright spots in a diffraction image, sometimes near or on top of a diffraction peak. Algorithmic removal of these speckles is therefore necessary to minimize spurious signals in time-resolved analysis.

Taking into these considerations, we use the following four steps as pre-processing protocols for each diffraction image: (i) speckle detection and removal, (ii) CCD dark count subtraction, (iii) intensity normalization, and (iv) image registration.

Speckle detection is based on the sharp footprint of muon peaks, whose width is at least two times smaller than any diffraction peak. To programmatically locate them, a Sobel filter is applied to the diffraction image [62], and candidate speckles are selected based on an intensity threshold in the filtered image. These candidates are further sieved by three criteria. First, the connected area of a candidate speckle in the Sobel image needs to fall within a preset bound. This largely eliminates single-pixel false positives or actual diffraction peaks that have a large area. Second, each speckle candidate is fitted with a 2D Gaussian function to determine its width, which needs to be smaller than a preset threshold. This step further eliminates very intense diffraction peaks that are mistaken for muons. Third, an absolute intensity threshold near the camera saturation value is applied to the original diffraction image, which captures the remaining speckles that may be missed by the Sobel filter. Finally, the detected speckles have their intensity values replaced by neighboring pixels via a bilinear interpolation procedure.

For CCD dark count subtraction, the counts are determined near the corner regions of the sensor array, which are not covered by the circular phosphor screen. For intensity normalization, assuming the total electron counts within the solid angle covered by the detector is conserved, the normalization value is the total intensity captured by the CCD. This assumption is not strictly correct because the (000) peak is typically blocked to protect the sensor and intensities at large scattering wavevectors are not captured. However, these are higher order considerations, and the described normalization procedure works well empirically, especially for correcting the large intensity drift shown in Fig. 3.15c. For image registration, all diffraction images are aligned to the very first image taken during an experimental run. This step utilizes a multi-resolution image pyramid and minimizes the mean square intensity difference between the reference and the test image [63]. The alignment is restricted to translations along the two in-plane axes to correct for the beam position drift.

3.4.2 Diffraction Peak and Lattice Constant Characterizations

For single-crystalline samples, photoinduced signals are reflected in changes in the peak profile, including its intensity, width, and momentum position (Fig. 3.9). For our keV UED setup, this profile is mostly set by the Lorentzian lineshape of the raw electron beam (Fig. 3.6). Hence, these three quantities are extracted by fitting integrated line cuts along the x- and y-axis of the CCD image to a Lorentzian function. A Gaussian function is used for the MeV data presented in this dissertation due to the different electron beam profile. We note that the extracted changes are typically not sensitive to the choice of the peak fitting function. To model the momentum-dependent diffuse scattering, a linear background is also added. In the literature, a pseudo-Voigt profile is sometimes used to separate different peak broadening mechanisms, such as the formation of uncorrelated domains (Lorentzian) and instrument resolution broadening (Gaussian) [64–66]. However, given the electron beam profile and the momentum resolution of the setup, we find that a pseudo-Voigt function may result in data overfitting and the interpretation of relative contributions from the Lorentzian and the Gaussian parts may not be clear.

Although a typical electron diffraction pattern has poorer momentum resolution compared to XRD (Sect. 3.1.2), a large number of peaks in an electron diffraction image allow us to accurately determine changes in the projected lattice constant in the detector plane. This is accomplished by a weighted regression procedure that takes into consideration all peak positions and their associated uncertainties obtained from the peak profile fitting. Specifically, let $\mathbf{b}_1$ and $\mathbf{b}_2$ denote the projections of the reciprocal lattice unit vectors, and let $\mathbf{r}_0 = (x_0, y_0)$ represent the coordinates of the (000) peak in the diffraction pattern. Given the fitted location of the i-th peak $\mathbf{r}_{i=1,\ldots,n} = (x_i, y_i)$ with in-plane Miller indices $\mathbf{m}_i = (H_i, K_i)$, we have

$$\begin{bmatrix} \mathbf{r}_0 \\ \mathbf{b}_1 \\ \mathbf{b}_2 \end{bmatrix} = \left(M^T W M\right)^{-1} M^T W \begin{bmatrix} \mathbf{r}_1 \\ \vdots \\ \mathbf{r}_n \end{bmatrix}, \quad \text{where } M = \begin{bmatrix} 1 & \mathbf{m}_1 \\ \vdots & \vdots \\ 1 & \mathbf{m}_n \end{bmatrix}. \tag{3.13}$$

Here, W is a diagonal matrix, whose diagonal entries correspond to the reciprocal of the position uncertainty in the i-th peak. This procedure has enabled us to detect lattice parameter changes down to 0.02% for the MeV UED setup and to 0.002% for our keV setup, corresponding to $\leq$0.1 pm for typical solids.

3.4.3 Phenomenological Model of Temporal Evolution

The photoinduced evolution of various quantities, such as diffraction peak intensity and width, may be fitted to a phenomenological model that helps to quantify the various timescales involved. Unless otherwise specified, a single-exponential relaxation function, $\Delta f(t)$, is used in this dissertation to describe photoinduced changes. It takes the following form [67–69]:

$$\Delta f(t) = \left[\frac{1}{2}\left(1 + \mathrm{Erf}\left(\frac{2\sqrt{2}(t - t_0)}{w}\right)\right) \cdot \left(I_\infty + I_0 e^{-(t-t_0)/\tau}\right)\right] * g(w_0, t). \tag{3.14}$$

In this model, w represents the intrinsic system response time to photoexcitation, which depends on the microscopic degrees of freedom and differs significantly between electrons, spins, and lattice ions (Sect. 1.1.3). I_0 represents the maximum system response, which grows with pump laser fluence. I_∞ denotes the value of Δf at long time delays, when the system reaches a quasi-equilibrium thermal state. τ is the characteristic relaxation time to the quasi-equilibrium, which reveals important information about the relaxation mechanism. t_0 is associated with the relative arrival time of pump and probe pulses, and it is the time delay when Δf reaches $(I_0 + I_\infty)/2$.

The effect of the finite pulse width in both pump and probe branches is taken into account by convolving the terms in the square brackets with a normalized Gaussian

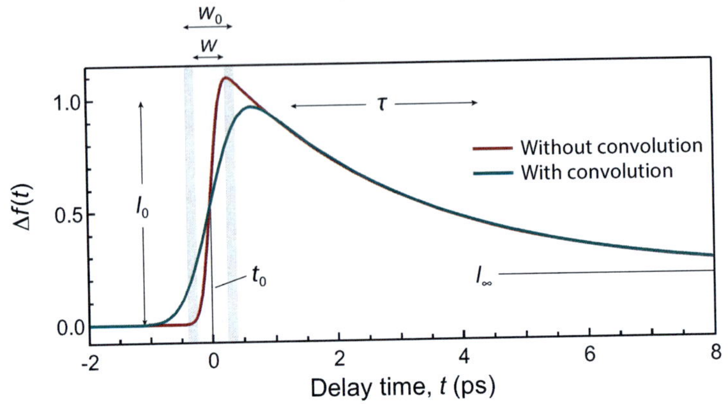

Fig. 3.16 Fitting function of temporal evolutions. A plot of Eq. (3.14) with and without the Gaussian convolution. The parameters used for the plot are $t_0 = 0\,\text{ps}$, $\tau = 3\,\text{ps}$, $w = 0.5\,\text{ps}$, $w_0 = 0.8\,\text{ps}$, $I_0 = 1$, and $I_\infty = 0.2$

pump–probe cross-correlation function $g(w_0, t)$, where w_0 denotes the FWHM. A plot of the fitting function with and without the convolution is shown in Fig. 3.16.

References

1. Freundlich, M.M.: Origin of the electron microscope. Science **142**, 185–188 (1963)
2. Eckert, M.: Max von Laue and the discovery of X-ray diffraction in 1912. Ann. Phys. (Berl.) **524**, A83–A85 (2012)
3. Glaeser, R.M.: Electron crystallography of biological macromolecules. Annu. Rev. Phys. Chem. **36**, 243–275 (1985)
4. Mannebach, E.M., Li, R., Duerloo, K.-A., Nyby, C., Zalden, P., et al.: Dynamic structural response and deformations of monolayer MoS_2 visualized by femtosecond electron diffraction. Nano Lett. **15**, 6889–6895 (2015)
5. Lin, M.-F., Kochat, V., Krishnamoorthy, A., Bassman, L., Weninger, C., et al.: Ultrafast non-radiative dynamics of atomically thin $MoSe_2$. Nat. Commun. **8**, 1745 (2017)
6. Giret, Y., Naruse, N., Daraszewicz, S.L., Murooka, Y., Yang, J., et al.: Determination of transient atomic structure of laser-excited materials from time-resolved diffraction data. Appl. Phys. Lett. **103**, 253107 (2013)
7. Fu, F., Liu, S., Zhu, P., Xiang, D., Zhang, J., et al.: High quality single shot ultrafast MeV electron diffraction from a photocathode radio-frequency gun. Rev. Sci. Instrum. **85**, 083701 (2014)
8. Weathersby, S.P., Brown, G., Centurion, M., Chase, T.F., Coffee, R., et al.: Mega-electron-volt ultrafast electron diffraction at SLAC National Accelerator Laboratory. Rev. Sci. Instrum. **86**, 073702 (2015)
9. Zhu, P., Zhu, Y., Hidaka, Y., Wu, L., Cao, J., et al.: Femtosecond time-resolved MeV electron diffraction. New J. Phys. **17**, 063004 (2015)
10. Ji, F., Durham, D.B., Minor, A.M., Musumeci, P., Navarro, J.G., et al.: Ultrafast relativistic electron nanoprobes. Commun. Phys. **2**, 54 (2019)

11. Kim, H.W., Vinokurov, N.A., Baek, I.H., Oang, K.Y., Kim, M.H., et al.: Towards jitter-free ultrafast electron diffraction technology. Nat. Photon. **14**, 245–250 (2019)
12. Sokolowski-Tinten, K., Li, R.K., Reid, A.H., Weathersby, S.P., Quirin, F., et al.: Thickness-dependent electron–lattice equilibration in laser-excited thin bismuth films. New J. Phys. **17**, 113047 (2015)
13. Chase, T., Trigo, M., Reid, A.H., Li, R., Vecchione, T., et al.: Ultrafast electron diffraction from non-equilibrium phonons in femtosecond laser heated Au films. Appl. Phys. Lett. **108**, 041909 (2016)
14. Stern, M.J., René de Cotret, L.P., Otto, M.R., Chatelain, R.P., Boisvert, J.-P., et al.: Mapping momentum-dependent electron-phonon coupling and nonequilibrium phonon dynamics with ultrafast electron diffuse scattering. Phys. Rev. B **97**, 165416 (2018)
15. René de Cotret, L.P., Pöhls, J.-H., Stern, M.J., Otto, M.R., Sutton, M., et al.: Time- and momentum-resolved phonon population dynamics with ultrafast electron diffuse scattering. Phys. Rev. B **100**, 214115 (2019)
16. Zheng, J.C., Zhu, Y., Wu, L., Davenport, J.W.: On the sensitivity of electron and X-ray scattering factors to valence charge distributions. J. Appl. Cryst. **38**, 648–656 (2005)
17. Morrison, V.R., Chatelain, R.P., Tiwari, K.L., Hendaoui, A., Bruhács, A., et al.: A photoinduced metal-like phase of monoclinic VO_2 revealed by ultrafast electron diffraction. Science **346**, 445–448 (2014)
18. Otto, M.R., René de Cotret, L.P., Valverde-Chavez, D.A., Tiwari, K.L., Émond, N., et al.: How optical excitation controls the structure and properties of vanadium dioxide. Proc. Natl. Acad. Sci. U. S. A. **116**, 450–455 (2019)
19. Li, J., Li, J., Sun, K., Wu, L., Li, R., et al.: Concurrent probing of electron-lattice dephasing induced by photoexcitation in 1*T*-TaSeTe using ultrafast electron diffraction. Phys. Rev. B **101**, 100304 (2020)
20. Gulde, M., Schweda, S., Storeck, G., Maiti, M., Yu, H.K., et al.: Ultrafast low-energy electron diffraction in transmission resolves polymer/graphene superstructure dynamics. Science **345**, 200–204 (2014)
21. Storeck, G., Vogelgesang, S., Sivis, M., Schäfer, S., Ropers, C.: Nanotip-based photoelectron microgun for ultrafast LEED. Struct. Dyn. **4**, 044024 (2017)
22. Gliserin, A., Walbran, M., Krausz, F., Baum, P.: Sub-phonon-period compression of electron pulses for atomic diffraction. Nat. Commun. **6**, 8723 (2015)
23. Kasmi, L., Kreier, D., Bradler, M., Riedle, E., Baum, P.: Femtosecond single-electron pulses generated by two-photon photoemission close to the work function. New J. Phys. **17**, 033008 (2015)
24. Sciaini, G., Miller, R.J.D.: Femtosecond electron diffraction: heralding the era of atomically resolved dynamics. Rep. Prog. Phys. **74**, 096101 (2011)
25. Siwick, B.J., Dwyer, J.R., Jordan, R.E., Miller, R.J.D.: Ultrafast electron optics: propagation dynamics of femtosecond electron packets. J. Appl. Phys. **92**, 1643–1648 (2002)
26. Stupakov, G., Penn, G.: Self field of a relativistic beam. In: Classical Mechanics and Electromagnetism in Accelerator Physics, pp. 137–150. Springer, Cham (2018)
27. Kealhofer, C., Schneider, W., Ehberger, D., Ryabov, A., Krausz, F., et al.: All-optical control and metrology of electron pulses. Science **352**, 429–433 (2016)
28. Chatelain, R.P., Morrison, V.R., Godbout, C., Siwick, B.J.: Ultrafast electron diffraction with radio-frequency compressed electron pulses. Appl. Phys. Lett. **101**, 081901 (2012)
29. Zhao, L., Wang, Z., Lu, C., Wang, R., Hu, C., et al.: Terahertz streaking of few-femtosecond relativistic electron beams. Phys. Rev. X **8**, 21061 (2018)
30. Li, R.K., Hoffmann, M.C., Nanni, E.A., Glenzer, S.H., Kozina, M.E., et al.: Terahertz-based subfemtosecond metrology of relativistic electron beams. Phys. Rev. Accel. Beams **22**, 012803 (2019)
31. Qi, F., Ma, Z., Zhao, L., Cheng, Y., Jiang, W., et al.: Breaking 50 femtosecond resolution barrier in MeV ultrafast electron diffraction with a double bend achromat compressor. Phys. Rev. Lett. **124**, 134803 (2020)

32. Wang, Y., Gedik, N.: Electron pulse compression with a practical reflectron design for ultrafast electron diffraction. IEEE J. Sel. Top. Quantum Electron. **18**, 140–147 (2012)
33. Aidelsburger, M., Kirchner, F.O., Krausz, F., Baum, P.: Single-electron pulses for ultrafast diffraction. Proc. Natl. Acad. Sci. U. S. A. **107**, 19714–19719 (2010)
34. Mulcahy, M.J., Bolin, P.C.: Handbook of Vacuum Insulation. Technical Report, Ion Physics Corp., Burlington (1971)
35. Kumar, V.: Understanding the focusing of charged particle beams in a solenoid magnetic field. Am. J. Phys. **77**, 737–741 (2009)
36. Edelstein, D.C., Romney, R.B., Scheuermann, M.: Rapid programmable 300 ps optical delay scanner and signal-averaging system for ultrafast measurements. Rev. Sci. Instrum. **62**, 579–583 (1991)
37. Kwong, K.F., Yankelevich, D., Chu, K.C., Heritage, J.P., Dienes, A.: 400-Hz mechanical scanning optical delay line. Opt. Lett. **18**, 558–560 (1993)
38. Sucha, G.: Rapid scanning time delays for ultrafast measurement systems. In: Fermann, M.E., Galvanauskas, A., Sucha, G. (eds.): Ultrafast Lasers, pp. 419–472. CRC Press, Boca Raton (2003)
39. Shen, X., Li, R.K., Lundström, U., Lane, T.J., Reid, A.H., et al.: Femtosecond mega-electron-volt electron microdiffraction. Ultramicroscopy **184**, 172–176 (2018)
40. Lu, D., Baek, D.J., Hong, S.S., Kourkoutis, L.F., Hikita, Y., et al.: Synthesis of freestanding single-crystal perovskite films and heterostructures by etching of sacrificial water-soluble layers. Nat. Mater. **15**, 1255–1260 (2016)
41. Eichberger, M., Krumova, M., Berger, H., Demsar, J.: Sample preparation methods for femtosecond electron diffraction experiments. Ultramicroscopy **127**, 9–13 (2013)
42. Hu, J., Vanacore, G.M., Cepellotti, A., Marzari, N., Zewail, A.H.: Rippling ultrafast dynamics of suspended 2D monolayers, graphene. Proc. Natl. Acad. Sci. U. S. A. **113**, E6555–E6561 (2016)
43. Han, T.-R.T., Zhou, F., Malliakas, C.D., Duxbury, P.M., Mahanti, S.D., et al.: Exploration of metastability and hidden phases in correlated electron crystals visualized by femtosecond optical doping and electron crystallography. Sci. Adv. **1**, e1400173 (2015)
44. Cremons, D.R., Plemmons, D.A., Flannigan, D.J.: Femtosecond electron imaging of defect-modulated phonon dynamics. Nat. Commun. **7**, 11230 (2016)
45. Harb, M., Enquist, H., Jurgilaitis, A., Tuyakova, F.T., Obraztsov, A.N., et al.: Phonon-phonon interactions in photoexcited graphite studied by ultrafast electron diffraction. Phys. Rev. B **93**, 104104 (2016)
46. Adrian, M., Senftleben, A., Morgenstern, S., Baumert, T.: Complete analysis of a transmission electron diffraction pattern of a MoS_2-graphite heterostructure. Ultramicroscopy **166**, 9–15 (2016)
47. Novoselov, K.S., Mishchenko, A., Carvalho, A., Castro Neto, A.H.: 2D materials and van der Waals heterostructures. Science **353**, aac9439 (2016)
48. Frisenda, R., Navarro-Moratalla, E., Gant, P., Pérez De Lara, D., Jarillo-Herrero, P., et al.: Recent progress in the assembly of nanodevices and van der Waals heterostructures by deterministic placement of 2D materials. Chem. Soc. Rev. **47**, 53–68 (2018)
49. Castellanos-Gomez, A., Buscema, M., Molenaar, R., Singh, V., Janssen, L., et al.: Deterministic transfer of two-dimensional materials by all-dry viscoelastic stamping. 2D Mater. **1**, 011002 (2014)
50. Bohr, J.: Adhesive tape exfoliation: why it works for graphene. EPL **109**, 58004 (2015)
51. Nair, R.R., Blake, P., Grigorenko, A.N., Novoselov, K.S., Booth, T.J., et al.: Fine structure constant defines visual transparency of graphene. Science **320**, 1308 (2008)
52. Nishiyama, H., Suga, M., Ogura, T., Maruyama, Y., Koizumi, M., et al.: Atmospheric scanning electron microscope observes cells and tissues in open medium through silicon nitride film. J. Struct. Biol. **172**, 191–202 (2010)
53. Deshpande, S.V., Gulari, E., Brown, S.W., Rand, S.C.: Optical properties of silicon nitride films deposited by hot filament chemical vapor deposition. J. Appl. Phys. **77**, 6534–6541 (1995)

54. Megí, Molist, P., Pombal, M.A.: Atlas of Plant and Animal Histology: Ultramicrotome. http://mmegias.webs.uvigo.es (2019). Accessed: 14 Aug 2020
55. Kum, H.S., Lee, H., Kim, S., Lindemann, S., Kong, W., et al.: Heterogeneous integration of single-crystalline complex-oxide membranes. Nature **578**, 75–81 (2020)
56. Bakaul, S.R., Serrao, C.R., Lee, M., Yeung, C.W., Sarker, A., et al.: Single crystal functional oxides on silicon. Nat. Commun. **7**, 10547 (2016)
57. Paskiewicz, D.M., Sichel-Tissot, R., Karapetrova, E., Stan, L., Fong, D.D.: Single-crystalline $SrRuO_3$ nanomembranes: a platform for flexible oxide electronics. Nano Lett. **16**, 534–542 (2016)
58. Zhang, Y., Shen, L., Liu, M., Li, X., Lu, X., et al.: Flexible quasi-two-dimensional $CoFe_2O_4$ epitaxial thin films for continuous strain tuning of magnetic properties. ACS Nano **11**, 8002–8009 (2017)
59. Ji, D., Cai, S., Paudel, T.R., Sun, H., Zhang, C., et al.: Freestanding crystalline oxide perovskites down to the monolayer limit. Nature **570**, 87–90 (2019)
60. René de Cotret, L.P., Otto, M.R., Stern, M.J., Siwick, B.J.: An open-source software ecosystem for the interactive exploration of ultrafast electron scattering data. Adv. Struct. Chem. Imag. **4**, 11 (2018)
61. Groom, D.: Cosmic rays and other nonsense in astronomical CCD imagers. Exp. Astron. **14**, 45–55 (2002)
62. Gonzalez, R.C., Woods, R.E.: Digital Image Processing, 2nd edn. Addison-Wesley, Boston (1992)
63. Thevenaz, P., Ruttimann, U.E., Unser, M.: A pyramid approach to subpixel registration based on intensity. IEEE Trans. Image Process. **7**, 27–41 (1998)
64. Fultz, B., Howe, J.: Transmission Electron Microscopy and Diffractometry of Materials. Springer, Berlin/Heidelberg (2013)
65. Feng, Y., Wang, J., Jaramillo, R., van Wezel, J., Haravifard, S., et al.: Order parameter fluctuations at a buried quantum critical point. Proc. Natl. Acad. Sci. U. S. A. **109**, 7224–7229 (2012)
66. Konstantinova, T., Wu, L., Abeykoon, M., Koch, R.J., Wang, A.F., et al.: Photoinduced dynamics of nematic order parameter in FeSe. Phys. Rev. B **99**, 180102 (2019)
67. Prasankumar, R.P., Taylor, A.J.: Optical Techniques for Solid-State Materials Characterization. CRC Press, Boca Raton (2011)
68. Hellmann, S., Rohwer, T., Kalläne, M., Hanff, K., Sohrt, C., et al.: Time-domain classification of charge-density-wave insulators. Nat. Commun. **3**, 1069 (2012)
69. Moore, R.G., Lee, W.S., Kirchman, P.S., Chuang, Y.D., Kemper, A.F., et al.: Ultrafast resonant soft X-ray diffraction dynamics of the charge density wave in $TbTe_3$. Phys. Rev. B **93**, 024304 (2016)

Chapter 4
Dynamics of Complex Order Parameter After Photoexcitation

Abstract Upon photoexcitation, a symmetry-broken state can undergo a nonequilibrium phase transition through different pathways from those in thermal equilibrium. Determining the mechanism underlying these photoinduced phase transitions (PIPTs) has been a long-standing issue in the study of condensed matter systems, but many details in the ultrafast, non-adiabatic regime still remain to be clarified. Here, we study the light-induced melting of a unidirectional charge density wave (CDW) in $LaTe_3$. Using three time-resolved probes – including ultrafast electron diffraction (UED), time- and angle-resolved photoemission spectroscopy (tr-ARPES), and transient reflectivity – we independently track the amplitude and phase dynamics of the complex CDW order parameter. We find that a fast recovery of the CDW amplitude ($\sim$1 ps) is followed by a slower reestablishment of phase coherence. This longer timescale is dictated by light-induced topological defects: long-range order is inhibited and is only restored when the defects annihilate. Our results offer a framework for understanding PIPTs by identifying the generation of defects as a governing mechanism.

Illustration of a defect disrupting a unidirectional charge density wave

A. Zong, *Emergent States in Photoinduced Charge-Density-Wave Transitions*, Springer Theses, https://doi.org/10.1007/978-3-030-81751-0_4

4.1 Non-adiabatic Symmetry-Breaking Transitions

4.1.1 Emergence of Topological Defects

The understanding of equilibrium phase transitions caused by spontaneous symmetry breaking is a hallmark achievement of twentieth-century physics. When these transitions are induced by adiabatically cooling from a disordered to an ordered phase, they are marked by a diverging correlation length and correlation time of equilibrium fluctuations at the transition temperature, T_c [1]. Much less is understood about non-adiabatic transitions, or quenches, where fluctuations are not expected to exhibit a diverging correlation length and time, preventing the onset of long-range order. This absence of critical behavior is often linked to the creation of topological defects in the ordered phase. The conventional framework for treating non-adiabatic transitions is the Kibble–Zurek theory [2, 3]. As the system is quenched through a phase transition from a disordered state, topological defects are generated as a result of the simultaneous emergence of the ordered phase in disconnected regions of space, a process analogous to the formation of cosmic strings in the early universe (Fig. 4.1). While such a picture is supported by experiments, for example, in liquid crystals [4, 5], ^{3}He [6, 7], and multiferroics [8], it has so far seen only limited experimental support in a broad new class of non-adiabatic transitions induced in ordered systems by photoexcitation [9, 10].

PIPTs present a unique platform whereby non-adiabatic transitions can be studied. They have emerged as an intense research field in recent decades [13] as a

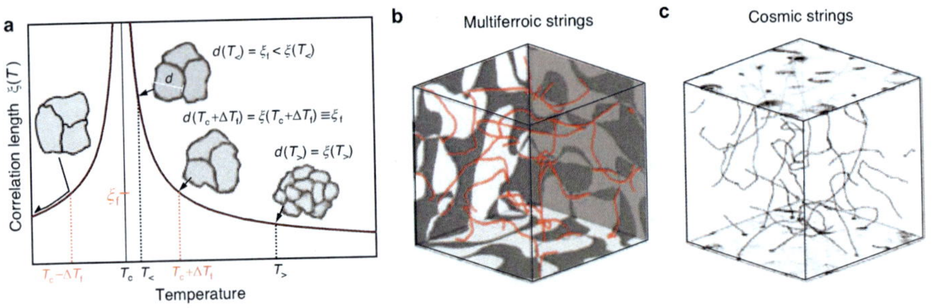

Fig. 4.1 Topological defects in a non-adiabatic phase transition. (**a**) The Kibble–Zurek mechanism for domain formation through a fast quench across T_c. At high temperature ($T_> > T_c + \Delta T_f$), the size of correlated regions (d) is the same as the correlation length ξ. Below the freeze-out temperature $T_f = T_c + \Delta T_f$, the lateral size of the fluctuating regions (indicated by the fuzzy boundaries in the schematics) can no longer match the diverging correlation length due to the fast quench. As the system cools down below $T_c - \Delta T_f$, frozen domains (indicated by the solid boundaries) of a characteristic size $\xi_f \equiv \xi(T_c + \Delta T_f)$ form. (**b**) Ferroelectric domains in $YMnO_3$ (patches with contrasting grayscale colors) measured by piezoresponse force microscopy, superimposed with simulated 3D strings that mark the intersections of different domain orientations. (**c**) Simulated cosmic strings in the early universe. (Adapted from Ref. [11] (**a**); adapted with permission from Ref. [8], Springer Nature Ltd (**b**); Ref. [12], APS (**c**))

consequence of the technological advances offered by ultrafast lasers. During these transitions, the initial state appearing immediately after photoexcitation, from which order recovers, is far from equilibrium. Moreover, topological defects in this case are not necessarily generated through a complete melting of the broken symmetry phase, but may also arise within the ordered state as a result of spatially localized absorption of high-energy photons.

4.1.2 CDW as a Platform for Multimodal Investigations

Materials that exhibit a unidirectional incommensurate CDW are well-suited for investigating PIPTs. Topological defects in these systems, such as dislocations, have been classified theoretically [14, 15] and are thought to play a negligible role in an equilibrium metal-to-CDW transition [16]. Indeed, if a sample is adiabatically cooled below T_c, a resolution-limited diffraction peak appears [17–19]. This observation indicates that the phase coherence extends macroscopically without impedance from topological defects, which, when present, reduce the correlation length and disrupt long-range order. By contrast, previous studies on PIPTs in unidirectional CDW systems have hinted at the existence of topological defects [9, 10], but more direct probes are needed to elucidate how their presence affects the order parameter dynamics.

In this work, we use three different time-resolved probes introduced in Sect. 1.3.1 to gain insight into the light-driven phase transition kinetics in a paradigmatic CDW system, rare-earth tritellurides. In each probe, an incident pump pulse perturbs or melts the CDW, and a delayed probe pulse is utilized to measure the ensuing dynamics of the relevant observable. We employ UED to probe the long-range density correlations [20], while using transient reflectivity and tr-ARPES to track the CDW gap amplitude [9, 21–25]. Transient reflectivity has the advantage that it possesses the highest temporal resolution and signal-to-noise ratio among the probes used, enabling us to additionally investigate the coherent response from collective excitations. The main benefit of tr-ARPES lies in its energy and momentum resolution; hence, it can directly probe the relevant gap dynamics. As tr-ARPES is a surface sensitive probe [26], the transient reflectivity measurements are essential in providing a bulk-sensitive view of the CDW amplitude dynamics.

It is instructive to estimate the probe depth of these three techniques. For UED measurements, in our transmission geometry, all layers of the sample (10–30 nm) contribute to the diffraction pattern. For transient reflectivity, where the probe photon used has a wavelength of 690 nm (1.80 eV), the penetration depth is 47 nm, calculated using the method described in Sect. 4.6.1. For tr-ARPES, at the 10.75 eV photon energy used, the inelastic electron mean free path near the Fermi surface is $\lesssim$2 nm [26], which explains why it is a surface sensitive probe. Each of the three techniques hence provides a unique perspective of the PIPT, allowing us to gain a comprehensive view.

The material we chose to focus on is the first member of the rare-earth tritelluride family, $LaTe_3$. As introduced in Sect. 2.3, it is a unidirectional CDW compound with a simple phase diagram [19], providing a clean platform to explore the effect of photoexcitation. Its layered structure, whose b-axis lies out-of-plane [27], makes it susceptible to CDW order that forms below an estimated transition temperature of 670 K, with an associated gap of $2\Delta \approx 750$ meV [28]. Because of a small in-plane anisotropy in the material, the CDW forms solely along the crystallographic c-axis, with an incommensurate wavevector $\mathbf{q}_0 \approx \frac{2}{7}\mathbf{c}^*$, where $\mathbf{c}^*$ is the reciprocal unit vector [29]. The high value of T_c ensures that, in the course of the PIPT, the transient lattice temperature is maintained below T_c despite laser-induced heating. We will come back to this heating effect in Sect. 4.2.3.

4.2 Temporal Evolution of Diffraction Peaks

We first describe the UED experiments, which monitor the structural modulation through the intensity and width of diffraction peaks. These experiments, carried out in a transmission geometry, are sensitive to both the amplitude and phase coherence of the CDW [15, 30]. Figure 4.2a shows an equilibrium electron diffraction cut along the (3 0 L) line at room temperature. Superlattice peaks, characteristic of CDW formation, are indicated by arrows. They appear in pairs at $\pm\mathbf{q}_0$ relative to the much brighter lattice Bragg peaks, signifying the presence of long-range density-wave order.

4.2.1 Dynamics of CDW Satellite Peaks

Following photoexcitation at $t = 0$, the integrated intensity of the superlattice peak initially decreases within $\sim$1 ps (Fig. 4.2b), a timescale limited by the temporal resolution of our setup [31]. The intensity then recovers to a quasi-equilibrium value. Meanwhile, the peak width (measured by FWHM) also increases by several times its equilibrium value and subsequently decreases (Fig. 4.2b inset and c). The peak broadening observed here is a signature of a loss of long-range order, which, in turn, requires the appearance of topological defects in high concentrations [20, 32], a non-trivial consequence of the PIPT. The peak width does not relax fully within the time window examined, and this partial recovery is attributed to a residual defect density. In principle, the presence of crystallographic disorder can also cause a linewidth broadening during a phase transition [33, 34]. However, this scenario is ruled out by the width of the structural Bragg peaks, which is resolution-limited. In the dataset corresponding to Fig. 4.2b, the maximum value of the peak width implies a CDW correlation length of less than $\sim$10 crystallographic unit cells (blue squares in Fig. 4.2c, d). Based on these estimates, we calculate that for every two

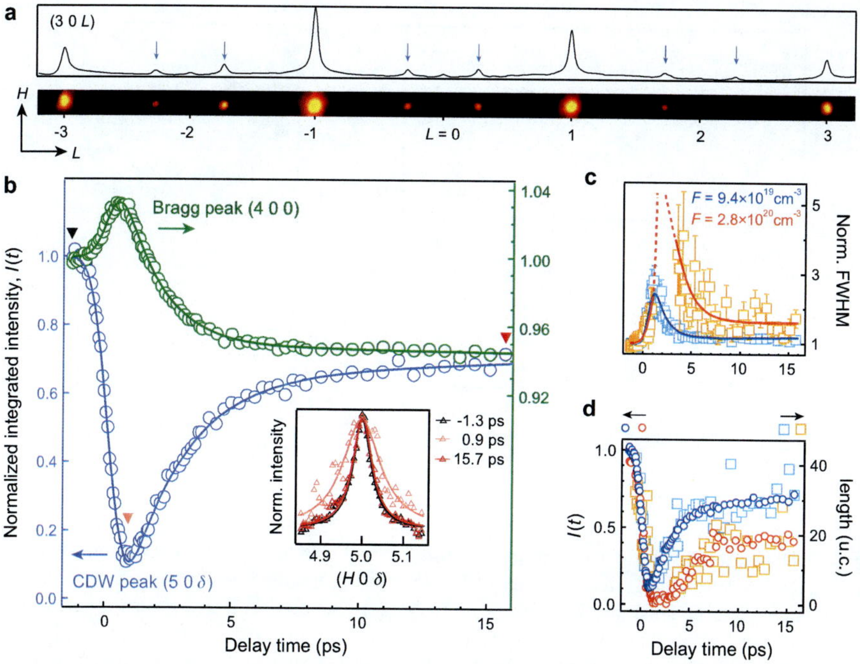

Fig. 4.2 Time evolution of electron diffraction after photoexcitation. (**a**) Electron diffraction cut along (3 0 L). Superlattice peaks are indicated by arrows. The line cut is obtained by integrating the colored strip along the H direction. (**b**) Time evolution of integrated intensities, $I(t)$, of the (4 0 0) Bragg peak and the (5 0 δ) superlattice peak after photoexcitation with an excitation density of 9.4×10^{19} cm^{-3}. Intensities are normalized to values before the arrival of the light pulse. *Inset*, snapshots of the superlattice peak at selected time delays, indicated by the triangles in the main panel. The transient broadening is isotropic along both H and L directions, and normalized line profiles shown are along H, from which full-width at half maximum (FWHM) is computed by fitting to a Lorentzian function (solid curves). (**c**) The time evolution of the superlattice peak width, normalized to values before photoexcitation, showing significant broadening through the CDW transition. Error bars represent one standard deviation in the Lorentzian fittings. Solid curves in **b** and **c** are fits to Eq. (3.14), while dashed lines in the orange curve are extrapolated to regions where the peak vanishes. (**d**) Time evolution of the integrated intensity of the (5 0 δ) superlattice peak (circles) overlaid onto the CDW correlation length (squares), showing good agreement. Excitation densities are the same as in **c**. Missing orange squares correspond to the time range where the CDW peaks are indistinguishable from the background and the width cannot be reliably extracted. u.c., crystallographic unit cell

photons absorbed, approximately one defect/anti-defect pair is created. We explain the details of this estimate in Sect. 4.6.2.

We next study how the time evolution of the superlattice peak changes with excitation density, F, where F is quoted in terms of absorbed photons per unit volume, calculated in Sect. 4.6.1. Beyond a critical value, $F_c \approx 2.0 \times 10^{20}$ cm^{-3}, the CDW melts, as the peak becomes indistinguishable from noise after photoexcitation (Fig. 4.3a, c). We estimate that the critical excitation density, F_c, corresponds to

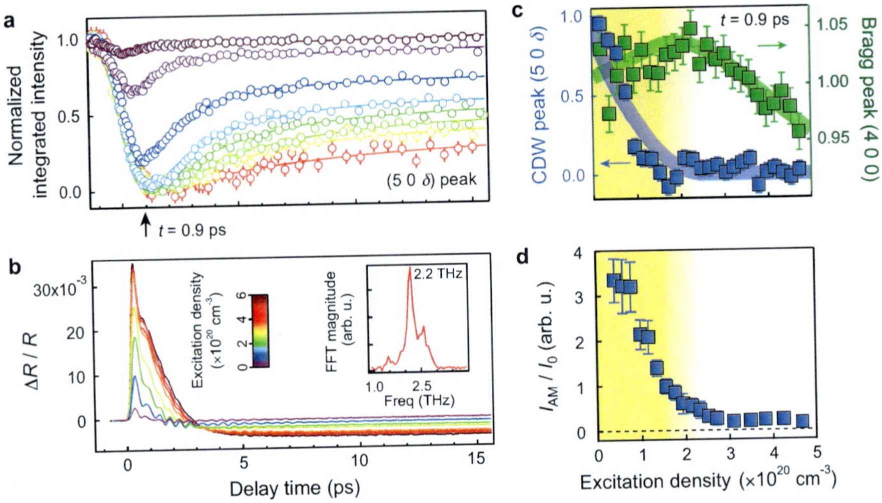

Fig. 4.3 Dependence of diffraction peaks and optical reflectivity on excitation density. (**a**) Time evolution of integrated intensity of the (5 0 δ) superlattice peak upon photoexcitation at different excitation densities. The colorscale is the same as used in **b**. Error bars are obtained from the standard deviation of noise prior to photoexcitation. Solid curves are fits to Eq. (3.14). The arrow indicates the time delay at which the intensity is plotted against the excitation density in **c**. (**b**) Transient reflectivity as a function of delay time at different excitation densities. *Inset*, Fourier transform of the oscillatory component measured at an excitation density of $4.1 \times 10^{19}\ \text{cm}^{-3}$. (**c**) The normalized (4 0 0) Bragg and (5 0 δ) superlattice peak integrated intensities at a time delay of $t = 0.9$ ps as a function of excitation density. Curves are guides to the eye. Error bars represent one standard deviation of statistical variation in intensity values. (**d**) Amplitude mode intensity, I_{AM}, as a function of excitation density. I_{AM} has been normalized against the maximum of the incoherent part, I_0 in Eq. (3.14), in the transient reflectivity trace. Error bars are obtained from one standard deviation in the fittings for I_0. In **c**,**d**, yellow-shaded regions denote the low excitation regime where the CDW has yet to melt completely

a defect every ~6 crystallographic unit cells, a length scale below which it is no longer appropriate to define the CDW with the wavevector $\mathbf{q}_0 \approx \frac{2}{7}\mathbf{c}^*$.

To understand how the CDW is reestablished after photoexcitation, we focus on the recovery timescale of the integrated intensity and linewidth shown in Fig. 4.2c, d. Most significantly, the characteristic time it takes for the peak to recover to quasi-equilibrium increases with excitation density. It should be noted that the recovery timescales are similar for the integrated intensity and the correlation length (Fig. 4.2d). This connection can be understood if we consider the ways by which the superlattice peak intensity can be reduced: (i) a suppression of the CDW amplitude; (ii) the excitation of phase modes (phasons) [30, 35]; (iii) a decrease in out-of-plane CDW correlation length [15], which we were not able to access in the transmission geometry of our experiment; and (iv) scattering from defect cores, which redistributes intensity across the entire Brillouin zone [20]. The latter two factors are controlled by the concentration of topological defects. The increased population of phasons originates partially from the temperature rise due to the laser pulse, but can also stem from defect motion [9]. These factors suggest that the dynamics of topological defects are intimately tied to the recovery of the superlattice peaks. Therefore, we infer that it is the reestablishment of CDW phase coherence that dominates the recovery timescale of the superlattice peak.

4.2.2 *Dynamics of Lattice Bragg Peaks*

The CDW amplitude—the first factor in the above list—recovers on a quicker timescale than the phase coherence, as we demonstrate in the following. For this purpose, we examine the structural Bragg peaks. As we recall from Sect. 2.2.1, their dynamics reflect the response of the CDW amplitude, whereas superlattice peaks additionally retain information about the phase coherence. In Fig. 4.2b, we show that when the CDW is suppressed, the Bragg peak first intensifies; subsequently, it weakens as the CDW amplitude recovers and the Debye–Waller factor dominates. This initial intensification signifies a reduction of the CDW amplitude, as distorted atoms return to their high symmetry positions.

These two competing contributions—transient reduction in the CDW amplitude and the Debye–Waller factor—make the analysis of Bragg peak intensity more complicated. The latter is also present in the superlattice peak, but its role is negligible compared to the suppression of the CDW order. The two contributions can be modeled by fitting the Bragg peak time trace to a modified version of Eq. (3.14)

$$\Delta I_{\text{Bragg}}(t) = \left[\frac{1}{2}\left(1 + \text{Erf}\left(\frac{2\sqrt{2}(t-t_0)}{w}\right)\right) \cdot \left(I_0 e^{-(t-t_0)/\tau_{\text{recovery}}}\right) + \Theta(t) I_{\text{DW}}\left(1 - e^{-t/\tau_{\text{DW}}}\right)\right] * g(w_0, t)\,, \quad (4.1)$$

where the first term in the square brackets represents the contribution due to a change in the CDW amplitude. The second term represents the Debye–Waller factor with a characteristic time τ_{DW} and a loss of intensity $I_{\text{DW}} < 0$. $\Theta(t)$ is the Heaviside step function to ensure the second term is non-positive. A representative fit is shown in Fig. 4.4b and the same analysis was performed for different excitation densities, shown in Fig. 4.4a. As expected, the Debye–Waller contribution scales linearly with the excitation density, as demonstrated in Fig. 4.4c.

Figure 4.4d shows a comparison of the CDW recovery time, τ_{recovery}, between the Bragg and superlattice peaks. These timescales are extracted from fitted curves in Figs. 4.3a and 4.4a. The CDW recovery timescale based on the Bragg peak is consistently shorter, and follows a different trend as a function of excitation density after the melting threshold, $F_c \approx 2.0 \times 10^{20}\,\text{cm}^{-3}$. Therefore, the Bragg peak enhancement disappears on a quicker timescale than it takes for the superlattice peak to regain its intensity. This discrepancy suggests that the CDW amplitude and phase coherence recover at different rates, an important characteristic of PIPTs that will be validated by further experiments using transient reflectivity and tr-ARPES.

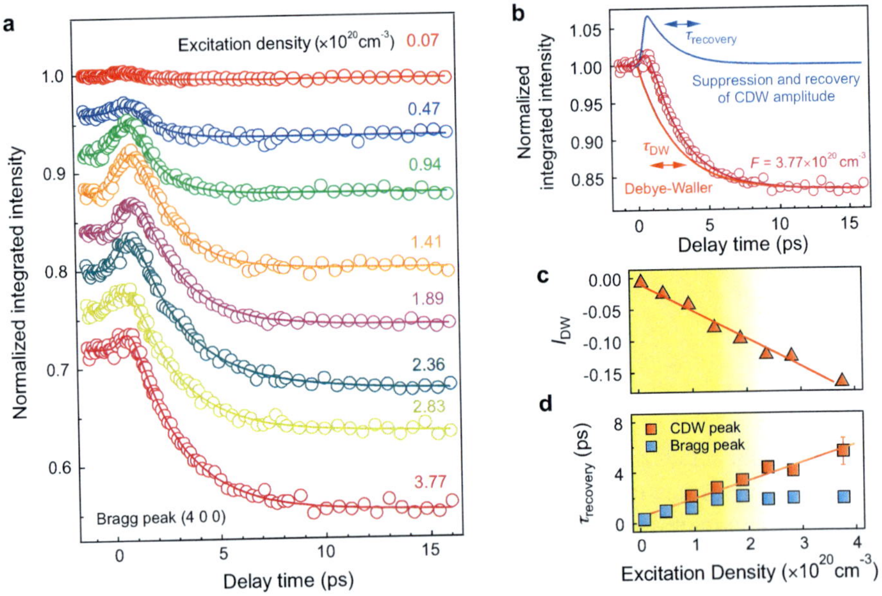

Fig. 4.4 Relaxation timescale of lattice Bragg peaks. (**a**) Time evolution of integrated intensity of the (4 0 0) Bragg peak at different excitation densities, with fitted curves superimposed. Each trace is vertically offset by −0.04 for clarity. (**b**) A curve fitting to Eq. (4.1) for an excitation density of 3.77×10^{20} cm^{-3}. Blue and orange curves represent two fitted components that account for the change in the CDW amplitude and the Debye–Waller factor, respectively. (**c**) Loss of Bragg peak intensity due to the Debye–Waller factor, I_{DW}, as a function of excitation density. The solid line is a linear fit. (**d**) Recovery timescales extracted from Bragg peaks (blue squares), compared with timescales from superlattice peaks (orange squares, the same as in Fig. 4.8a). The yellow-shaded regions in **c** and **d** are the same as those in Fig. 4.3, where the CDW is only partially suppressed

4.2.3 Effect of Temperature on Diffraction Peak Dynamics

Laser-induced heating is known to be significant in thin film samples used in UED where thermal dissipation is effectively limited to 2D. Therefore, the laser repetition rate was kept low for most diffraction measurements, between 0.5–5 kHz. Nevertheless, there is still a finite heating effect and it is necessary to estimate the steady-state and transient lattice temperatures and to investigate whether the recovery timescales of the CDW depend on the sample temperature.

The steady-state temperature was evaluated using the temperature-dependent superlattice peak intensity. Empirically, the intensity shows a BCS-like temperature dependence [19]. To obtain the steady-state temperature, the delay stage was positioned before time-zero (i.e. before photoexcitation), and the superlattice peak intensity was monitored with and without the pump beam present. Based on the value of T_c [28], the peak intensity in the absence of the pump beam, and the peak intensity with the pump beam present, we estimated the steady-state temperature of the sample using the BCS formula [15]. The results are plotted in Fig. 4.5a, b.

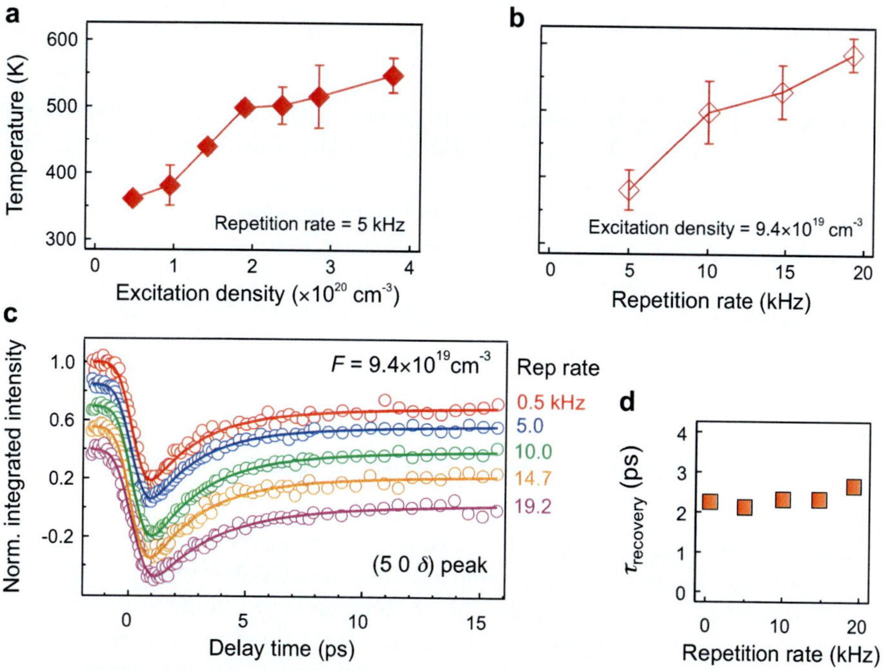

Fig. 4.5 Laser-induced heating and the effect on CDW peak recovery. (**a**) Estimated steady-state temperature of the sample as a function of excitation density at a fixed 5 kHz laser repetition rate. (**b**) Estimated steady-state temperature of the sample as a function of repetition rate at a fixed excitation density of 9.4×10^{19} cm^{-3}. Error bars in **a** and **b** are obtained by using two distinct superlattice peaks to estimate the temperature. (**c**) Time evolution of the integrated intensity of the (5 0 δ) superlattice peak at several laser repetition rates with a fixed excitation density of 9.4×10^{19} cm^{-3}. Each trace has been vertically offset by -0.15 for clarity. Solid curves are fits to Eq. (3.14). (**d**) Recovery time constants extracted from the fits in **c**, showing minimal dependence on the repetition rate and hence steady-state laser heating

As the sample temperature increases with both excitation density and laser repetition rate, it is necessary to verify that the results presented in the preceding sections are not due to an increased steady-state temperature. To show this, we compare the time evolution of the superlattice peak at a fixed excitation density but at a series of different repetition rates. These results are shown in Fig. 4.5c, d, where the recovery timescales are similar across a wide range of steady-state temperatures (Fig. 4.5b), in contrast to the phenomenology observed in Fig. 4.3a. Hence, the increase in the recovery time with a higher excitation density is not due to an elevated sample temperature from steady-state laser heating.

While the above arguments consider the effect of steady-state temperature, it is also necessary to obtain an estimate of the transient rise in the lattice temperature shortly after photoexcitation. An upper bound of such an increase is given by

$$\Delta T = \frac{Q}{mC}, \quad (4.2)$$

where ΔT is the change in the transient temperature, Q is the absorbed heat from a single laser pulse, m is the illuminated mass calculated using the laser spot size and penetration depth (estimated in Sect. 4.6.1), and C is the specific heat [27]. For instance, at the highest excitation density used in the experiment, 3.77×10^{20} cm^{-3}, the estimated transient increase in the lattice temperature is ~60 K. This calculation ensures that sum of the steady-state and the transient increase in temperature remains below $T_c \approx 670$ K [28], indicating that the observed photoinduced CDW melting is nonthermal in nature.

4.3 CDW Amplitude Dynamics from Transient Reflectivity

Comparing the recovery timescales extracted from Bragg and CDW peaks (Fig. 4.4d), we inferred that the amplitude of the CDW order parameter is reestablished before any long-range phase coherence. To confirm this picture, we turn to transient reflectivity measurements, which can probe the fast evolution of CDW amplitude [9, 21–23] with a finer temporal resolution and higher signal-to-noise ratio (see Sect. 1.3.1). In Fig. 4.3b, we present the transient reflectivity results where two components in each trace are visible: an incoherent response arising from the excitation and relaxation of quasiparticles, and an oscillatory coherent response predominantly from the 2.2 THz CDW amplitude mode [23]. The negative tail at long delay times results from laser-induced heating as the system reaches a quasi-equilibrium thermal state. If we compare the time traces between reflectivity and diffraction (Fig. 4.3a, b), there is one striking observation: The CDW recovery timescale extracted from the incoherent response in transient reflectivity is much quicker (up to ~1 ps), in contrast to that of the superlattice peak, which reaches 5.5 ps for the largest value of excitation density (Fig. 4.8a). As the incoherent response is known to be a sensitive probe of the CDW gap size [9, 21–23], we again infer that the amplitude recovers on a faster timescale than the phase coherence.

4.3.1 Correspondence of Excitation Densities

It is important to ensure that the excitation densities are consistent for a fair comparison of the timescales between the transient reflectivity and the UED measurements. In particular, the sample geometries are different between these two experiments: A thin film was used in diffraction while a cleaved surface was used for reflectivity. In this section, we check such consistency by comparing the critical value, F_c.

From Fig. 4.3a, above $F_c \approx 2.0 \times 10^{20}$ cm^{-3}, the superlattice peak vanishes into the background. This is summarized in the blue curve in Fig. 4.3c. In contrast to the behavior of the CDW peak, the response of the Bragg peak exhibits a downturn in intensity beyond F_c, which is shown in the green curve in Fig. 4.3c. This trend can be understood by noting that the Debye–Waller factor starts to dominate the Bragg peak response at high excitation densities, as the CDW has fully melted.

A similar critical excitation density is observed in the coherent amplitude mode response in the transient reflectivity measurement. We analyzed the coherent component by subtracting a single-exponential fit to Eq. (3.14) before Fourier transforming the remaining trace after 1 ps. A typical plot of the Fourier transformed trace is presented in the inset of Fig. 4.3b, where the 2.2 THz amplitude mode dominates the spectrum [23]. We present in Fig. 4.3d the amplitude mode oscillation magnitude, I_{AM}, which is taken from the peak height in the Fourier transformed spectrum. To compare I_{AM} across different excitation densities, it is normalized by the peak in the incoherent response, I_0 in Eq. (3.14). Figure 4.3d clearly illustrates a similar critical excitation density, F_c, for CDW melting: The amplitude mode oscillations weaken as regions with short-range CDW correlations dephase and become incoherent. Such correspondence of F_c with the UED measurement therefore justifies the comparison of recovery timescales between the two techniques.

4.4 Tracking the CDW Gap with Photoemission

To further investigate the amplitude restoration in a momentum- and energy-resolved fashion, we use tr-ARPES to probe the gap dynamics. Figure 4.6a shows a segment of the Fermi surface excited with a density above F_c. Following photoexcitation, spectral weight fills in the gapped portions of the Fermi surface and a normal-state Fermi surface is partially restored around 250 fs after the arrival of the pump pulse. To get some insights into the changes in the band dispersion, we examine the evolution of a particular momentum cut, $\mathbf{k}_{\parallel}$, as indicated by the yellow arrow in Fig. 4.6a. Snapshots of the cut at various time delays are shown in Fig. 4.6b–f. The Te $5p_z$ band above E_F (upper CDW band) is populated upon photoexcitation by quasiparticles decaying from higher energy. However, the CDW gap, Δ, remains. With increasing pump–probe delay, the edge of the upper CDW band moves toward E_F while spectral weight increases in the gapped region below E_F, indicating a reduction of the gap size, which is near its minimum at 250 fs. Beyond 400 fs, the upper CDW band edge starts to move away from E_F, and the spectral weight in the gapped region is then depleted, signaling a recovery of the gap. By 5 ps, the resulting band structure is essentially the same as the one before photoexcitation, but at an elevated lattice temperature.

This sequence of events can be summarized in Fig. 4.6g, which shows the time evolution of the momentum-integrated intensity of a representative gapped region (orange box in Fig. 4.6a), where the Fermi level, E_F, lies approximately at the

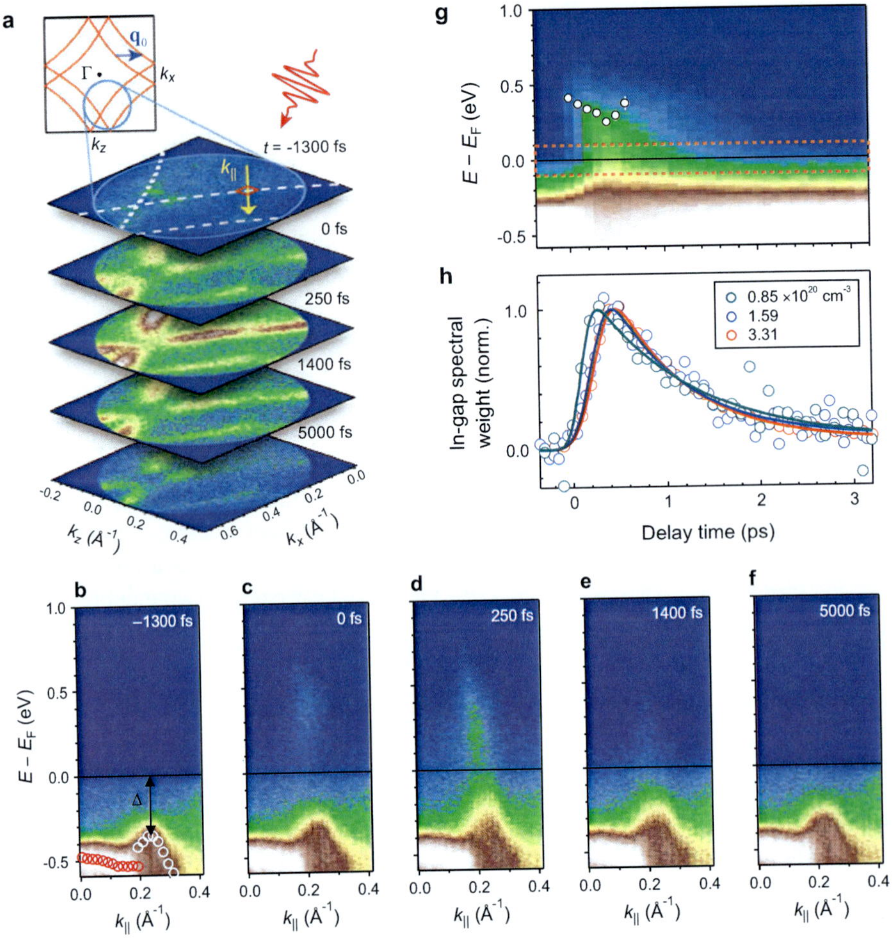

Fig. 4.6 **Photoemission spectra showing CDW gap dynamics.** (**a**) *Top*, tight-binding plot of the normal-state Fermi surface in the first Brillouin zone. The circle shows the probed part of the Fermi surface and the arrow marks the CDW wavevector $\mathbf{q}_0$. *Bottom*, a section of the Fermi surface through the photoexcitation process with an excitation density of 3.31×10^{20} cm^{-3}. Intensities are integrated over ± 10 meV around E_F. Cuts along $k_\parallel$ (yellow arrow) are shown in (**b**)–(**f**). In **b**, circles are fitted dispersions of Te $5p_z$ (white) and $5p_y$ (red) bands [36], where a gap $\Delta = 0.36(2)$ eV opens at the $5p_z$ band and causes it to back-bend. (**g**) Time evolution of the gapped region highlighted by the orange box in the $t = -1300$ fs slice of **a**. Fitted positions of the upper CDW band edge are marked by white dots. (**h**) Time evolution of the in-gap spectral weight obtained by integrating the intensity inside the orange box in **g**. Three traces correspond to different excitation densities, all normalized between 0 and 1 and fitted to Eq. (3.14)

center of the CDW gap (see Sect. 2.3.2 and Fig. 2.10c). The white dots in Fig. 4.6g highlight the fitted positions of the upper CDW band edge using a Lorentzian profile multiplied by the Fermi–Dirac distribution at each delay step [25]. The band edge of the lower CDW band cannot be reliably determined by the same fitting procedure due to the strong intensity of the Te $5p_y$ band, which lies in close proximity (red circles in Fig. 4.6b). Beyond 600 fs, the edge of the upper CDW band is difficult to determine as it is quickly depopulated. Even above the threshold excitation density, this band edge fitting suggests that a nonzero CDW gap remains through all time delays. This observation is further confirmed by a lack of spectral weight in certain parts of the Fermi surface, shown in Fig. 4.6a. Such a residual CDW gap has been observed before [25], and we interpret it as indicative of the presence of short-range CDW correlations [33, 37]. Notably, the existence of local correlations is consistent with the transient reflectivity data in Fig. 4.3b, d, where the amplitude mode oscillations do not vanish completely even above F_c.

To characterize the gap recovery more quantitatively, we plot in Fig. 4.6h the time evolution of in-gap spectral weight obtained by integrating the intensity over an energy window of ± 0.1 eV around E_F (orange box in Fig. 4.6g). After fitting the recovery with an exponential decay, the time constant obtained is less than 1.1 ps for all excitation densities measured, consistent with the transient reflectivity results (Fig. 4.8a). We thus identify $\sim$1 ps as the characteristic timescale for the restoration of the CDW amplitude, whereas the phase coherence takes up to 5.5 ps to recover at the highest excitation density measured (Fig. 4.8a).

4.4.1 *Effect of Temperature on CDW Gap Dynamics*

The tr-ARPES data presented in the preceding paragraphs was taken at 15 K, as opposed to the transient reflectivity and UED traces, which were taken at room temperature. This is because the energy resolution of the ARPES spectra is significantly improved at low temperature due to reduced Fermi–Dirac broadening. To verify that the results were not dependent on the temperature at which the measurements were performed, we took a control set of tr-ARPES spectra at room temperature. As shown in Fig. 4.7, the base temperature did not affect the recovery time of the CDW gap. This is expected because the transition temperature of $LaTe_3$ is $\sim$670 K [28], so the CDW order parameter amplitude at 295 K is very close to its value at 15 K [19, 28, 38].

4.5 Defect-Mediated Order Parameter Recovery

The CDW recovery timescales from the UED, tr-ARPES, and transient reflectivity measurements are summarized in Fig. 4.8a, showing two distinct trends. To rationalize the two decoupled time timescales, we sketch our interpretations in Fig. 4.8b–e. In this picture, the laser pulse excites energetic quasiparticles that create topological defects as they relax and dissipate energy. In the illustration, these defects are depicted as CDW dislocations [14, 32, 35]. In the meantime, the CDW gap and hence the amplitude of the CDW order parameter decreases, while the long-range

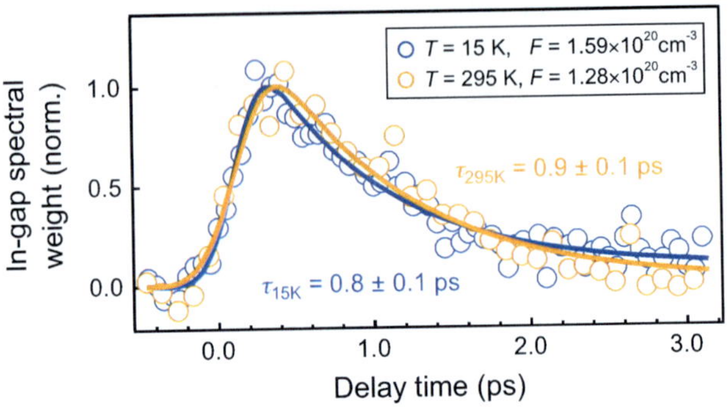

Fig. 4.7 Effect of sample temperature on photoemission dynamics. Time evolution of normalized in-gap spectral weight taken at two different base temperatures and similar excitation densities, showing negligible difference in the recovery timescale. The spectral weight is obtained by integrating the intensity over ±0.1 eV around E_F, the orange box in Fig. 4.6g. Solid curves are fits to Eq. (3.14)

phase coherence is also suppressed or destroyed (Fig. 4.8c). Within ~1 ps, the CDW amplitude recovers to quasi-equilibrium. Meanwhile, the long-range phase coherence is not fully restored (Fig. 4.8d). It takes several more picoseconds or longer, depending on the excitation density, for the defects to annihilate and for the CDW phase coherence to set in (Fig. 4.8e).

An important implication of Fig. 4.8a is that the phase coherence takes longer to recover with a higher concentration of topological defects (orange line). While pinpointing the exact reason behind this relationship requires a detailed theoretical treatment, here we limit ourselves to mentioning two possible mechanisms. If the recovery is determined by the annihilation of pairs of two-dimensional topological defects, then the presence of a large number of defects disturbs the coupling between CDWs in adjacent planes. Hence, this renormalized out-of-plane coupling reduces the restoring force that brings together defect/anti-defect pairs. A related possibility is that larger concentrations of photoinduced defects renormalize down the effective T_c [1] to a value close to that of the laser-heated sample in UED (Fig. 4.5a), and as a result, the restoration of the CDW order exhibits critical slowing-down associated with the proximity to the renormalized T_c [1].

The synergy of the three time-resolved probes used in this work has provided a uniquely comprehensive view of the photoinduced phase transition in a symmetry-broken state. Across the multiple techniques, a consistent picture is obtained where a quick recovery of the CDW amplitude is followed by a slower restoration of phase coherence. Topological defects are preeminent in this regard, inhibiting the reestablishment of long-range order in the nonequilibrium setting. A transition driven by photoexcitation, where topological defects are generated in the ordered phase, therefore represents a distinctive framework under which non-adiabatic transitions can be instigated. Rapid improvements in the temporal resolution of UED may further allow the destruction of the ordered state to be investigated in detail in subsequent work [39]. The results presented in this paper pave the way to future

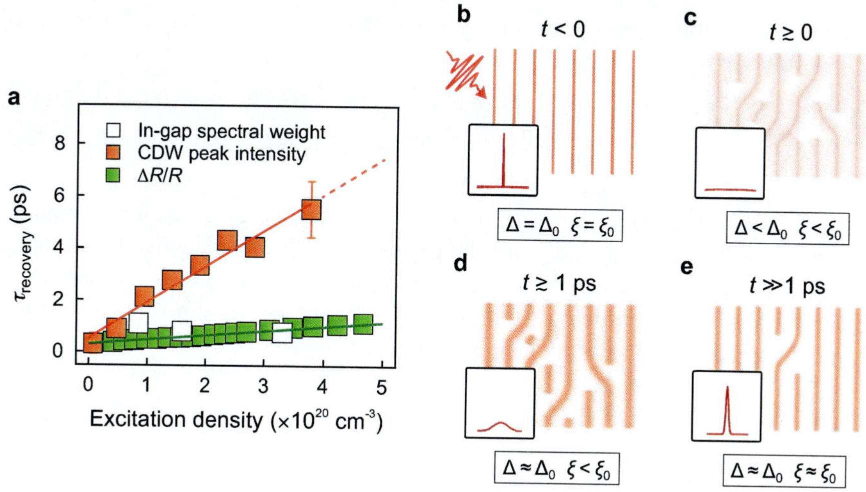

Fig. 4.8 Summary of CDW recovery timescales and dynamics. (**a**) Characteristic recovery times of the amplitude and phase coherence as a function of excitation density. Transient reflectivity and tr-ARPES track the amplitude dynamics. The recovery timescale of the phase coherence is quantified by the diffraction intensity instead of the peak width due to the better signal-to-noise ratio (Fig. 4.2d). Error bars, when larger than the symbol size, denote one standard deviation of fits to Eq. (3.14). Lines are fits to the data; the dashed segment is extrapolated. **b–e** Schematic illustration of the CDW evolution after photoexcitation. In each image, the unidirectional charge density modulation is depicted as stripes in real space. Stripe brightness indicates the strength of the CDW amplitude and smearing represents phase excitations. A cartoon of the CDW diffraction peak is presented in the lower left corner. Δ and ξ denote CDW amplitude and correlation length, respectively; Δ_0 and ξ_0 are values at equilibrium. (**b**) Before photoexcitation, the CDW amplitude is large and the CDW is long-range ordered. The corresponding superlattice diffraction is represented by a narrow-width peak. (**c**) Following photoexcitation, the CDW amplitude is suppressed and topological defects are formed. These effects lead to a reduction in the integrated intensity and a broadening of the peak width. (**d**) After $\sim$1 ps, the CDW amplitude is largely restored, while defects persist. The diffraction peak remains broad due to the presence of these topological defects. (**e**) Many defects annihilate at a further time delay though a nonzero defect concentration remains. The superlattice peak significantly narrows as the phase coherence sets in

studies on nonequilibrium defect-mediated transitions and the optical manipulation of topological defects in other ordered states of matter.

4.6 Supplementary

4.6.1 Calculation of Excitation Density

We presented the pump pulse fluence in terms of excitation density, F, to ensure consistency across the different setups, which use pump lasers with different wavelengths: 1038 nm for UED, 780 nm for transient reflectivity, and 720 nm for tr-ARPES. Excitation density values are calculated in terms of absorbed photons per unit volume:

$$F = \frac{E}{\hbar\omega} \cdot (1 - R) \cdot \frac{1}{\pi(w_x/2)(w_y/2)d}, \tag{4.3}$$

where E is the incident laser pulse energy, $\hbar\omega$ is the energy of a single photon, R is the reflectivity, w_x and w_y are the effective laser spot sizes on the sample (measured at FWHM), which take into consideration the incident angle, and d is the penetration depth at which the light *intensity* decays to $1/e$ of its original value. From the Lambert–Beer law, the penetration depth is given by [40]

$$d = \frac{c}{\kappa\omega}, \tag{4.4}$$

where c is the speed of light in vacuum, κ is the imaginary part of the complex refractive index $N = n + i\kappa$. If we are given the values of reflectivity R and the real part of optical conductivity σ_1, κ can be determined by simultaneously solving

$$R = \left|\frac{N-1}{N+1}\right|^2 = \frac{(n-1)^2 + \kappa^2}{(n+1)^2 + \kappa^2}, \tag{4.5}$$

$$\kappa = \frac{\sigma_1}{2\epsilon_0 n\omega}, \tag{4.6}$$

where ϵ_0 is the vacuum permittivity. Using R and σ_1 values at the relevant wavelengths [41, 42], the penetration depths are calculated to be 44 nm, 44 nm, and 46 nm for the 1038 nm (1.19 eV), 780 nm (1.59 eV), and 720 nm (1.72 eV) pump wavelengths, respectively. Using these values, we are able to obtain an excitation density that holds a more consistent comparison across the different time-resolved techniques, and the consistency has been experimentally verified in Sect. 4.3.1.

4.6.2 Estimating CDW Correlation Length and Defect Density

In Fig. 4.2d, we plotted the time evolution of the superlattice peak intensity and the CDW correlation length. At high excitation density, the recovery timescales of both quantities are significantly slower compared to the case at low excitation density. Here, we detail the quantitative analysis that estimates the CDW correlation length, ξ, from the peak width. The correlation length ξ is computed as [43]

$$\xi = (\mathrm{FWHM}^2 - w_0^2)^{-1/2}, \tag{4.7}$$

where w_0 is the instrumental resolution, taken to be the average width of the CDW peak before photoexcitation. Here, we assume that the CDW width is instrumental resolution-limited in equilibrium [19].

The knowledge of both excitation density (F) and CDW correlation length (ξ) enables us to establish a correspondence between the number of absorbed photons and that of created defects. Let n_p denote the number of defect/anti-defect *pairs* generated in one Te plane for each photon absorbed. Given the correlation length ξ, which is measured in crystallographic unit cells, $\frac{1}{2}\xi^{-2}$ defect/anti-defect pairs exist per Te plane per unit cell. On the other hand, assuming that the absorbed photons are evenly distributed among the four Te planes in each unit cell (Fig. 2.8a), we have $\frac{1}{4}FV$ photons per Te plane per unit cell, where V is the unit cell volume. Therefore,

$$n_p = \frac{2}{\xi^2 FV}, \tag{4.8}$$

from which we estimated in Sect. 4.2.1 that for every two photons absorbed, approximately one defect/anti-defect pair is generated.

References

1. Goldenfeld, N.: Lectures on Phase Transitions and the Renormalization Group. Westview, Boulder (1992)
2. Kibble, T.W.B.: Topology of cosmic domains and strings. J. Phys. A **9**, 1387–1398 (1976)
3. Zurek, W.H.: Cosmological experiments in condensed matter systems. Phys. Rep. **276**, 177–221 (1996)
4. Bowick, M.J., Chandar, L., Schiff, E.A., Srivastava, A.M.: The cosmological Kibble mechanism in the laboratory: String formation in liquid crystals. Science **263**, 943–945 (1994)
5. Chuang, I., Durrer, R., Turok, N., Yurke, B.: Cosmology in the laboratory: Defect dynamics in liquid crystals. Science **251**, 1336–1342 (1991)
6. Ruutu, V.M.H., Eltsov, V.B., Gill, A.J., Kibble, T.W.B., Krusius, M., et al.: Vortex formation in neutron-irradiated superfluid ^{3}He as an analogue of cosmological defect formation. Nature **382**, 334–336 (1996)
7. Bäuerle, C., Bunkov, Y.M., Fisher, S.N., Godfrin, H., Pickett, G.R.: Laboratory simulation of cosmic string formation in the early Universe using superfluid ^{3}He. Nature **382**, 332–334 (1996)
8. Spaldin, N.A.: Multiferroics: From the cosmically large to the subatomically small. Nat. Rev. Mater. **2**, 17017 (2017)
9. Yusupov, R., Mertelj, T., Kabanov, V.V., Brazovskii, S., Kusar, P., et al.: Coherent dynamics of macroscopic electronic order through a symmetry breaking transition. Nat. Phys. **6**, 681–684 (2010)
10. Mertelj, T., Kusar, P., Kabanov, V.V., Giraldo-Gallo, P., Fisher, I.R., et al.: Incoherent topological defect recombination dynamics in $TbTe_3$. Phys. Rev. Lett. **110**, 156401 (2013)
11. Griffin, S.M., Lilienblum, M., Delaney, K.T., Kumagai, Y., Fiebig, M., et al.: Scaling behavior and beyond equilibrium in the hexagonal manganites. Phys. Rev. X **2**, 041022 (2012)
12. Bevis, N., Hindmarsh, M., Kunz, M., Urrestilla, J.: CMB polarization power spectra contributions from a network of cosmic strings. Phys. Rev. D **76**, 043005 (2007)
13. Nasu, K. (ed.): Photoinduced Phase Transitions. World Scientific, Singapore (2004)

14. Lee, P.A., Rice, T.M.: Electric field depinning of charge density waves. Phys. Rev. B **19**, 3970–3980 (1979)
15. Grüner, G.: Density Waves in Solids. Addison-Wesley, Boston (1994)
16. Hoesch, M., Bosak, A., Chernyshov, D., Berger, H., Krisch, M.: Giant Kohn anomaly and the phase transition in charge density wave $ZrTe_3$. Phys. Rev. Lett. **102**, 086402 (2009)
17. Girault, S., Moudden, A.H., Pouget, J.P.: Critical X-ray scattering at the Peierls transition of the blue bronze. Phys. Rev. B **39**, 4430–4434 (1989)
18. Holt, M., Zschack, P., Hong, H., Chou, M.Y., Chiang, T.-C.: X-ray studies of phonon softening in $TiSe_2$. Phys. Rev. Lett. **86**, 3799–3802 (2001)
19. Ru, N., Condron, C.L., Margulis, G.Y., Shin, K.Y., Laverock, J., et al.: Effect of chemical pressure on the charge density wave transition in rare-earth tritellurides $R\mathrm{Te}_3$. Phys. Rev. B **77**, 035114 (2008)
20. Fultz, B., Howe, J.: Transmission Electron Microscopy and Diffractometry of Materials. Springer, Berlin/Heidelberg (2013)
21. Kabanov, V.V., Demsar, J., Podobnik, B., Mihailovic, D.: Quasiparticle relaxation dynamics in superconductors with different gap structures: Theory and experiments on $YBa_2Cu_3O_{7-\delta}$. Phys. Rev. B **59**, 1497–1506 (1999)
22. Demsar, J., Biljaković, K., Mihailovic, D.: Single particle and collective excitations in the one-dimensional charge density wave solid $K_{0.3}MoO_3$ probed in real time by femtosecond spectroscopy. Phys. Rev. Lett. **83**, 800–803 (1999)
23. Yusupov, R.V., Mertelj, T., Chu, J.H., Fisher, I.R., Mihailovic, D.: Single-particle and collective mode couplings associated with 1- and 2-directional electronic ordering in metallic $R\mathrm{Te}_3$ (R=Ho, Dy, Tb). Phys. Rev. Lett. **101**, 24602 (2008)
24. Schmitt, F., Kirchmann, P.S., Bovensiepen, U., Moore, R.G., Rettig, L., et al.: Transient electronic structure and melting of a charge density wave in $TbTe_3$. Science **321**, 1649–1652 (2008)
25. Rettig, L., Cortés, R., Chu, J.-H., Fisher, I.R., Schmitt, F., et al.: Persistent order due to transiently enhanced nesting in an electronically excited charge density wave. Nat. Commun. **7**, 10459 (2016)
26. Seah, M.P., Dench, W.A.: Quantitative electron spectroscopy of surfaces: A standard data base for electron inelastic mean free paths in solids. Surf. Interface Anal. **1**, 2–11 (1979)
27. Ru, N., Fisher, I.R.: Thermodynamic and transport properties of YTe_3, $LaTe_3$, $CeTe_3$. Phys. Rev. B **73**, 033101 (2006)
28. Hu, B.F., Cheng, B., Yuan, R.H., Dong, T., Wang, N.L.: Coexistence and competition of multiple charge-density-wave orders in rare-earth tritellurides. Phys. Rev. B **90**, 085105 (2014)
29. DiMasi, E., Aronson, M.C., Mansfield, J.F., Foran, B., Lee, S.: Chemical pressure and charge-density waves in rare-earth tritellurides. Phys. Rev. B **52**, 14516–14525 (1995)
30. Overhauser, A.W.: Observability of charge-density waves by neutron diffraction. Phys. Rev. B **3**, 3173–3182 (1971)
31. Moore, R.G., Lee, W.S., Kirchman, P.S., Chuang, Y.D., Kemper, A.F., et al.: Ultrafast resonant soft X-ray diffraction dynamics of the charge density wave in $TbTe_3$. Phys. Rev. B **93**, 024304 (2016)
32. Vogelgesang, S., Storeck, G., Horstmann, J.G., Diekmann, T., Sivis, M., et al.: Phase ordering of charge density waves traced by ultrafast low-energy electron diffraction. Nat. Phys. **14**, 184–190 (2018)
33. Chatterjee, U., Zhao, J., Iavarone, M., Di Capua, R., Castellan, J.P., et al.: Emergence of coherence in the charge-density wave state of $2H$-$NbSe_2$. Nat. Commun. **6**, 6313 (2015)
34. Kogar, A., de la Pena, G.A., Lee, S., Fang, Y., Sun, S.X.-L., et al.: Observation of a charge density wave incommensuration near the superconducting dome in Cu_xTiSe_2. Phys. Rev. Lett. **118**, 027002 (2017)
35. Lee, W.S., Chuang, Y.D., Moore, R.G., Zhu, Y., Patthey, L., et al.: Phase fluctuations and the absence of topological defects in a photo-excited charge-ordered nickelate. Nat. Commun. **3**, 838 (2012)

36. Brouet, V., Yang, W.L., Zhou, X.J., Hussain, Z., Ru, N., et al.: Fermi surface reconstruction in the CDW state of $CeTe_3$ observed by photoemission. Phys. Rev. Lett. **93**, 126405 (2004)
37. Lee, P.A., Rice, T.M., Anderson, P.W.: Fluctuation effects at a Peierls transition. Phys. Rev. Lett. **31**, 462–465 (1973)
38. Moore, R.G., Brouet, V., He, R., Lu, D.H., Ru, N., et al.: Fermi surface evolution across multiple charge density wave transitions in $ErTe_3$. Phys. Rev. B **81**, 073102 (2010)
39. Morimoto, Y., Baum, P.: Diffraction and microscopy with attosecond electron pulse trains. Nat. Phys. **14**, 252–256 (2018)
40. Dressel, M., Grü Electrodynamics of Solids: Optical Properties of Electrons in Matter. Cambridge University Press, Cambridge (2002)
41. Sacchetti, A., Degiorgi, L., Giamarchi, T., Ru, N., Fisher, I.R.: Chemical pressure and hidden one-dimensional behavior in rare-earth tri-telluride charge-density wave compounds. Phys. Rev. B **74**, 125115 (2006)
42. Pfuner, F., Degiorgi, L., Chu, J.-H., Ru, N., Shin, K.Y., et al.: Optical properties of the charge-density-wave rare-earth tri-telluride compounds: A view on $PrTe_3$. Physica B **404**, 533–536 (2009)
43. Laulhé, C., Huber, T., Lantz, G., Ferrer, A., Mariager, S.O., et al.: Ultrafast formation of a charge density wave state in $1T$-TaS_2: Observation at nanometer scales using time-resolved x-ray diffraction. Phys. Rev. Lett. **118**, 247401 (2017)

Chapter 5
Dynamical Slowing-Down in an Ultrafast Transition

Abstract Complex systems, which consist of a large number of interacting constituents, often exhibit universal behavior near a phase transition, such as a slowdown of certain dynamical observables. This phenomenon, known as critical slowing-down, is well studied in thermodynamic phase transitions but is less understood in highly nonequilibrium settings, where the time it takes to traverse the phase boundary becomes comparable to the timescale of dynamical fluctuations. Using transient optical spectroscopy and femtosecond electron diffraction, we studied a photoinduced phase transition in a model charge density wave (CDW) compound, $LaTe_3$. We observed that it takes the longest time to suppress the order parameter at the threshold photoexcitation density, where the CDW transiently vanishes. This finding can be captured by generalizing the time-dependent Landau theory to a system far from equilibrium. Our understanding of dynamical slowing-down may offer insights into other general principles behind nonequilibrium transitions in many-body systems.

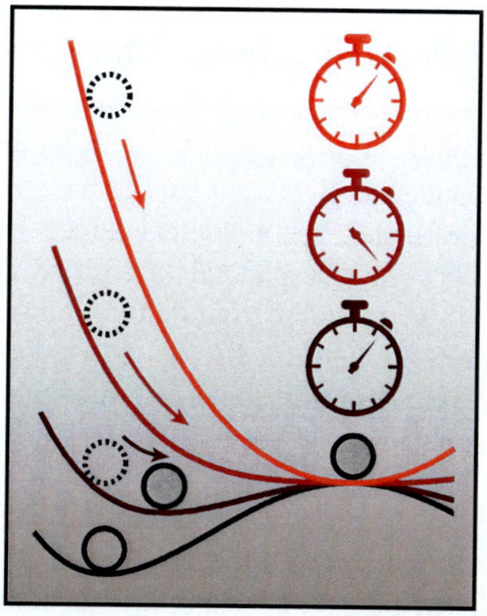

Cartoon of the melting dynamics of an electronic crystal. The melting takes the longest time (middle clock) when the order, depicted as a ball, first vanishes transiently

A. Zong, *Emergent States in Photoinduced Charge-Density-Wave Transitions*, Springer Theses, https://doi.org/10.1007/978-3-030-81751-0_5

5.1 Critical Slowing-Down

5.1.1 Equilibrium Phase Transitions

In a second-order symmetry-breaking phase transition, the spatial extent of fluctuating regions diverges close to the critical temperature, T_c (Fig. 5.1a). Correspondingly, the relaxation time of these fluctuations tends to infinity, a phenomenon known as critical slowing-down [5, 6]. The phenomenology of slowing dynamics near a critical point is much more general: it has been observed in first-order transitions [7, 8], glasses [9, 10], dynamical systems [11], and even microbial communities [12]. Its common occurrence makes it a robust signature of phase transitions in a vast array of complex systems [13].

Close to equilibrium, critical slowing-down has been well characterized in condensed matter systems. Theoretically, it is described by a dynamical critical exponent, whose value depends on the dynamic universality class [6]. Experimentally, the evidence comes from a vanishing rate of change in the order parameter close to T_c, with early reports in refs. [3, 7, 14] (Fig. 5.1c, d). While these measurements probe the slowing dynamics in the time domain, it can be observed in the frequency domain as well. For example, inelastic neutron scattering has revealed a narrowing quasi-elastic peak along the energy axis as T_c is approached (Fig. 5.1b), indicating a suppressed relaxation rate of critical fluctuations [2, 15, 16]. Moreover, if there is a collective mode associated with the phase transition, the mode softening in the vicinity of T_c (Fig. 5.1e) is also taken as a signature of critical slowing-down [17].

5.1.2 Nonequilibrium Photoinduced Transitions

For symmetry-breaking phase transitions in a highly nonequilibrium setting, the dynamics are much less understood. Recent studies have found important features in nonequilibrium transitions, such as topological defects, which are absent in their equilibrium counterparts [18–21]. Despite the differences, a slowdown in dynamics is thought to carry over to systems far from equilibrium. For example, in a rapid quench into a broken-symmetry state, the Kibble–Zurek theory suggests that critical slowing-down plays a central role in domain formation: as the phase boundary is traversed at a faster rate than the system can respond, spatially disconnected regions may adopt distinct configurations of the same degenerate ground state [22]. Characteristic domain structures in liquid crystals have indeed been observed [23, 24], providing indirect evidence for the slowdown.

To study the dynamics in a nonequilibrium setting, charge density wave (CDW) transitions instigated by an intense femtosecond laser pulse provide an accessible platform with well-controlled tuning parameters. A suite of time-resolved probes can track the evolution of electronic and lattice orders after strong photoexcitation,

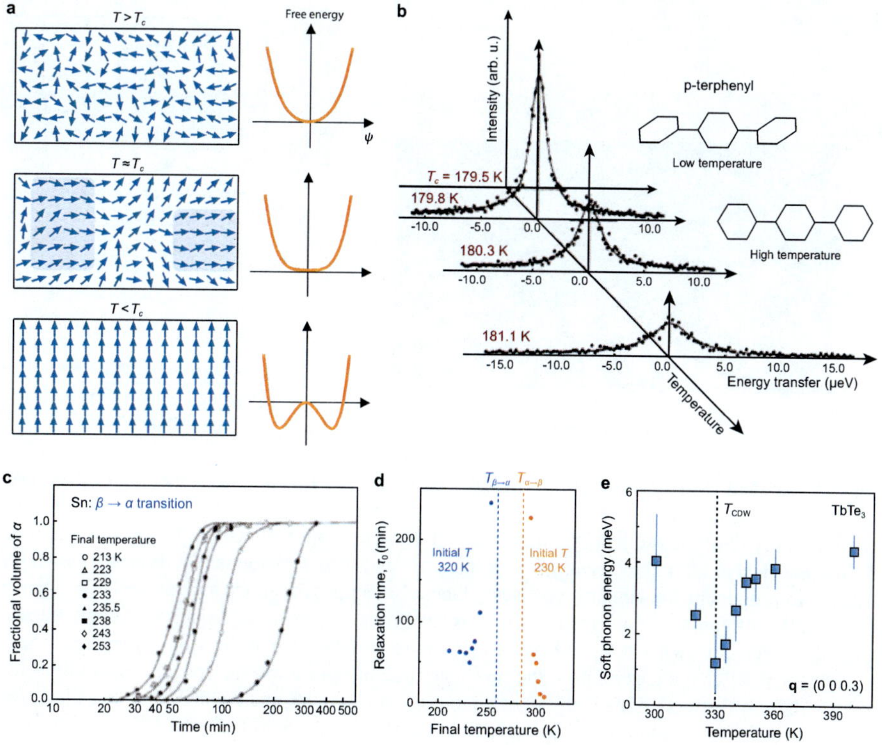

Fig. 5.1 Critical slowing-down in equilibrium phase transitions. (**a**) Schematics of a second-order ferromagnetic phase transition and the associated Landau free energy diagrams. Close to T_c, critical fluctuations exhibit a diverging correlation length (shaded areas) and a diverging relaxation time. (**b**) Neutron scattering intensity of a molecular crystal, p-terphenyl, at wavevector $(\frac{5}{2}, \frac{1}{2}, 0)$. The energy-resolved diffraction displays a narrowing quasi-elastic peak as its T_c of the antiferrodistortive structural transition is approached from the above. This transition occurs as the planar phenyl rings at high temperature become twisted upon cooling (*inset*), leading to unit cell doubling in the a-b plane [1]. From the energy linewidth, the lifetimes of orientational fluctuation of the phenyl rings are estimated to be 11.5 ns (179.8 K), 5.7 ns (180.3 K), and 2.5 ns (181.1 K). (**c**) Time taken for the first-order phase transition between the high-temperature white tin (β) and the low-temperature gray tin (α). The plot shows the time-dependent volume fraction of α-Sn, determined from neutron diffraction, when fine-powder samples initially kept at 320 K (β-Sn) were brought to various final temperatures. (**d**) Relaxation times of $\alpha \to \beta$ and $\beta \to \alpha$ transitions extracted from time-dependent volume faction curves, which are fitted to the Avrami equation, $1 - \exp[-(t/\tau_0)^n]$. (**e**) Soft phonon frequency near the CDW wavevector in $TbTe_3$ across its T_{CDW}, determined by inelastic X-ray scattering. As phonon frequency approaches zero at T_{CDW}, it takes longer for atoms to move back to their high-symmetry positions. (Adapted with permission from Ref. [2], IOP Publishing (**b**); Ref. [3], Elsevier (**c**, **d**); Ref. [4], APS (**e**))

as we have seen in Chap. 4. These probes can also offer insights into the critical behavior, if present, during the phase transition. Immediately after photoexcitation, a coherently excited CDW amplitude mode was observed to soften transiently [18], hinting at critical slowing-down (Fig. 5.2a). Right below T_c, a diverging relaxation time back to equilibrium was interpreted as another signature [8, 25] (Fig. 5.2b).

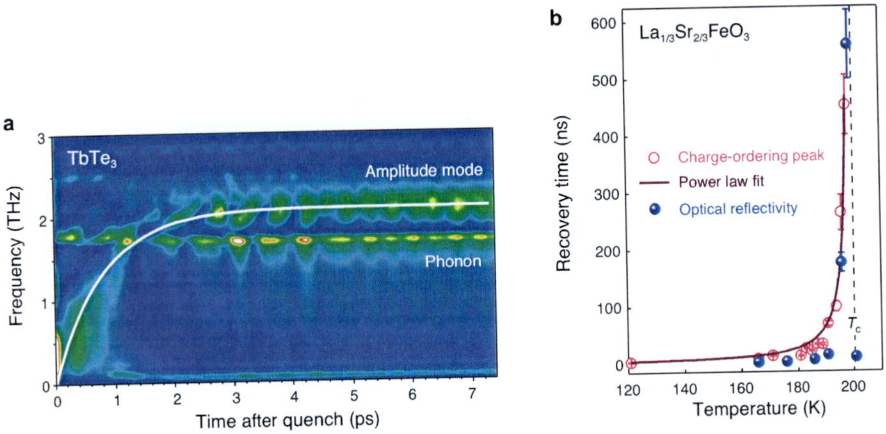

Fig. 5.2 Critical slowing-down in photoinduced phase transitions. (**a**) Softening of the CDW amplitude mode (AM) in $TbTe_3$ after a photoinduced quench. The image shows the Fourier transform spectra of the oscillatory components of transient optical reflectivity, which is measured at different delays after the arrival of a strong laser pulse that momentarily destroys the CDW order. Two prominent frequencies were seen, labeled as amplitude mode and phonon. The former significantly softens near time zero when the CDW order is the most suppressed (white curve), suggesting critical slowing-down. Measurements were performed at 15 K, deep in the CDW phase of $TbTe_3$. (**b**) After photoexcitation, the recovery time of both transient optical reflectivity (blue balls) and X-ray diffraction intensity of the charge-ordering peak (pink circles) exhibits a diverging trend near the charge-ordering transition temperature of a $La_{1/3}Sr_{2/3}FeO_3$ thin film. Solid curve is a fit of the X-ray data to a power law $(1 - T/T_c)^{-\delta}$, yielding $\delta = 1.25 \pm 0.10$. (Adapted with permission from Ref. [18], Springer Nature Ltd (**a**); Ref. [8] (**b**))

However, observables in previous studies, such as amplitude mode frequency or quasiparticle relaxation time across the spectroscopic gap, are only well defined in the broken-symmetry state [26–28]. To demonstrate slowing dynamics in the vicinity of a nonequilibrium phase transition, ideally one would measure an increased timescale near the phase boundary compared to both ordered and disordered states.

In this chapter, we circumvent this obstacle by focusing on a different observable during the photoinduced melting of a CDW: the time taken to suppress the condensate. Note that we have shifted our focus from the order parameter recovery discussed in Chap. 4 to its initial photoinduced melting analyzed here. With increasing photoexcitation densities, the perturbed system will enter one of the two transient states, where the long-range CDW is either partially or completely suppressed [20]. The two states are separated by the threshold excitation density, F_{melt}, where the spatially coherent condensate first vanishes completely. Through transient reflectivity and time-resolved diffraction measurements at different excitation densities, we will show that it takes the maximum time to suppress the CDW right at F_{melt}, indicating dynamical slowing-down near the boundary between the two transient states. Here, we use *dynamical* slowing-down to emphasize the highly nonequilibrium nature of the system and to distinguish it from *critical* slowing-down commonly defined in a second-order phase transition in equilibrium [5, 6].

5.2 Timescale for Photoinduced CDW Suppression

To study the timescale of photoinduced CDW suppression, our material of interest is again the paradigmatic CDW system, $LaTe_3$ [29]. Like other rare-earth tritellurides, $LaTe_3$ possesses a quasi-2D structure (Fig. 2.8a) and develops a unidirectional CDW with wavevector $\mathbf{q}_{\mathrm{CDW}}$ along the c-axis below $T_c \approx 670\,\mathrm{K}$ [30]. In equilibrium, the CDW transition is characterized by the appearance of satellite peaks in a diffraction pattern as well as gap openings at certain parts of the Fermi surface connected by $\mathbf{q}_{\mathrm{CDW}}$ (Fig. 2.10b).

5.2.1 *Initial System Response from Multiple Probes*

Upon the arrival of a strong femtosecond laser pulse, the CDW order is transiently suppressed [20, 31, 32]. We first establish the timescale for this process by performing ultrafast electron diffraction (UED) and transient optical spectroscopy (TOS), which reveal how the lattice and electrons respond to intense photoexcitation, respectively. Unlike the keV UED setup we used in Chap. 4, in which the initial response after photoexcitation is limited by the instrument temporal resolution, here we employ an MeV UED setup [33] that has a much improved time resolution, as determined by THz electron streaking (Sect. 5.5.2). Previous measurements from time- and angle-resolved photoemission spectroscopy (tr-ARPES) [20] and time-resolved X-ray diffraction (tr-XRD) [32] are also included to obtain a comprehensive and consistent view of the ultrafast melting process.

While UED and tr-XRD track the evolution of CDW satellite peaks at characteristic wavevector $\mathbf{q}_{\mathrm{CDW}}$ (Fig. 5.3a), TOS and tr-ARPES probe the change in the spectroscopic gap (Fig. 5.3b). Despite the different observables, the initial response that corresponds to CDW melting proceeds with a similar timescale, denoted by τ in Fig. 5.3c. The rising edges across the four techniques in Fig. 5.3c all span a time interval of $\tau \approx 400\,\mathrm{fs}$, with variations arising from the different temporal resolutions in each setup (Sect. 5.5.1) and different photoexcitation densities used (Fig. 5.5a). The agreement among structural and electronic probes suggests the presence of strong electron–phonon coupling in this system. Notably, the value of 400 fs is on the same scale as the period of the 2.2-THz CDW amplitude mode [28], further indicating the vital role of lattice vibrations in the formation of the charge order [34].

5.2.2 *CDW Suppression Time for Varying Excitation Densities*

Among the four techniques discussed, TOS possesses the best temporal resolution and signal-to-noise ratio. Hence, it enables us to more quantitatively investigate the

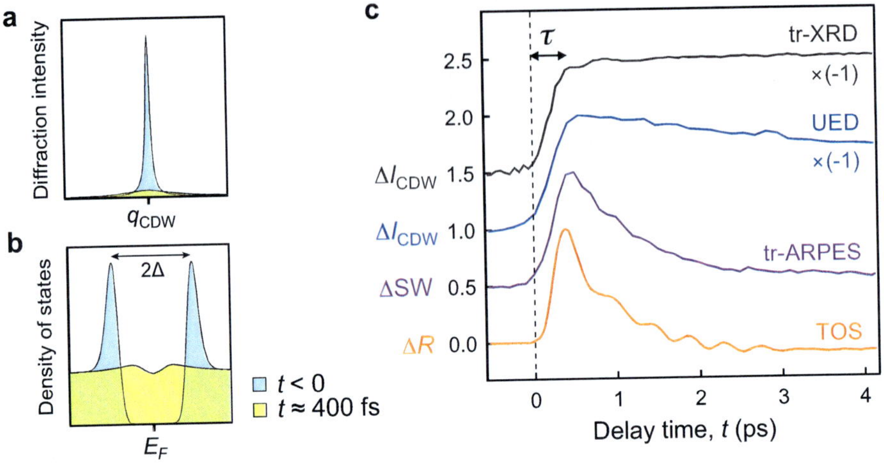

Fig. 5.3 Photoinduced CDW melting probed by multiple time-resolved techniques. (**a**, **b**), Schematics of the superlattice peak and density of states before (blue) and after (yellow) photoexcitation. (**c**) Transient response of superlattice peak intensity (ΔI_{CDW}), in-gap spectral weight (ΔSW), and reflectivity (ΔR) probed by corresponding time-resolved techniques. All traces are normalized between 0 and 1 and vertically offset for clarity. ΔI_{CDW} is inverted for easier comparison. All traces are measured in $LaTe_3$ except for tr-XRD, which measures $TbTe_3$ [32], a similar compound in the same rare-earth tritelluride family with a lower T_c

timescale of CDW suppression, τ, as we vary the laser excitation density, F, quoted in terms of absorbed photon number per unit volume [20]. Figure 5.4a shows the temporal evolution of the transient reflectivity, $\Delta R/R$, across a large range of F. The trace from Fig. 5.3c is overlaid at the corresponding F. The data presented was taken using a probe photon energy of 1.80 eV (690 nm). As I will explain in Sect. 5.5.3, this probe wavelength is selected among the white light supercontinuum because it is the energy most sensitive to the dynamics of the CDW gap.

If we zoom into the initial change in the transient reflectivity (Fig. 5.4d, e), we find that the rise time is slightly longer when the excitation density is approximately 2×10^{20} cm^{-3}. To quantitatively extract a characteristic timescale, we fit $\Delta R/R$ traces at each excitation density to Eq. (3.14), which provides a phenomenological model involving a single-exponential decay. In Fig. 5.4b, we present an example cut (red curve) at $F = 5.81 \times 10^{20}$ cm^{-3}, together with the fitting result (black curve). The fitting shows poor agreement. In particular, the kink around 0.6 ps in the $\Delta R/R$ trace leads to significant discrepancies between the data and the fitted curve from 1 to 4 ps. To improve the fitting procedure, we hence performed a global fit for traces at all excitation densities using a two-component phenomenological model, as detailed in Sect. 5.3.1. An example fit is presented in Fig. 5.4c, showing excellent agreement. We attribute the two components to quasiparticle excitations in different parts of the Brillouin zone (see Sect. 5.3.1 for details), and we extract the CDW melting time (τ) from the rise time of the first component (Fig. 5.4c, blue dashed curve).

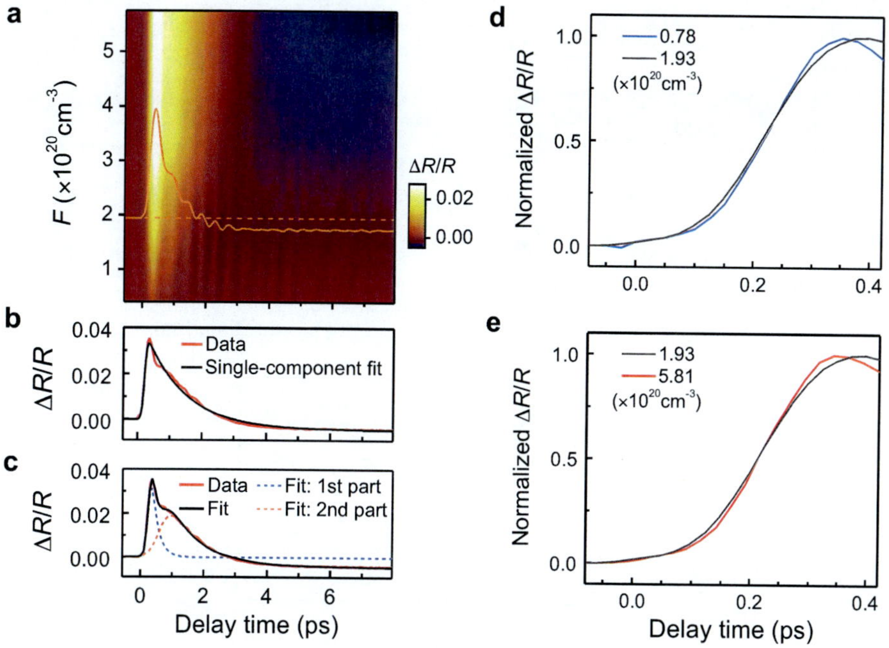

Fig. 5.4 **Transient reflectivity at different excitation densities.** (**a**) $\Delta R/R$ traces at different excitation densities, F, expressed in terms of absorbed photons per unit volume. A particular $\Delta R/R$ cut is overlaid at the excitation density indicated by the dashed line. It is the same trace shown in Fig. 5.3c. (**b**) $\Delta R/R$ trace at $F = 5.81 \times 10^{20}$ cm^{-3}, with a single-component fit to Eq. (3.14), showing poor agreement from 1 to 4 ps. (**c**) The same $\Delta R/R$ trace as in **b**, but with a fit to a two-component decay using Eq. (5.1), showing excellent agreement. (**d**, **e**) Selected traces of $\Delta R/R$ in the CDW melting time window, featuring a maximum melting timescale when F reaches the threshold value of $F_{\text{melt}} \approx 2 \times 10^{20}$ cm^{-3}, as exemplified by a less steep slope of the initial rise. All traces are normalized between 0 and 1 for easier comparison

Remarkably, the rise time τ displays a non-monotonic trend as a function of excitation density (Fig. 5.5a, orange circles), with a maximum at $\sim 2 \times 10^{20}$ cm^{-3} (black arrow). To confirm this observation, we similarly track the suppression of superlattice peaks using UED at various excitation densities (Fig. 5.5a, blue diamonds and Fig. 5.5c). Despite significantly larger errors due to lower signal-to-noise ratio and poorer temporal resolution compared to the TOS measurements, the initial timescale in the UED experiments suggests the same non-monotonic behavior in τ. We further note an independent measurement on $SmTe_3$ [35], a CDW compound in the same family as $LaTe_3$, that demonstrates a similar fluence-dependent trend in the initial system response (Fig. 5.5b).

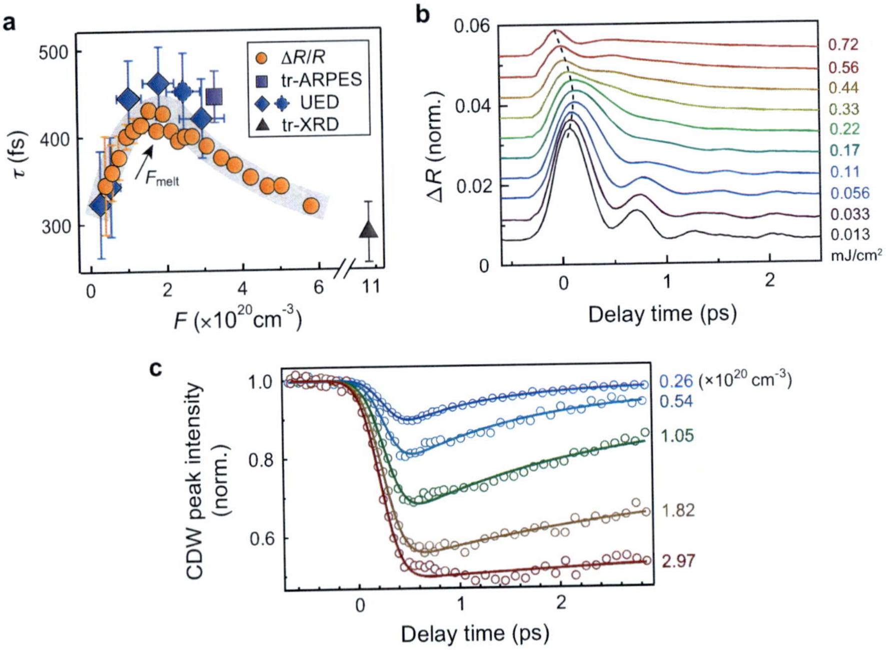

Fig. 5.5 CDW suppression time across the threshold pump fluence. (**a**) CDW suppression time, τ, as a function of excitation density, F. Gray curve is a guide to the eye, highlighting the maximum τ at F_{melt} as indicated by the arrow. Data points for tr-ARPES and tr-XRD correspond to curves in Fig. 5.3c; data points for $\Delta R/R$ correspond to traces in Fig. 5.4a. For the UED data, the corresponding trace of the dashed diamond is shown in Fig. 5.3c, while the rest, measured on a separate sample, are shown in **c**. Error bars indicate uncertainties in curve fittings and in the instrumental temporal resolution. (**b**) Differential reflectivity of $SmTe_3$ at 800 nm normalized by the incident pump fluence, which is labeled for each trace. Dashed curve is a guide to the eye, highlighting the non-monotonic trend of the initial rise time. (**c**) Initial suppression of the superlattice peak intensity measured by MeV UED, where the excitation density is labeled for each trace. Intensities are normalized to values before photoexcitation. Solid curves are fits to Eq. (3.14) and the FWHM of the erf function is taken as the CDW suppression time (τ) in panel **a**. The non-vanishing intensity beyond the melting threshold arises from background intensity in the diffraction pattern and non-uniform illumination of all layers of the sample due to a finite pump laser penetration depth. (Adapted with permission from Ref. [35], APS (**b**))

5.3 Signatures of the Threshold Excitation Density

To associate this non-monotonic behavior in τ with dynamical slowing-down during photoinduced CDW melting, we next establish that the melting proceeds the slowest precisely at the threshold excitation density when the CDW in the illuminated sample volume is just fully destroyed, namely, $F_{\text{melt}} \approx 2 \times 10^{20}$ cm^{-3}. We make three observations in this regard. First, the value 2×10^{20} cm^{-3} corresponds to the point where the superlattice peak is observed to be maximally suppressed in UED measurements [20], suggesting that τ indeed peaks at the threshold excitation

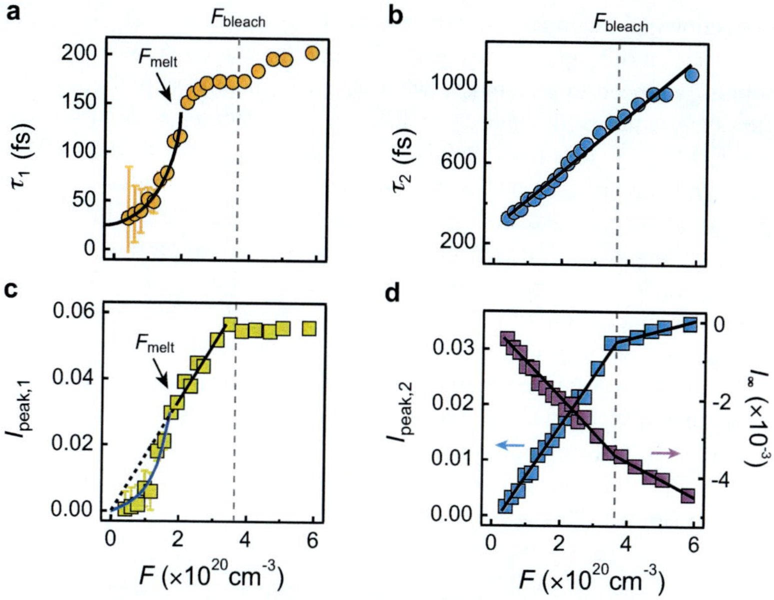

Fig. 5.6 Excited quasiparticle dynamics across the threshold pump fluence. (**a**) Relaxation time of excited quasiparticles across the CDW gap (τ_1), showing a steeply increasing trend near F_{melt} (black arrow), an indicator of the CDW gap closing. Black curve is a guide to the eye. (**b**) Relaxation of excited quasiparticles in the ungapped part of the Brillouin zone (τ_2), showing a linear increase with excitation density. (**c**) Initial excited quasiparticle population, $I_{\text{peak},1}$, showing a super-linear dependence on F below F_{melt} (arrow) and a plateau beyond F_{bleach} (vertical dashed line). Black line is a linear fit with extrapolation (dashed) to zero. Blue curve is a fit to Eq. (5.2). (**d**) Excited quasiparticle population in the ungapped momentum region, represented by $I_{\text{peak},2}$ (left axis), and lattice heating, represented by I_∞ (right axis), both showing a kink at F_{bleach} without any salient features at F_{melt}. All quantities were obtained by fitting $\Delta R/R$ traces in Fig. 5.4a to Eq. (5.1). Error bars, if larger than the marker size, represent uncertainties in curve fittings

density. Second, the time for the initial fast relaxation in transient reflectivity displays a steeply increasing trend at F_{melt} (Fig. 5.6a). This is attributed to a vanishing energy gap at the Fermi level when the CDW is completely suppressed, which limits the decay rate of excited quasiparticles [26–28]. Third, the maximum reflectivity change, $I_{\text{peak},1}$, also displays distinct behavior below and above F_{melt} (Fig. 5.6c). Below F_{melt}, the presence of a CDW gap modifies the transient population of excited quasiparticles, resulting in a super-linear $I_{\text{peak},1}$ as a function of excitation density (Fig. 5.6c, blue curve; see Sect. 5.3.2). Beyond F_{melt}, the excited quasiparticle population is directly proportional to the excitation density (Fig. 5.6c, black line). Above an even higher value F_{bleach}, the peak reflectivity $I_{\text{peak},1}$ plateaus (Fig. 5.6c, vertical dashed line) due to quasiparticle bleaching. It is worth emphasizing that at F_{melt}, the lattice temperature stays below T_c at all time delays after photoexcitation [20], reaffirming that the observed CDW melting is nonthermal in nature and is not a result of transient lattice heating above the transition temperature.

In the following sections, we substantiate the observations in the preceding paragraph via a more detailed analysis of the transient reflectivity traces. First and foremost, we need to understand what physical quantity is probed by $\Delta R/R$. Unlike the superlattice peak intensity in diffraction measurements or in-gap spectral weight in tr-ARPES, optical reflectivity is not a direct gauge of the CDW order parameter. Typically, in a gapped system, the value of transient reflectivity is taken to be proportional to the excited quasiparticle density [26–28], which in turn is sensitive to the gap size. For example, clear oscillations are present in ΔR traces (Figs. 5.3c, 5.4a, and 5.10), with a dominant contribution from the CDW amplitude mode [28]—the modulation of the gap magnitude. Based on this sensitivity of ΔR to the gap size as well as the consistency of the initial timescale in Figs. 5.3c and 5.5a across techniques, we take the initial rise time (τ) in transient reflectivity as the time needed for the amplitude of the CDW order parameter to be maximally suppressed. The value of τ is well separated from any electron–electron scattering timescale ($\leq$100 fs) [26, 36] and represents a simultaneous population of excited quasiparticles and renormalization of the gap, which occur self-consistently.

5.3.1 *Quasiparticle Dynamics Across the Excitation Threshold*

In addition to the initial rise time in $\Delta R/R$, its maximum height and relaxation time also reveal valuable information about the excited quasiparticle and CDW dynamics. For a quantitative analysis, we adopt the following phenomenological model for fitting $\Delta R/R$ traces,

$$\begin{aligned} \Delta R(t)/R(t) = \Bigg\{ & \left[\frac{1}{2}\left(1+\mathrm{Erf}\left(2\sqrt{2}(t-t_0)/\tau\right)\right) I_{\mathrm{peak},1}\, e^{-(t-t_0)/\tau_1}\right] \\ & + \left[\frac{1}{2}\left(1+\mathrm{Erf}\left(2\sqrt{2}(t-t_1)/\tau'\right)\right)\left(I_\infty + I_{\mathrm{peak},2}\, e^{-(t-t_1)/\tau_2}\right)\right]\Bigg\} * g(w_0, t). \end{aligned} \tag{5.1}$$

Before discussing the meaning of the fitting parameters, we first describe the procedure and present the result. All traces shown in Fig. 5.4a are included in a global fitting algorithm, where τ' and t_1 are constrained to be the same for all excitation densities to limit the number of free parameters. Their fitted values are $\tau' = 1.078(7)$ and $t_1 = 0.763(3)$ ps. Other important parameters as a function of excitation density are shown in Fig. 5.6. For all fittings, we use a Gaussian kernel, $g(w_0, t)$, to model the intrinsic instrument temporal resolution and w_0 represents the FWHM of the Gaussian.

To understand the physical picture behind the two components, which are represented by the two $[\cdot]$ terms in Eq. (5.1), we first compare their relaxation times, τ_1 and τ_2, shown in Fig. 5.6a, b. While τ_1 shows an abrupt change at F_{melt}, τ_2 is featureless across the threshold excitation density. This suggests that the first component is more sensitive to the CDW gap, as the order parameter transiently vanishes at F_{melt}. Indeed, the steeply increasing trend highlighted by the black curve in Fig. 5.6a is consistent with previous reports of suppressed quasiparticle relaxation in similar CDW systems close to T_c [26–28]. Given that the Fermi surface is only partially gapped in the CDW state and the gap size is highly anisotropic in the Brillouin zone [37], we associate the first $[\cdot]$ term with quasiparticle dynamics in the gapped region, while the second with that in the metallic region. We note that the smoothly increasing τ_2 as a function of excitation density is consistent with phonon-mediated quasiparticle relaxation in metallic systems [36].

This interpretation of $\Delta R/R$ traces is further supported by the comparison between $I_{\mathrm{peak},1}$ (Fig. 5.6c) and $I_{\mathrm{peak},2}$ (Fig. 5.6d, left). At the threshold F_{melt}, $I_{\mathrm{peak},1}$ changes from a super-linear to a linear trend, while $I_{\mathrm{peak},2}$ displays no distinctive feature. As will be explained in Sect. 5.3.2, the super-linear trend in $I_{\mathrm{peak},1}$ is consistent with the excited quasiparticle population in a gapped band structure, where the transient gap size is dependent on the excitation density. On the other hand, the quasiparticle dynamics in the ungapped part of the Brillouin zone is less affected by the changing CDW gap, so there is a lack of feature in $I_{\mathrm{peak},2}$ at F_{melt}.

With this interpretation in mind, our primary subject of the current study, τ, represents the time to modify the excited quasiparticle population in the gapped region. Quantitatively, τ in the transient reflectivity measurement is consistent with the initial response time in other time-resolved techniques (Figs. 5.3c and 5.5a). Therefore, gap renormalization and quasiparticle excitation in the gapped region occur self-consistently. Compared to τ, the rise time of the second component, τ', is considerably longer. This slower rise is accompanied by the relaxation of the first component (Fig. 5.6b), suggesting quasiparticle scattering across different parts of the Brillouin zone.

Another important parameter, I_∞, represents the transient reflectivity at long delay time relative to the probed time window, indicating laser-induced heating. By plotting I_∞ and $I_{\mathrm{peak},i=1,2}$ as a function of photoexcitation density (Fig. 5.6c, d), we further observe a kink at $F \approx 3.6 \times 10^{20}$ cm^{-3}. This value of excitation density corresponds to approximately 0.2 absorbed photon per unit cell of $LaTe_3$. We interpret this value as F_{bleach}, where bands near the Fermi level within the energy of the probe photon are depleted. It thus accounts for the plateau feature in $I_{\mathrm{peak},1}$ (Fig. 5.6c), where the population of excited quasiparticles is saturated.

5.3.2 Excited Quasiparticle Population

In the previous section, we identified the critical excitation density, F_{melt}, with the location of a kink in the maximum population of excited quasiparticles in the gapped region, $I_{\mathrm{peak},1}$ (Fig. 5.6c). This is the threshold point where a super-linear trend of $I_{\mathrm{peak},1}$ turns into a linear behavior, evident in the fitted blue curve and the black line in Fig. 5.6c. In this section, we expand the discussion on the super-linear trend below F_{melt}.

At low excitation density where the CDW is not fully suppressed, a phonon-mediated decay channel of quasiparticles results in a maximum transient quasiparticle population of [27, 28]

$$I_{\text{peak},1} = I_0 F \frac{(\Delta + k_B T/2)^{-1}}{1 + \gamma \sqrt{2k_B T/\pi\Delta}\, e^{-\Delta/k_B T}}, \tag{5.2}$$

where I_0 is a constant of proportionality, F is the excitation density, k_B is the Boltzmann constant, $T = 300$ K is the sample temperature, and Δ is the CDW gap size. Here, γ is a material specific parameter accounting for the phonon modes that participate in the relaxation of the quasiparticles [27, 28]. Strictly speaking, Eq. (5.2) only applies to a system with an isotropic gap [27], but empirically it works well for the rare-earth tritelluride series where the gap size is momentum-dependent [28].

In the fit of Fig. 5.6c at $F < F_{\text{melt}}$, we assume the simplest case where the gap size Δ depends on F linearly,

$$\Delta = \Delta_0(1 - F/F_{\text{melt}}), \tag{5.3}$$

where $\Delta_0/k_B = 4062$ K is the equilibrium value of the gap at 300 K [30]. The only free parameters in the curve fitting of Fig. 5.6c are I_0 and γ, where the fitted value of $\gamma = 2.8 \pm 0.5$. The fit reproduces the super-linear trend, in particular near $F = 0$. When F approaches F_{melt}, Eq. (5.2) is no longer applicable because it predicts a vanishing $I_{\text{peak},1}$ as Δ is reduced to zero. The discrepancy arises from the assumption in Eq. (5.2) that Δ is constant in time but Δ is in fact time-dependent. Therefore, the fit presented in Fig. 5.6c is restricted to a range below F_{melt}.

5.4 The Slowdown During CDW Suppression

We have established two important points so far: (i) the initial rise time (τ) in the transient reflectivity data corresponds to the time taken to suppress the CDW order; (ii) upon varying photoexcitation density, τ peaks at F_{melt}, a threshold point where the CDW order is maximally suppressed. To interpret these observations, we draw some parallels between the present nonequilibrium study and its equilibrium counterparts. At equilibrium, when the temperature is close to T_c, time-domain measurements of the order parameter indicate a reduced rate of change, which signifies critical slowing-down [3, 7, 14] (Fig. 5.1c, d). Here, we use photoexcitation density in lieu of temperature as the tuning parameter, and we extend the timescale to the femtosecond regime. Similarly, we interpret the maximum value of τ at exactly the threshold excitation density as a signature of dynamical slowing-down in this ultrafast phase transition.

To understand how a slowdown in dynamics can be extended to a regime far from equilibrium, we again make reference to the established framework of symmetry-breaking transitions in equilibrium, which are parameterized by an order parameter

ψ. On a phenomenological level, we consider the standard Landau potential (Fig. 5.1a), $\mathcal{W}(\psi) = -\alpha|\psi|^2 + \frac{\beta}{2}|\psi|^4$, which gives the simplest description of the second-order CDW transition in $LaTe_3$ [29]; here, α, β are the usual Landau coefficients. To see the slowdown near T_c, the typical treatment is to solve the time-dependent Landau equation [6],

$$\frac{\partial \psi}{\partial t} = -\Gamma \frac{\delta \mathcal{W}}{\delta \psi} + \eta(t), \tag{5.4}$$

where Γ is a phenomenological parameter and $\eta(t)$ is the Langevin noise term representing fluctuations. Close to T_c where the order parameter ψ is small and the free energy $\mathcal{W}(\psi)$ develops a flat bottom (Fig. 5.1a, middle), the relaxation time of long-wavelength fluctuations in ψ can be shown to approach infinity, which is the origin of critical slowing-down [6].

5.4.1 *Time-Dependent Landau Theory*

We now generalize this treatment to a highly nonequilibrium regime. A CDW transition is associated with both a modulation of electron density, described by the electronic order parameter ψ_e, and a periodic lattice distortion, described by the lattice order parameter ψ_l. If we are only interested in equilibrium properties, we can limit ourselves to considering either of these orders. However, as mobile electrons and heavy ions may have very different dynamics, we need to consider the orders separately to capture the nonequilibrium state. Our analysis is based on the Landau free energy functional [6, 38–40]:

$$\mathcal{W} = -\alpha_e|\psi_e|^2 + \frac{\beta}{2}|\psi_e|^4 - \zeta(\psi_e\psi_l^* + \psi_e^*\psi_l) + \alpha_l|\psi_l|^2. \tag{5.5}$$

Here, α_e, α_l, and β are the model parameters; the $\zeta(\cdot)$ term represents electron–phonon coupling. This model implies that (i) non-linearity (the $|\psi_e|^4$ term) occurs due to electron–electron effects, (ii) the two orders are linearly coupled, and (iii) the lattice potential has a parabolic form originating from the elastic force. This potential can be straightforwardly generalized to non-homogeneous cases by including spatial derivatives, to cases where non-linearity is due to phononic effects, or to cases where more than one phonon mode is coupled to ψ_e. Despite its simplicity, this model contains all required physical ingredients and has a small number of free parameters. In addition, it has well described a number of recent experiments related to the CDW dynamics [18, 38, 40]. Below, we also assume that photoexcitation acts as a homogeneous energy quench, allowing us to fix the order parameters to be real and to focus on the dynamical properties of the CDW amplitude.

The next step is to formulate the dynamical equations. The electron–electron interaction defines the fastest timescale ($\tau_e < 100\,\text{fs}$) in the system; on the phononic timescale, electrons instantly adjust themselves to the local value of the Landau potential. Hence, we impose overdamped dynamics on ψ_e, as done in the equilibrium time-dependent Landau formalism [6]. On the other hand, the heavy ions behave more like classical objects. This allows us to introduce the following equations of motion:

$$\frac{d\psi_e}{dt} \propto -\frac{\delta W}{\delta \psi_e}, \tag{5.6}$$

$$\frac{d^2\psi_l}{dt^2} \propto -\frac{\delta W}{\delta \psi_l}. \tag{5.7}$$

In normalized variables $x \equiv \psi_e(t)/\psi_e(T=0)$ and $y \equiv \psi_l(t)/\psi_l(T=0)$, Eqs. (5.6)–(5.7) read

$$\tau_e \frac{dx}{dt} - \alpha(t)x + x^3 + \zeta_0(x-y) = 0, \tag{5.8}$$

$$\frac{1}{\omega_0^2}\frac{d^2y}{dt^2} + (y-x) = 0. \tag{5.9}$$

Equations (5.8)–(5.9) imply that the effective force acting on the electronic (or lattice) order is $f_x \propto \alpha x - x^3 - \zeta_0(x-y)$ (or $f_y \propto x - y$). ζ_0 is the coupling strength between the electrons and the lattice; τ_e is the electron–electron scattering time that characterizes the relaxation within the electron subsystem; ω_0 is the unrenormalized phonon frequency at $\mathbf{q}_{\mathrm{CDW}}$; $\alpha(t)$ is the phenomenological Landau parameter that takes the value of $\alpha_0 = (T_c - T_{\mathrm{env}})/T_c$ at equilibrium, where $T_{\mathrm{env}} = 300\,\text{K}$ is the environment temperature. The time-dependence of $\alpha(t)$ is associated with the relaxation of excited quasiparticles after the laser pulse.

To model the photoexcitation event, we impose the following dynamics on the Landau potential,

$$\alpha(t) = \alpha_0 - \Theta(t)\kappa F e^{-t/\tau_0}, \tag{5.10}$$

where $\Theta(t)$ is the Heaviside step function, F is the excitation density in the unit of photons/cm^3, and κ is a constant of proportionality for normalizing F. τ_0 is the intrinsic thermal relaxation time of the quasiparticles after photoexcitation. For numerical calculations, we set $\tau_0 = 0.3\,\text{ps}$, taken from the transient reflectivity relaxation time at a small excitation density far below T_c [18, 20, 28]. Note that in Eq. (5.10), we neglect the temperature dependence of $\alpha(t)$ arising from transient lattice heating after photoexcitation. This is because based on heat capacity estimates, we expect a maximum lattice temperature rise of $\Delta T_{\max} \lesssim 100\,\text{K}$ [20], which is small compared to the high transition temperature $T_c \approx 670\,\text{K}$ [30] for our room temperature measurements.

To determine the values of τ_e and ζ_0, we consider small fluctuations around the equilibrium state, $x(t) = x_{\rm eq} + \delta x(t)$, $y(t) = y_{\rm eq} + \delta y(t)$, where $x_{\rm eq} = y_{\rm eq} = \sqrt{\alpha_0}$. For small δx and δy, we can linearize Eqs. (5.8)–(5.9),

$$\tau_e \frac{d\delta x}{dt} + 2\alpha_0 \delta x + \zeta_0(\delta x - \delta y) = 0, \tag{5.11}$$

$$\frac{1}{\omega_0^2}\frac{d^2\delta y}{dt^2} + (\delta y - \delta x) = 0. \tag{5.12}$$

Using the ansatz $\delta x = ae^{\lambda t}$ and $\delta y = be^{\lambda t}$, we obtain a cubic equation for the eigenmode λ,

$$\lambda^3 + \lambda^2 \frac{2\alpha_0 + \zeta_0}{\tau_e} + \lambda\omega_0^2 + \frac{2\alpha_0\omega_0^2}{\tau_e} = 0. \tag{5.13}$$

The solution to Eq. (5.13) defines the CDW amplitude mode frequency $\omega_{\rm AM} = {\rm Im}\,\lambda$ and its damping $\gamma_{\rm AM} = -{\rm Re}\,\lambda$. Therefore, if one fixes the values of $\omega_{\rm AM}$, $\gamma_{\rm AM}$, and ω_0, one can solve for τ_e and ζ_0. Based on the AM measured in transient reflectivity (Fig. 5.10b), the position and width of the AM peak give $\omega_{\rm AM} = 2.2 \cdot (2\pi)$ THz and $\gamma_{\rm AM} = 0.2 \cdot (2\pi)$ THz. From previous Raman measurements, we expect $\omega_0 = 3.25 \cdot (2\pi)$ THz [41, 42]. These experimental parameters lead to an electron–electron scattering time $\tau_e = 27$ fs and electron–phonon coupling strength $\zeta_0 = 1.35$. It is worth noting that the numerically solved value of τ_e is below 100 fs and is expected for $LaTe_3$—a metallic system without strong electron–electron correlation effects [26, 36].

Using $x(t = 0) = y(t = 0) = \sqrt{\alpha_0}$ as the initial condition for Eqs. (5.8)–(5.10), we track the dynamics of the order parameter—both electronic and lattice parts—after photoexcitation for different excitation densities (Fig. 5.7). For weak

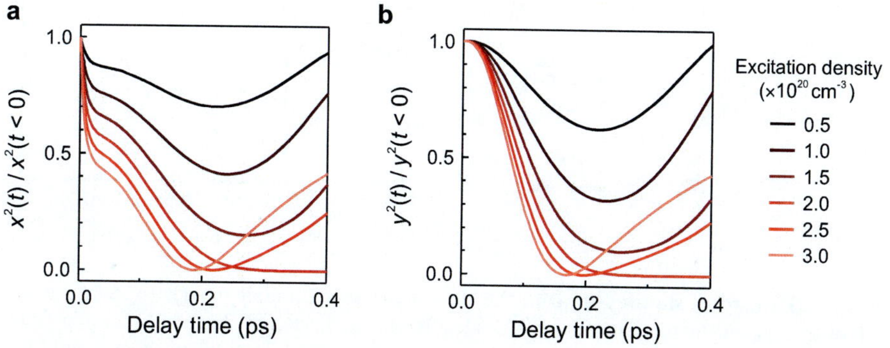

Fig. 5.7 Simulated dynamics of the CDW order parameter amplitude. (**a**) and (**b**) represent the electronic and lattice parts of the order parameter, respectively, both normalized to their values before photoexcitation ($t < 0$). Different curves denote different excitation densities, F, in Eq. (5.10)

excitation, the order parameter ($x^2(t)$ or $y^2(t)$) is slightly suppressed and then quickly recovers to the initial value. Increasing the excitation density results in a more pronounced suppression until the critical point is reached at F_c, where $x^2(t)$ and $y^2(t)$ asymptotically approach zero for the first time. It is interesting to observe that at even higher excitation density, the order parameters pass through zero in finite time. We associate this peculiarity with the fact that at F_c, the system is indeed in the critical regime where the Landau potential becomes flat (or equivalently, the effective forces $f_x \approx 0$ *and* $f_y \approx 0$), and, as such, we expect an effect of dynamical slowing-down. From Fig. 5.7a, the slowdown can be quantified by tracking the position of the first minimum in $x^2(t)$, as plotted in Fig. 5.8b. The result is similar if $y^2(t)$ is used instead of $x^2(t)$. We observe that indeed at F_c the time needed to reach the minimum order parameter amplitude diverges. In the simulation, we adjusted κ to match F_c with the experimentally determined F_{melt}.

We summarize the time-dependent Landau formalism by highlighting the physical picture presented in Fig. 5.8a. Unlike the equilibrium case where microscopic fluctuations, captured by $\eta(t)$ in Eq. (5.4), initiate the dynamics, the temporal evolution in the nonequilibrium case is driven by a femtosecond laser pulse, where a coherently excited CDW amplitude mode plays an instrumental role [34]. The pulse modifies the free energy functional [α_e in Eq. (5.5)] and sets off the order parameter to seek a new global minimum. Though we draw the free energy as fixed curves, it should be noted that it evolves dynamically according to Eq. (5.10). As the dynamics of ψ_e and ψ_l closely follow each other, we use a single circle to denote the order parameter in Fig. 5.8a. At the critical excitation density, F_{melt}, beyond which the order parameter vanishes transiently, the time taken to suppress the order is the

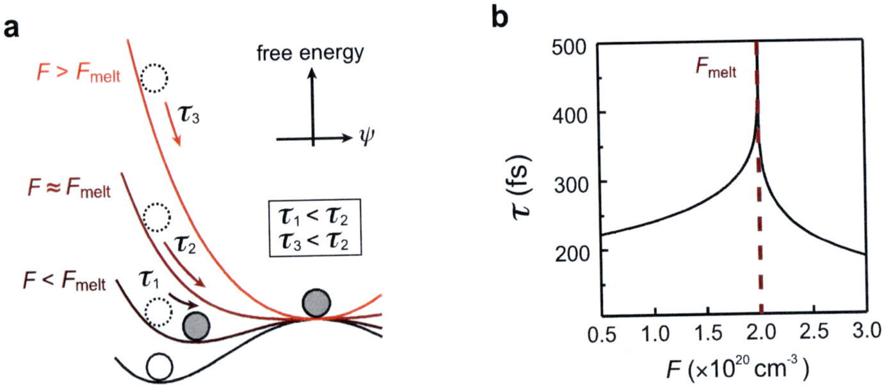

Fig. 5.8 Dynamical slowing-down in the time-dependent Landau theory. (**a**) Schematic of CDW order parameter (ψ) dynamics in a Landau free energy landscape. The solid circle represents ψ before photoexcitation and dashed ones are nonequilibrium ψ in an impulsively altered free energy, whose subsequent evolution, which is not drawn, is described by Eq. (5.10). Filled circles represent ψ when CDW is transiently suppressed, either partially or completely. Colored curves are snapshots of transient free energies at different excitation densities right after laser pulse incidence. At F_{melt}, ψ first reaches zero in its temporal evolution. (**b**) Calculated CDW suppression time as a function of excitation density, based on the time-dependent Landau simulation in Fig. 5.7a

longest ($\tau_2 > \tau_1, \tau_3$ in Fig. 5.8a). Similar to the equilibrium situation, the slow evolution reflects a transiently flat potential landscape when the order parameter is close to zero, which leads to its reduced rate of change. In the calculation for Fig. 5.8b, it is worth emphasizing that we have no adjustable parameter except a constant (κ) that converts a dimensionful F in the experiment to a dimensionless quantity in the computation. In addition, the CDW suppression time, τ, falls under a similar range of magnitudes as observed in the experiment. This timescale is associated with the frequency of the CDW amplitude mode (ω_{AM}) and the unrenormalized frequency of the phonon that strongly couples to the CDW (ω_0), indicating the instrumental role of phonons in mediating the ultrafast transition.

There is one key difference between the calculated and measured trend of τ: the latter lacks a sharp divergence at the threshold excitation density. We attribute this rounding of the divergence to the presence of temporally or spatially varying perturbations on the system, such as photoinduced topological defects [18–21] or additional phonons coupled to the CDW order [28, 32, 42], which are not considered in our minimal model. In the Landau picture, they disrupt the flat potential energy landscape when the order parameter approaches zero, which is required for the diverging behavior. Furthermore, the divergence only happens in a very narrow window of excitation densities (Fig. 5.8b), which makes experimental detection challenging as any small uncertainties or fluctuations in the pulse energy can smear the singularity.

It is worth mentioning that there is a kink in the dynamics of the electronic order in Fig. 5.7a at $\sim$30 fs, which is associated with the short electron–electron scattering time τ_e. For the dynamics of the lattice order, no such kink is present in Fig. 5.7b because the characteristic lattice response time is $\sim 2\pi/\omega_0 \approx 0.3\,\mathrm{ps} \gg \tau_e$.

There are a few ingredients that can be added to the model to improve the quantitative agreement with the measurements. For example, $x(t)$ and $y(t)$ may vary spatially, mimicking photoinduced topological defects that smear the divergence in τ. In addition, the relaxation of quasiparticles is represented by a fixed τ_0, which is a simplification expected only to work in a limited range of excitation densities. An F-dependent τ_0 may better represent the actual experiments. Lastly, only one phonon mode is considered in the current simulation, though other phonons have been shown to couple to the CDW order. Their incorporation may make the model more realistic. In Chap. 7, we will consider some of these extensions in a time-dependent $O(N)$ model.

In conclusion, two different time-resolved probes are used to systematically study the ultrafast melting of a CDW instigated by an intense laser pulse. We have experimentally demonstrated the phenomenon of dynamical slowing-down, manifested as the longest time it takes to suppress the CDW at the threshold excitation density in the nonequilibrium phase transition. The agreement in timescale across techniques and with theoretical simulation by time-dependent Landau equations highlights the important role of phonons in this photoinduced transition. Despite complexities involved in phase transitions far from equilibrium, the observation of slowing dynamics in this setting pinpoints a robust commonality for us to understand nonequilibirum phenomena of more intricate systems.

5.5 Supplementary

5.5.1 Experimental Parameters

MeV Ultrafast Electron Diffraction

The experiment was carried out on an exfoliated thin flake of $LaTe_3$ (Fig. 5.9a) at room temperature, prepared using the method detailed in Sect. 3.3.1. The measurement was performed in the MeV UED setup in the Accelerator Structure Test Area facility at SLAC National Laboratory [33, 43] using a 3.1-MeV electron bunch oper-

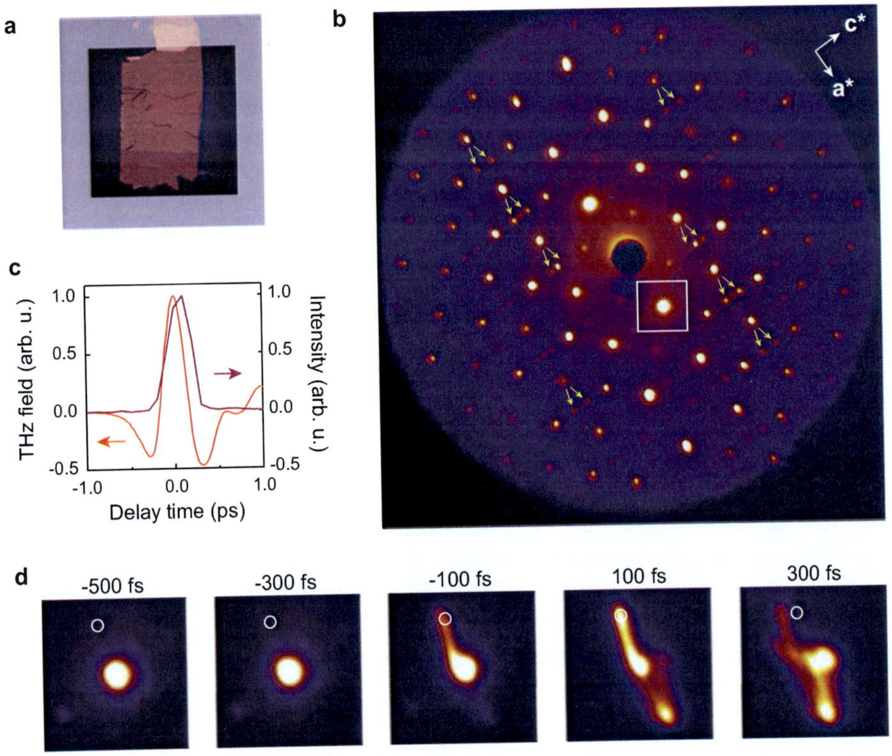

Fig. 5.9 Estimating UED temporal resolution from THz streaking. (**a**) Optical image of the UED sample, exfoliated in an inert gas environment onto a 10-nm-thick silicon nitride window. (**b**) Full room temperature diffraction pattern of $LaTe_3$ in the $(H\ 0\ L)$ plane with 3.1 MeV incident electron kinetic energy. Pairs of arrows indicate selected superlattice peaks whose intensity averages yield the curve shown in Fig. 5.3c. These peaks are selected for optimal signal-to-noise as their intensity exceeds a preset threshold. White square denotes the region of interest (ROI) of the (2 0 0) Bragg peak in **d**. (**c**) THz field strength $E(t)$ (left axis) from electro-optic sampling and intensity $I(t)$ (right axis) integrated over the circular ROI shown in **d**. Both quantities are normalized for easier comparison. The initial rise of $I(t)$ results from THz-induced electron deflection. (**d**) Snapshots of the (2 0 0) Bragg peak at selected time delays, showing electron deflection due to the THz field

ating at 180 Hz. The electron beam size on the sample was approximately 90 μm by 90 μm (FWHM), much smaller than the 800-nm (1.55-eV) pump laser beam from a commercial Ti:sapphire laser (Vitara and Legend Elite HE, Coherent Inc.).

At this pump laser wavelength, the penetration depth ($1/e$ of intensity) is 44 nm, calculated from static reflectivity and optical conductivity in refs. [44, 45]. This depth is comparable to the sample thickness, so not all layers are uniformly photoexcited even though all layers equally contribute to the electron diffraction pattern at 3.1 MeV beam energy. We therefore estimate an effective excitation density F based on the recovery timescale of the superlattice peak, which is shown to have an approximately linear dependence on F in the range probed [20]. The error bars in the estimated F shown in Fig. 5.5a are thus derived from errors in the fitted recovery timescale and the uncertainty in the quasi-linear relationship.

Transient Optical Spectroscopy

The experimental setup was detailed in Sect. 1.3.1 with an optical pump at 780 nm (1.59 eV) and a white light probe ranging from 500 to 700 nm (2.48 to 1.77 eV). As determined from the pump–probe cross-correlation, the overall temporal resolution was determined to be 70 fs. Unless otherwise specified, the data presented selects a probe wavelength of 690 nm (1.80 eV) as it is the wavelength most sensitive to the CDW gap dynamics in the available spectral range (see Sect. 5.5.3). The sample was held at room temperature throughout the measurement.

5.5.2 *Temporal Resolution of the MeV UED*

The investigation of the initial system response after photoexcitation requires the knowledge of the instrumental temporal resolution. Unlike transient optical spectroscopy where the resolution can be extracted from a pump–probe cross-correlation experiment, it is less straightforward to measure in UED. One common method is to perform UED on a reference sample, where the initial dynamics are known to be fast [33]. However, if the temporal resolution is comparable to the sample response time, a sample-independent method is preferred to disentangle the resolution effect from the intrinsic response.

For this purpose, we determined the temporal resolution by streaking the electron bunch with an intense THz field, from which the electron profile was extracted [46, 47]. The THz pulse (∼650 kV/cm) was generated by a DSTMS (4-N,N-dimethylamino-4'-N'-methyl-stilbazolium 2,4,6-trimethylbenzenesulfonate) crystal with the spectral peak at 2 THz. A typical THz electric field profile, $E(t)$, is shown in Fig. 5.9c (left axis), measured via electro-optic sampling with an 800-nm (1.55-eV), 80-fs pulse. As the single-cycle THz pulse and the electron bunch overlap spatially and temporally on the sample, the electrons are deflected sideways due to the strong net Lorentz force (Fig. 5.9d) [46]. Therefore, one can estimate the instrumental temporal resolution, in this case dominated by the contribution from the electron source, by observing the streaking pattern.

More specifically, as the THz pulse is incident on the sample, we monitor the temporal evolution of integrated intensity, $I(t)$, of a circular region of interest (ROI), whose diameter is the FWHM of the nearby Bragg peak (Fig. 5.9c, d). The finite width of the rising edge in $I(t)$ gives an upper bound of the electron bunch length, as the rising edge has additional contributions from a nonzero transverse bunch size and a finite streaking speed. It should be noted that we only use the initial rising edge of $I(t)$. This is because motions of the MeV electrons after 100 fs are more complex when electrons inside the metallic sample interact with the THz pulse and distort the local electric field (Fig. 5.9d). As the streaked electrons can span the entire detector segment from its original position to the extremum position in the ROI (Fig. 5.9d), the bunch width is therefore lower-bounded by the width of the rising edge in the THz pulse, $E(t)$ (Fig. 5.9c, left axis). By fitting the rising edges of $E(t)$ and $I(t)$ to an erf function, we determine the UED temporal resolution from the fitted FWHM to be between 280 and 325 fs. This uncertainty in the temporal resolution is propagated to the corresponding error bars in Fig. 5.5a.

5.5.3 *Probe Photon Energy in Transient Optical Spectroscopy*

The transient reflectivity measurement employed several different probe photon energies in the white light continuum (500 to 700 nm). As different photon energies are sensitive to different inter-band transitions and scattering cross sections, we selected the energy that is the most sensitive to the dynamics of the CDW amplitude. For this purpose, we compared the transient response at a fixed pump excitation density among several probe energies (Fig. 5.10a). The shown traces are normalized to a maximum of unity for easy comparison. In Fig. 5.10b, we plot the Fourier transformed spectra of the coherent oscillatory component in Fig. 5.10a. In our measurement geometry of parallel polarizations between the pump and the probe beams, the most prominent oscillation is the 2.2 THz A_{1g} CDW amplitude mode (AM). Among all probe energies, the 690 nm (1.80 eV) photon gives the highest AM peak (Fig. 5.10b), making it the most suitable energy for probing the CDW gap dynamics in our available spectral range.

The observed photon energy dependence can be rationalized if one examines the full optical conductivity spectrum, $\sigma_1(\omega)$, of $LaTe_3$ [44, 45]. In the range of our white light continuum, we access the high-frequency tail of the mid-infrared Lorentz harmonic oscillators in $\sigma_1(\omega)$, which compose the single particle peak of the CDW condensate. As the probe photon energy decreases, one moves closer to the center of the single particle peak. Therefore, the detection becomes more sensitive to a changing size of the CDW gap in the course of the photoinduced transition.

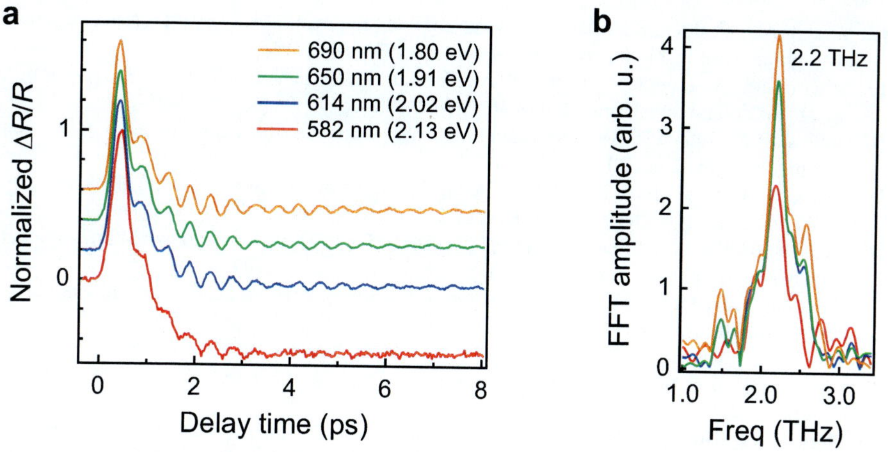

Fig. 5.10 Transient reflectivity with different probe photon wavelengths. (**a**) Normalized $\Delta R/R$ traces at several probe photon wavelengths. The pump photons are the same (780 nm, 1.59 eV) with an identical excitation density of $F = 1.2 \times 10^{20}$ cm^{-3}. Traces are vertically offset for clarity. (**b**) Fourier transform of the oscillatory component of the traces in **a**. The most prominent peak features the 2.2 THz amplitude mode of the CDW. Traces are zero padded before the Fourier transform

References

1. Baudour, J.L., Delugeard, Y., Cailleau, H.: Transition structurale dans les polyphényles. I. Structure cristalline de la phase basse température du *p*-terphényle à 113 K. Acta Cryst. **B32**, 150–154 (1976)
2. Cailleau, H., Heidemann, A., Zeyen, C.M.E.: Observation of critical slowing down at the structural phase transition in p-terphenyl by high-resolution neutron spectroscopy. J. Phys. C Solid State Phys. **12**, L411–L413 (1979)
3. Iizumi, M.: Real-time neutron diffraction studies of phase transition kinetics. Physica B+C **136**, 36–41 (1986)
4. Maschek, M., Rosenkranz, S., Heid, R., Said, A.H., Giraldo-Gallo, P., et al.: Wave-vector-dependent electron-phonon coupling and the charge-density-wave transition in $TbTe_3$. Phys. Rev. B **91**, 235146 (2015)
5. Collins, M.F.: Magnetic Critical Scattering. Oxford University Press, Oxford (1989)
6. Goldenfeld, N.: Lectures on Phase Transitions and the Renormalization Group. Westview, Boulder (1992)
7. Horie, Y., Fukami, T., Mase, S.: First order structural phase transition in $BaPb_{1-x}Bi_xO_3$ and the scaling law. Solid State Commun. **62**, 471–474 (1987)
8. Zhu, Y., Hoffman, J., Rowland, C.E., Park, H., Walko, D.A., et al.: Unconventional slowing down of electronic recovery in photoexcited charge-ordered $La_{1/3}Sr_{2/3}FeO_3$. Nat. Commun. **9**, 1799 (2018)
9. Souletie, J., Tholence, J.L.: Critical slowing down in spin glasses and other glasses: Fulcher versus power law. Phys. Rev. B **32**, 516–519 (1985)
10. Lasjaunias, J.C., Biljaković, K., Nad', F., Monceau, P., Bechgaard, K.: Glass transition in the spin-density wave phase of $(TMTSF)_2PF_6$. Phys. Rev. Lett. **72**, 1283–1286 (1994)
11. Strogatz, S.H.: Nonlinear Dynamics and Chaos: With Applications to Physics, Biology, Chemistry, and Engineering. CRC Press, Boca Raton (2018)

12. Veraart, A.J., Faassen, E.J., Dakos, V., van Nes, E.H., Lü, et al.: Recovery rates reflect distance to a tipping point in a living system. Nature **481**, 357–359 (2012)
13. Scheffer, M., Bascompte, J., Brock, W.A., Brovkin, V., Carpenter, S.R., et al.: Early-warning signals for critical transitions. Nature **461**, 53–59 (2009)
14. Collins, M.R., Teh, H.C.: Neutron-scattering observations of critical slowing down of an Ising system. Phys. Rev. Lett. **30**, 781–784 (1973)
15. Toudic, B., Cailleau, H., Lechner, R.E., Petry, W.: Direct observation of critical phenomena by incoherent neutron scattering. Phys. Rev. Lett. **56**, 347–350 (1986)
16. Press, W., Hüller, A., Stiller, H., Stirling, W., Currat, R.: Critical slowing down of orientational fluctuations in a plastic crystal. Phys. Rev. Lett. **32**, 1354–1356 (1974)
17. Niermann, D., Grams, C.P., Becker, P., Bohatý, L., Schenck, H., et al.: Critical slowing down near the multiferroic phase transition in $MnWO_4$. Phys. Rev. Lett. **114**, 037204 (2015)
18. Yusupov, R., Mertelj, T., Kabanov, V.V., Brazovskii, S., Kusar, P., et al.: Coherent dynamics of macroscopic electronic order through a symmetry breaking transition. Nat. Phys. **6**, 681–684 (2010)
19. Mertelj, T., Kusar, P., Kabanov, V.V., Giraldo-Gallo, P., Fisher, I.R., et al.: Incoherent topological defect recombination dynamics in $TbTe_3$. Phys. Rev. Lett. **110**, 156401 (2013)
20. Zong, A., Kogar, A., Bie, Y.-Q., Rohwer, T., Lee, C., et al.: Evidence for topological defects in a photoinduced phase transition. Nat. Phys. **15**, 27–31 (2019)
21. Trigo, M., Giraldo-Gallo, P., Clark, J.N., Kozina, M.E., Henighan, T., et al.: Ultrafast formation of domain walls of a charge density wave in $SmTe_3$. Phys. Rev. B **103**, 054109 (2021)
22. Zurek, W.H.: Cosmological experiments in condensed matter systems. Phys. Rep. **276**, 177–221 (1996)
23. Chuang, I., Durrer, R., Turok, N., Yurke, B.: Cosmology in the laboratory: Defect dynamics in liquid crystals. Science **251**, 1336–1342 (1991)
24. Bowick, M.J., Chandar, L., Schiff, E.A., Srivastava, A.M.: The cosmological Kibble mechanism in the laboratory: String formation in liquid crystals. Science **263**, 943–945 (1994)
25. Tomeljak, A., Schäfer, H., Städter, D., Beyer, M., Biljakovic, K., et al.: Dynamics of photoinduced charge-density-wave to metal phase transition in $K_{0.3}MoO_3$. Phys. Rev. Lett. **102**, 066404 (2009)
26. Demsar, J., Biljaković, K., Mihailovic, D.: Single particle and collective excitations in the one-dimensional charge density wave solid $K_{0.3}MoO_3$ probed in real time by femtosecond spectroscopy. Phys. Rev. Lett. **83**, 800–803 (1999)
27. Kabanov, V.V., Demsar, J., Podobnik, B., Mihailovic, D.: Quasiparticle relaxation dynamics in superconductors with different gap structures: Theory and experiments on $YBa_2Cu_3O_{7-\delta}$. Phys. Rev. B **59**, 1497–1506 (1999)
28. Yusupov, R.V., Mertelj, T., Chu, J.H., Fisher, I.R., Mihailovic, D.: Single-particle and collective mode couplings associated with 1- and 2-directional electronic ordering in metallic $R\mathrm{Te}_3$ (R=Ho, Dy, Tb). Phys. Rev. Lett. **101**, 24602 (2008)
29. Ru, N., Condron, C.L., Margulis, G.Y., Shin, K.Y., Laverock, J., et al.: Effect of chemical pressure on the charge density wave transition in rare-earth tritellurides $R\mathrm{Te}_3$. Phys. Rev. B **77**, 035114 (2008)
30. Hu, B.F., Cheng, B., Yuan, R.H., Dong, T., Wang, N.L.: Coexistence and competition of multiple charge-density-wave orders in rare-earth tritellurides. Phys. Rev. B **90**, 085105 (2014)
31. Schmitt, F., Kirchmann, P.S., Bovensiepen, U., Moore, R.G., Rettig, L., et al.: Transient electronic structure and melting of a charge density wave in $TbTe_3$. Science **321**, 1649–1652 (2008)
32. Moore, R.G., Lee, W.S., Kirchman, P.S., Chuang, Y.D., Kemper, A.F., et al.: Ultrafast resonant soft X-ray diffraction dynamics of the charge density wave in $TbTe_3$. Phys. Rev. B **93**, 024304 (2016)
33. Weathersby, S.P., Brown, G., Centurion, M., Chase, T.F., Coffee, R., et al.: Mega-electron-volt ultrafast electron diffraction at SLAC National Accelerator Laboratory. Rev. Sci. Instrum. **86**, 073702 (2015)

34. Hellmann, S., Rohwer, T., Kalläne, M., Hanff, K., Sohrt, C., et al.: Time-domain classification of charge-density-wave insulators. Nat. Commun. **3**, 1069 (2012)
35. Trigo, M., Giraldo-Gallo, P., Kozina, M.E., Henighan, T., Jiang, M.P. et al.: Coherent order parameter dynamics in $SmTe_3$. Phys. Rev. B **99**, 104111 (2019)
36. Prasankumar, R.P., Taylor, A.J.: Optical Techniques for Solid-State Materials Characterization. CRC Press, Boca Raton (2011)
37. Brouet, V., Yang, W.L., Zhou, X.J., Hussain, Z., Moore, R.G., et al.: Angle-resolved photoemission study of the evolution of band structure and charge density wave properties in $R\mathrm{Te}_3$ (R = Y, La, Ce, Sm, Gd, Tb, and Dy). Phys. Rev. B **77**, 235104 (2008)
38. Schäfer, H., Kabanov, V.V., Beyer, M., Biljakovic, K., Demsar, J.: Disentanglement of the electronic and lattice parts of the order parameter in a 1D charge density wave system probed by femtosecond spectroscopy. Phys. Rev. Lett. **105**, 066402 (2010)
39. Schaefer, H., Kabanov, V.V., Demsar, J.: Collective modes in quasi-one-dimensional charge-density wave systems probed by femtosecond time-resolved optical studies. Phys. Rev. B **89**, 045106 (2014)
40. Beaud, P., Caviezel, A., Mariager, S.O., Rettig, L., Ingold, G., et al.: A time-dependent order parameter for ultrafast photoinduced phase transitions. Nat. Mater. **13**, 923–927 (2014)
41. Eiter, H.-M., Lavagnini, M., Hackl, R., Nowadnick, E.A., Kemper, A.F., et al.: Alternative route to charge density wave formation in multiband systems. Proc. Natl. Acad. Sci. U.S.A. **110**, 64–69 (2013)
42. Lavagnini, M., Baldini, M., Sacchetti, A., Di Castro, D., Delley, B., et al.: Evidence for coupling between charge density waves and phonons in two-dimensional rare-earth tritellurides. Phys. Rev. B **78**, 201101 (2008)
43. Shen, X., Li, R.K., Lundström, U., Lane, T.J., Reid, A.H., et al.: Femtosecond mega-electron-volt electron microdiffraction. Ultramicroscopy **184**, 172–176 (2018)
44. Sacchetti, A., Degiorgi, L., Giamarchi, T., Ru, N., Fisher, I.R.: Chemical pressure and hidden one-dimensional behavior in rare-earth tri-telluride charge-density wave compounds. Phys. Rev. B **74**, 125115 (2006)
45. Pfuner, F., Degiorgi, L., Chu, J.-H., Ru, N., Shin, K.Y., et al.: Optical properties of the charge-density-wave rare-earth tri-telluride compounds: A view on $PrTe_3$. Physica B **404**, 533–536 (2009)
46. Li, R.K., Hoffmann, M.C., Nanni, E.A., Glenzer, S.H., Kozina, M.E., et al.: Terahertz-based subfemtosecond metrology of relativistic electron beams. Phys. Rev. Accel. Beams **22**, 012803 (2019)
47. Ofori-Okai, B.K., Hoffmann, M.C., Reid, A.H., Edstrom, S., Jobe, R.K., et al.: A terahertz pump mega-electron-volt ultrafast electron diffraction probe apparatus at the SLAC Accelerator Structure Test Area facility. J. Instrum. **13**, P06014 (2018)

Chapter 6
Light-Induced Charge Density Wave in $LaTe_3$

Abstract When electrons in a solid are excited by light, they can alter the free energy landscape and access phases of matter that are out of reach in thermal equilibrium. This accessibility becomes important in the presence of phase competition, when one state of matter is preferred over another by only a small energy scale that, in principle, is surmountable by the excitation. Here, we study $LaTe_3$, where a small lattice anisotropy in the *a*-*c* plane results in a unidirectional charge density wave (CDW) along the *c*-axis. Using ultrafast electron diffraction, we find that after photoexcitation, the *c*-axis CDW is weakened and subsequently, a different competing CDW along the *a*-axis emerges. The timescales characterizing the relaxation of this new CDW and the reestablishment of the original CDW are nearly identical, which points toward a strong competition between the two orders. The new CDW represents a nonequilibrium phase of matter with no equilibrium counterpart, and this study provides a route for discovering similar hidden states that are "trapped" in equilibrium.

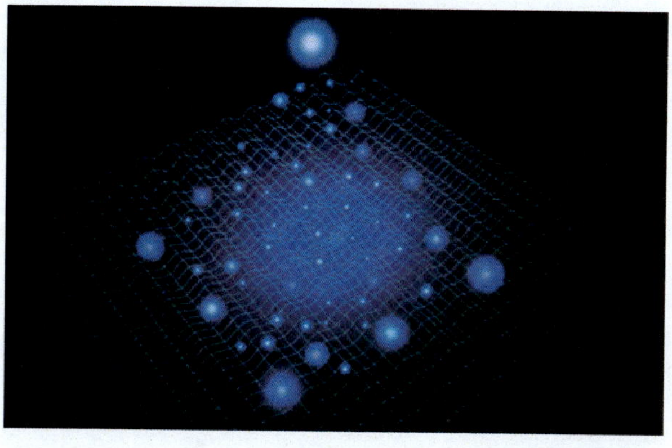

An artist's impression of a light-induced charge density wave (CDW). The wavy mesh represents the CDWs while glowing spheres represent photons

A. Zong, *Emergent States in Photoinduced Charge-Density-Wave Transitions*, Springer Theses, https://doi.org/10.1007/978-3-030-81751-0_6

6.1 Proximal Phases of Matter

A major theme in condensed matter physics is the relationship between proximal phases of matter, where one ordered ground state gives way to another as a function of some external parameter such as pressure, magnetic field, doping, or disorder. It is in such a neighborhood that we find colossal magnetoresistance in manganites [1] and unconventional superconductivity in heavy fermion, copper oxide, and iron-based compounds [2]. In these materials, the nearby ground states can affect one another in several ways. For example, phases can compete, impeding the formation of one state in place of another. This scenario is played out, for instance, in $La_{2-x}Ba_xCuO_4$ at $x = 1/8$, where the development of alternating charge and spin-ordered regions prevents the onset of superconductivity [3, 4]. On the other hand, fluctuations of an adjacent phase can help another be realized, such as in ^{3}He, where ferromagnetic spin fluctuations enable the atoms to form Cooper pairs and hence a p-wave superfluid [2, 5]. In more complicated situations, such as in manganites, nanoscale phase separation occurs, where local insulating antiferromagnetism coexists next to patches of metallic ferromagnetism, resulting in large magnetic and electrical responses to small perturbations [1]. In each case, the macroscopic properties of a material are heavily influenced by the nearby presence of different phases.

Intense light pulses have recently emerged as a tool to tune between neighboring broken-symmetry phases of matter [6–11]. Conventionally, light pulses are used to restore symmetry, but in certain cases symmetries can also be broken. For instance, exposing $SrTiO_3$ to MIR or THz radiation has led to ferroelectricity [6, 7], while ferromagnetism has been induced in a manganite with NIR light [9]. In this chapter, we will re-examine the rare-earth tritelluride compound, $LaTe_3$, where a unidirectional CDW phase is only present along the c-axis with no counterpart along the nearly equivalent, perpendicular a-axis. We show that femtosecond light pulses can be used to break translational symmetry and unleash an a-axis CDW. Using ultrafast electron diffraction (UED), we are able to visualize this process and track both order parameters simultaneously, gaining a unique perspective of the two CDWs in the time domain.

6.2 Two Density Waves in Rare-Earth Tritellurides

As we recall from Sect. 2.3, rare-earth tritellurides (RTe$_3$) possess a layered, quasi-tetragonal structure (Fig. 2.8a) with a slight in-plane anisotropy ($a \geq 0.997c$, see Fig. 2.9b) [12, 13], which leads to a preferred direction for the CDW order along the c-axis. The Fermi surface, which is similar for all RTe$_3$, arises from the nearly square-shaped Te sheets; the rare-earth atoms, with different radii, effectively serve to apply chemical pressure [14, 15]. The normal-state Fermi surface is depicted in Fig. 2.10b, along with the CDW wavevector parallel to the c-axis. Depending on the specific rare-earth element in $R\text{Te}_3$, some of the members display a CDW only along the c-direction while others have an additional CDW along the orthogonal a-direction (Fig. 2.8b). As one moves from lighter to heavier rare-earth elements, the transition temperature of the CDW along the c-axis, T_{c1}, decreases while that along the a-axis, T_{c2}, is first finite in TbTe_3 and increases with atomic number. This relationship strongly suggests that the two CDWs compete in equilibrium. The competition can be understood as follows: once the c-axis CDW forms, large portions of the Fermi surface open up a gap; the corresponding loss of states near the Fermi energy therefore inhibits the formation of the a-axis CDW [14].

6.2.1 Two Energy Gaps in ErTe_3

To better understand the relationship between the two orthogonal CDW orders in $R\text{Te}_3$, it is instructive to look at a member of this family with a heavy rare-earth element, for example, ErTe_3. Below its $T_{c2} = 159(5)$ K, both electron and X-ray diffractions have identified the satellite peaks along the a-axis in addition to the satellites along the c-axis [14, 16]. Similar to the c-axis CDW, the formation of this secondary CDW is accompanied by a secondary gap opening in the partially gapped Fermi surface at T_{c2} [17, 18]. Figure 6.1 shows the photoemission spectra of ErTe_3 taken at 15 K, deep inside the double-CDW phase. The Fermi surface is marked by two gaps at distinct momenta; the gap sizes, Δ_c and Δ_a, are associated with the c-axis and the a-axis CDW, respectively. Based on the relative magnitude of Δ_c and Δ_a (Fig. 6.1b, c), the c-axis CDW has a much larger amplitude than its a-axis counterpart, thus corroborating the trend seen in their respective transition temperatures. This relationship is also reflected in a much more intense satellite peak at wavevector q_c compared to q_a in diffraction measurements [16].

6.2.2 Hierarchy of Photoinduced CDW Dynamics in ErTe_3

The difference in the equilibrium strength between these two CDWs leads to their different responses after a photoexcitation event. Figure 6.2a displays a series of snapshots of the Fermi surface upon the arrival of an optical pump pulse at 720 nm. Both energy gaps transiently disappear, with the normal-state Fermi surface restored around 0.6 ps. At longer time delays, the c-axis CDW gap (Δ_c) reopens, while the a-axis CDW gap (Δ_a) remains largely filled even after 2.0 ps. Similar to our analysis of LaTe_3 in Chap. 4, we can obtain a more quantitative view by plotting the photoemission spectral weight within the CDW gaps as a function of pump-probe delay, shown in Fig. 6.2b. There are two prominent differences between the spectral weight evolutions in Δ_c and Δ_a: (i) the photoinduced closing of Δ_c is delayed by 100 fs compared to Δ_a, and (ii) as hinted by the Fermi surface snapshots, Δ_c recovers more quickly than Δ_a.

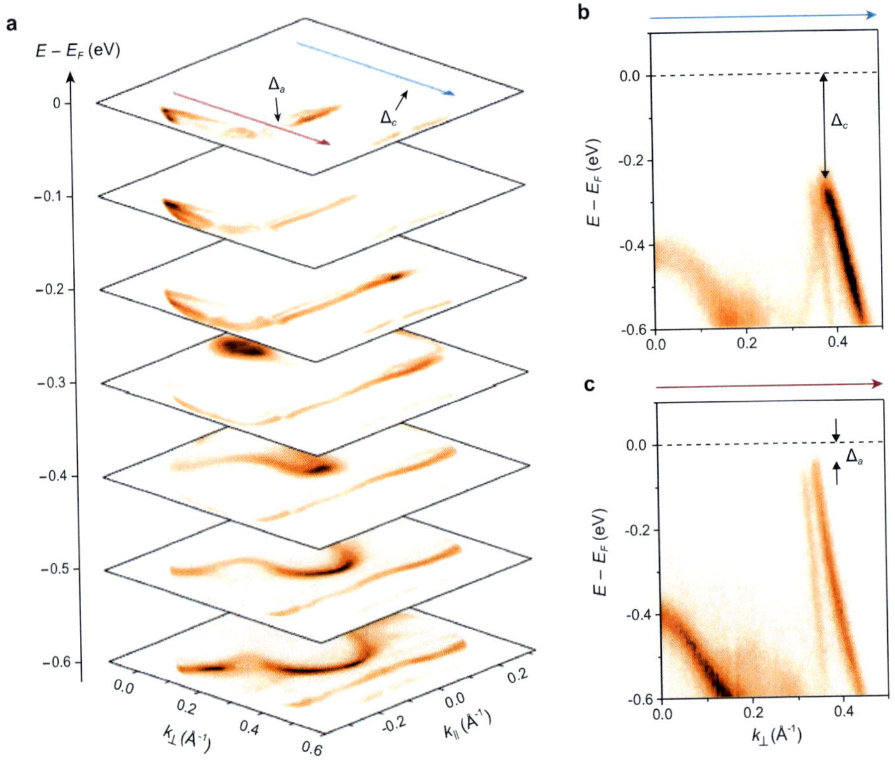

Fig. 6.1 **Two equilibrium CDW orders in $ErTe_3$.** (**a**) Constant energy maps of the photoemission spectra, taken at 15 K in the double-CDW phase. Each map is integrated over a ±10 meV window around the specified energy. The momentum locations of the two CDW gaps, Δ_c and Δ_a, are indicated on the Fermi surface. Following the convention in Ref. [18], $k_\perp$ and $k_\parallel$ are oriented at 45° with respect to the in-plane axes a and c. $k_\perp \equiv (k_z + k_a)/\sqrt{2}$; $k_\parallel \equiv (k_z - k_a)/\sqrt{2}$. (**b**,**c**) Electronic dispersions along $k_\perp$ at the corresponding cuts in **a**, indicated by the color-coded arrows. Respective CDW gaps are labeled. Data was acquired with 10.75 eV photon energy. The colorscale of each image is individually adjusted for better visualization

The relative sequence of events between the two CDW orders—for both photoinduced melting and subsequent recovery—suggests that the c-axis order is a prerequisite for the equilibrium a-axis order. This interpretation explains why the a-axis gap melts faster but recovers slower. From a microscopic point of view, the c-axis CDW leads to significant Fermi surface reconstruction, so the equilibrium a-axis order is a result of instabilities associated with the partially gapped Fermi surface, as opposed to the original normal-state Fermi surface. These instabilities are not present unless a long-range density wave is already formed along the c-axis, giving rise to a clear hierarchy of the two orthogonal CDWs in their ultrafast dynamics.

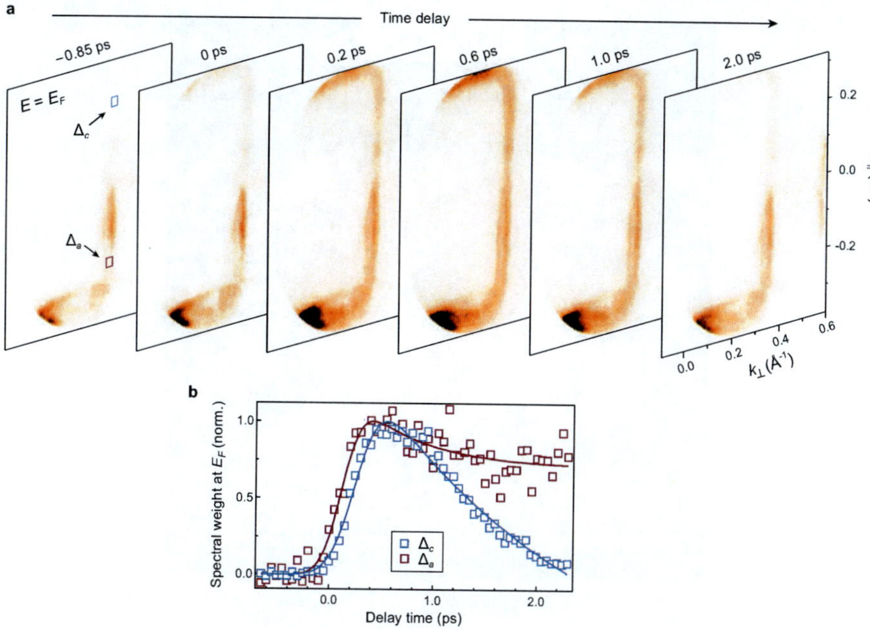

Fig. 6.2 Photoinduced evolution of the two CDW gaps in $ErTe_3$. (a) Snapshots of the Fermi surface following the excitation by a 720-nm (1.72-eV) pulse with an incident fluence of 0.7 mJ/cm^2. Both CDW gaps, Δ_c and Δ_a, transiently close, and Δ_c has a faster recovery. All panels share the same colorscale. (**b**) Evolutions of photoemission spectral weight at E_F within the momentum bounds specified by the square ROIs in **a** (0.025×0.025 Å^{-2}). The integration window along the energy axis is 25 meV. Spectral weights are normalized between 0 and 1. Curves are fits to Eq. (3.14), highlighting the slower suppression and faster recovery of Δ_c (blue) compared to Δ_a (purple)

6.3 Transient CDW in $LaTe_3$

When the a-axis CDW does not exist in equilibrium, such as in $LaTe_3$, one may wonder whether intense photoexcitation could disturb the hierarchy of CDWs between the two in-plane directions. To follow the temporal evolution of the CDW after light excitation, we used transmission ultrafast electron diffraction, which allows us to capture the $(H\ 0\ L)$ plane, with $(H\ K\ L)$ denoting the Miller indices. In the left panel of Fig. 6.3, we show a static diffraction pattern of $LaTe_3$ taken before the arrival of the pump laser pulse with 3.1 MeV electron kinetic energy. Satellite peaks (blue arrows) flanking the main Bragg peaks are observed only along the c-axis. These peaks are due to the existence of the equilibrium CDW. In the right panel, we show the diffraction pattern 1.8 ps after photoexcitation by an 80-fs, 800-nm (1.55-eV) laser pulse, which creates excitations across the single-particle gap and suppresses the CDW along the c-axis. As the equilibrium CDW is weakened, new peaks emerge along the a-direction (red arrows) independent of the pump laser polarization, a change that can also be visualized in the differential intensity plot in

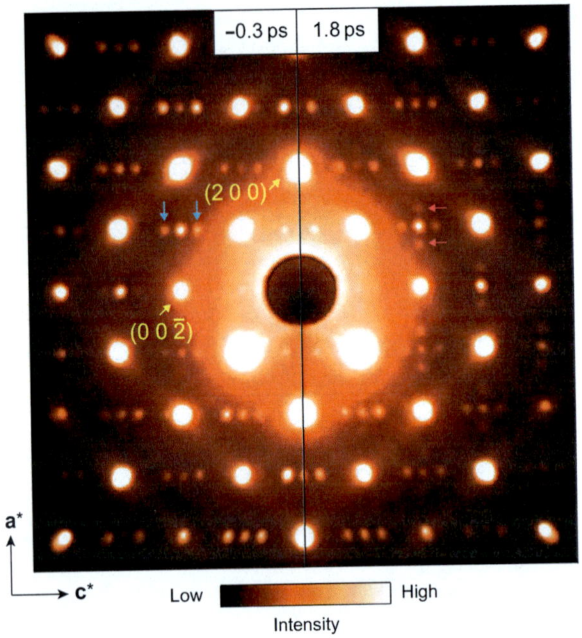

Fig. 6.3 Light-induced CDW observed by MeV ultrafast electron diffraction. Electron diffraction patterns before (left) and 1.8 ps after (right) photoexcitation with an 80-fs, 800-nm laser pulse, taken at 3.1 MeV electron kinetic energy. Blue and red arrows indicate the equilibrium CDW peaks along the c-axis and the light-induced CDW peaks along the a-axis, respectively. The right half is a mirror reflection of the left half, so they denote the same set of peaks in the diffraction pattern. Due to the transverse nature of atomic displacements, the a-axis CDW peaks are most visible along $\mathbf{c}^*$ and the opposite is true for the c-axis CDW peaks. $\mathbf{a}^* \equiv (2\pi/|\mathbf{a}|)\hat{\mathbf{a}}$ and $\mathbf{c}^* \equiv (2\pi/|\mathbf{c}|)\hat{\mathbf{c}}$ are reciprocal lattice unit (r.l.u.) vectors

Fig. 6.5a. Here, the appearance of a new lattice periodicity along the a-axis is clear, and we interpret these peaks as signaling the emergence of an out-of-equilibrium CDW.

These photoinduced CDW peaks were replicated in a separate UED setup with 26 keV electron kinetic energy. Compared to MeV electron diffraction, keV diffraction possesses significantly improved momentum resolution, but suffers from more background scattering (Fig. 6.4a), making the detection of weak intensities from the transient a-axis CDW more challenging. Nonetheless, the differential diffraction plot reveals clear peaks along the a-axis at 1.5 ps after photoexcitation, as shown in Fig. 6.4b. It is worth noting that the pump laser wavelengths are different in the two UED setups: 800 nm for MeV and 1038 nm for keV. Hence, the observed light-induced CDW is unlikely a result of any resonantly pumped interband transition.

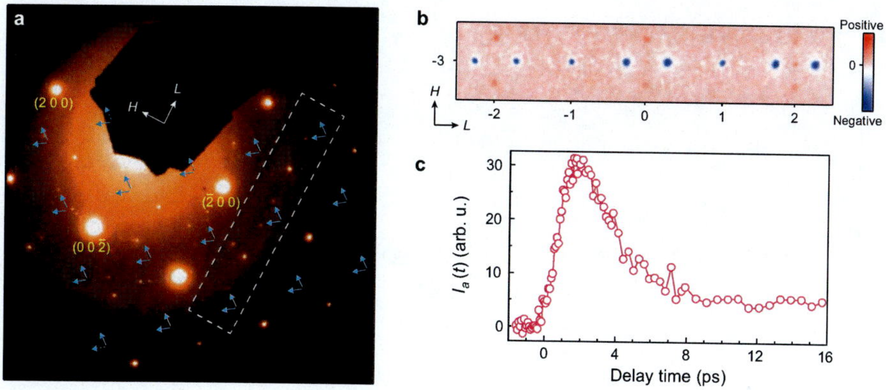

Fig. 6.4 Confirmation of photoinduced CDW by keV ultrafast electron diffraction. (**a**) Full diffraction pattern of $LaTe_3$ before photoexcitation, taken with 26 keV electron kinetic energy. Blue arrows indicate the equilibrium c-axis CDW peaks. Dashed rectangle denotes the region of interest (ROI) examined in **b**. (**b**) Differential intensity plot, focusing on the $H = -3$ row in **a** and showing photoinduced change at 1.5 ps compared to the pattern before photoexcitation. (**c**) Time evolution of integrated intensities of the transient a-axis CDW peak from keV UED. Peaks in multiple Brillouin zones are averaged for better signal-to-noise. The curve is vertically offset so that values before photoexcitation are averaged to zero. The incident pump laser fluence was 240 $\mu J/cm^2$

6.3.1 Temporal Evolution of the Photoinduced Order

This nonequilibrium CDW is ephemeral and only lasts for a few picoseconds. In Fig. 6.5b, we show the temporal evolution of the integrated intensity of the peaks along both the a- and c-axis. The intensity of the a-axis CDW peak reaches a maximum around 1.8 ps and then relaxes over the next couple of picoseconds to a quasi-equilibrium value. The residual intensity at long time delays is due to laser pulse-induced heating that causes a thermal occupation of phonons, which is shown in the diffuse scattering trace in Fig. 6.5b and as the overall red background in Fig. 6.5a. The intensity of the c-axis CDW peak shows the opposite behavior: It first reaches a minimum around 0.5 ps before recovering to a quasi-equilibrium. The initial decay of the c-axis CDW occurs markedly faster than the rise of the transient CDW. This is because the suppression of the equilibrium CDW involves a coherent motion of the lattice ions, whose timescale is tied to the period of the 2.2 THz CDW amplitude mode [19–21]. On the other hand, incoherent fluctuations dictate the ordering of the a-axis CDW, which occurs on a slower timescale. In Chap. 7, we will examine in more detail the initial formation of the a-axis order.

Despite the disparity in the initial timescales, the relaxation times are nearly identical and the overall intensity changes are perfectly anti-correlated, which suggests that these latter properties are governed by a single underlying mechanism. Figure 6.5b shows that one CDW forms at the cost of the other and the two recover back to quasi-equilibrium simultaneously, which, for this fluence, takes a characteristic time of $\tau_a \approx \tau_c \approx 1.7$ ps. The agreement in trends of both the

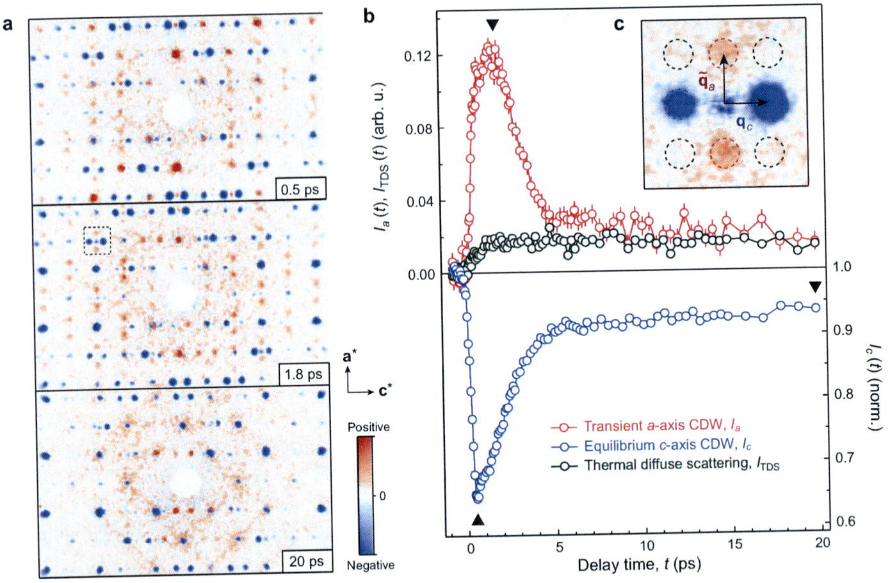

Fig. 6.5 Dynamics of the light-induced CDW. (**a**) Change in intensities with respect to the diffraction pattern before photoexcitation. Snapshots are taken at three select pump-probe time delays, as indicated by the triangles in **b**. (**b**) Time evolution of integrated intensities of the equilibrium c-axis CDW peak (I_c), the transient a-axis CDW peak (I_a), and the thermal diffuse scattering (I_{TDS}). Integration areas are marked by circles in **c** with corresponding colors. Peaks in multiple crystallographic Brillouin zones are averaged for improved signal-to-noise ratio. I_a and I_{TDS} are vertically offset to have their values zeroed prior to photoexcitation. I_c is normalized by its value before photoexcitation. Error bars represent the standard deviation of noise for $t < 0$. (**c**) Enlarged view of the dashed square in **a** at $t = 1.8$ ps. Each integration region has a diameter equal to 1.5 times the full-width-at-half-maximum (FWHM) of the equilibrium CDW peak. The incident pump laser fluence for all panels was 1.3 mJ/cm^2

intensities and the characteristic relaxation times is even more striking when we examine the data at different photoexcitation fluences. As shown in Fig. 6.6, for each fluence, the two CDWs reach anti-correlated extremum values and relax in almost perfect correspondence. Such a strong correlation in both the intensities and the relaxation timescales naturally points toward a phase competition in this nonequilibrium context where the transient CDW cannot exist once the equilibrium CDW recovers.

6.3.2 Anomalous Wavevector of the Transient CDW

From the example of $ErTe_3$ in Sect. 6.2.2, we learned that the equilibrium a-axis CDW gets suppressed faster and recovers slower after photoexcitation compared to its c-axis counterpart. This behavior is in contrast to our observation in $LaTe_3$,

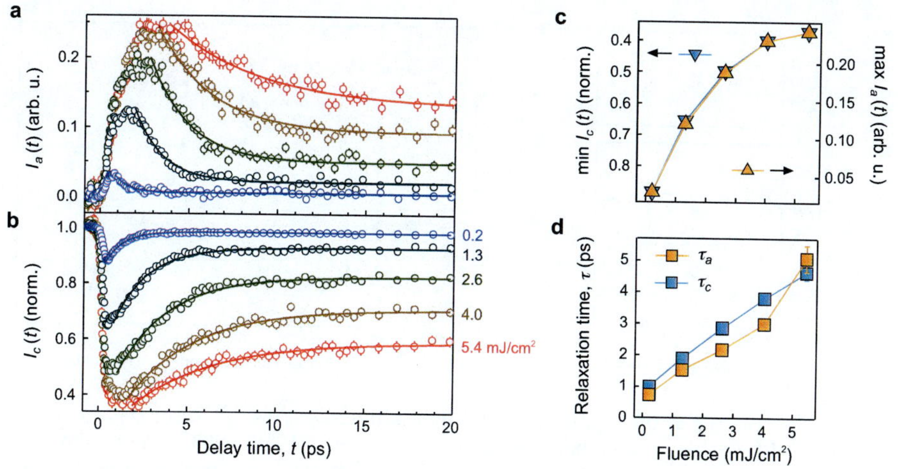

Fig. 6.6 Dependence of equilibrium and transient CDW peaks on pump laser fluence. (**a**,**b**) Time evolution of integrated intensities for the transient *a*-axis CDW peaks and the equilibrium *c*-axis CDW peaks, respectively. Each color denotes an incident fluence. Error bars are obtained from the standard deviation of noise prior to photoexcitation. Curves are single-exponential fits to the relaxation dynamics. In **b**, the intensity I_c does not transiently reach zero at high fluence because of background intensities in the diffraction pattern and non-uniform illumination of all layers of the sample due to a shorter pump laser penetration depth (44 at 800 nm wavelength) compared to the sample thickness ($\lesssim$60 nm) used in this case. (**c**) *Left*, minimum value of the integrated intensity for the equilibrium *c*-axis CDW peaks. *Right*, maximum value of the integrated intensity for the transient *a*-axis CDW peaks. The saturation at high fluence reflects a complete suppression of the *c*-axis order in the pumped volume. (**d**) Characteristic relaxation times at different fluences for the recovery of the *c*-axis CDW peaks (τ_c) and the disappearance of the *a*-axis CDW peaks (τ_a). The increasing trend reflects the longer time taken for topological defects in the *c*-axis CDW to disappear at higher fluence [19]. Error bars, if larger than the symbol size, denote one standard deviation in the corresponding single-exponential fits in **a**,**b**

where the light-induced *a*-axis CDW grows only when the *c*-axis order is weakened. To show that the photoinduced CDW in $LaTe_3$ is genuinely a nonequilibrium state, we may characterize its CDW wavevector and compare it to those in heavier RTe_3 compounds.

Despite the poor momentum resolution of MeV electron diffraction due to the short de Broglie wavelength ($\lambda = 0.35$ pm for electrons with 3.1 MeV kinetic energy) [22], the large number of peaks observed across multiple Brillouin zones yield a precise value of the CDW wavevector after statistical averaging. As the CDW wavevector is expressed in terms of the reciprocal lattice unit (r.l.u.), we measure the distance between an adjacent pair of CDW peaks in the diffraction pattern, normalized by the distance between the neighboring Bragg peaks. This normalization procedure minimizes distortion of the wavevector due to any misalignment of the electron beam with respect to the sample surface normal, or due to the slight curvature of the Ewald sphere.

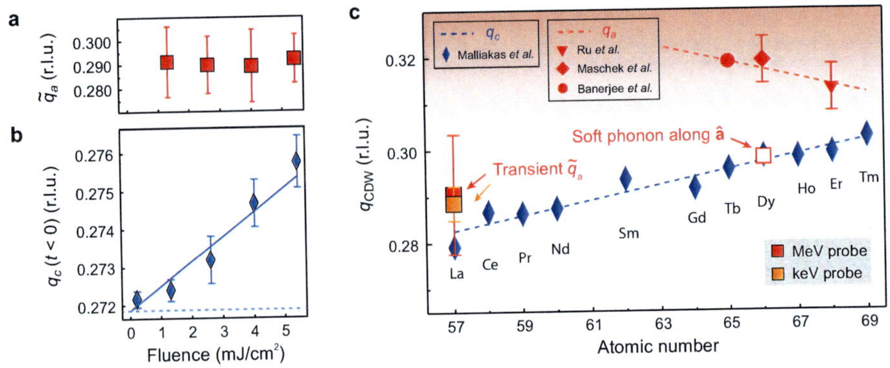

Fig. 6.7 Wavevector of the equilibrium and light-induced CDWs. (**a**) Wavevector $\tilde{q}_a$ of the transient CDW at different pump fluences. Error bars represent the half-width at half maximum of the Gaussian fit to a wavevector histogram of a large number of a-axis satellite peaks for each fluence. At 0.2 mJ/cm^2, the a-axis CDW peak is too weak to have its wavevector determined reliably. (**b**) Wavevector q_c of the equilibrium c-axis CDW peak taken before the arrival of the pump laser pulse at various fluences. Steady-state laser heating causes q_c to increase monotonically with fluence, consistent with previous reports [12, 14, 23]. Error bars represent one standard deviation among multiple peaks selected. The equilibrium value of q_c at the base temperature of 307 K is determined from the vertical intercept (dashed line) of the linear fit (solid line) at zero fluence. (**c**) Summary of CDW wavevectors across the rare-earth series; r.l.u.: reciprocal lattice unit. Values of q_c (blue diamonds) were taken from Ref. [12] at the highest temperature below T_{c1} (see Sect. 6.5.1 for a discussion of the temperature dependence of the wavevector). For q_a, only $TbTe_3$ [23], $DyTe_3$ [24], and $ErTe_3$ [25] were accurately measured by X-ray diffraction. White square denotes the calculated wavevector of the soft phonon along the a-axis, which was confirmed by inelastic X-ray scattering at T_{c1} [24]. Red (orange) square denotes the wavevector of the transient a-axis CDW, probed by time-resolved MeV (keV) electron diffraction. Dashed lines are guides to the eye, highlighting a monotonic trend for both q_a and q_c across the rare-earth elements. Shaded region represents extrapolated values of q_a for light rare-earth elements (La to Gd) if a bidirectional CDW were to form. Error bars, if larger than the marker size, denote reported uncertainty in the literature, or for $\tilde{q}_a$, the standard deviation of values obtained in multiple Brillouin zones and diffraction images

To determine the wavevector $\tilde{q}_a$ of the transient a-axis CDW,[1] we apply the above procedure at each time delay when the photoinduced peak can be distinguished from the background. The location of each peak is determined by fitting it to a Gaussian profile, and $\tilde{q}_a$ is computed for each adjacent pair. We find no observable time-dependence of $\tilde{q}_a$ beyond our experimental uncertainty. Hence, for each fluence, we plot a histogram of all $\tilde{q}_a$ extracted, fit it to a Gaussian distribution, and assign the Gaussian center and width as the value and uncertainty of $\tilde{q}_a$. Figure 6.7a shows the fluence dependence of $\tilde{q}_a$, which displays a constant trend within the uncertainties. Their average value is plotted as the red square in Fig. 6.7c. We could cross check this value by examining the keV diffraction data, which has a reduced error bar due to improved momentum resolution benefiting

[1] We use a tilde to denote the nonequilibrium value.

from a much longer de Broglie wavelength ($\lambda = 7.5$ pm for electrons with 26 keV kinetic energy). In this case, $\tilde{q}_a$ is computed from the statistical average of CDW pairs from 14 different Brillouin zones in the differential diffraction plot, and is indicated as the orange square in Fig. 6.7c. We see that $\tilde{q}_a$ values determined by both MeV and keV data are consistent with each other.

To confirm the validity of the above procedure to compute $\tilde{q}_a$, we apply a similar method to measure the wavevector of the equilibrium c-axis CDW, q_c. As q_c is known to change after photoexcitation [26] (Fig. 6.11), we focus on the diffraction images at delay time $t < 0$. Figure 6.7b summarizes $q_c(t < 0)$ for each laser fluence, where steady-state laser heating causes q_c to increase monotonically with fluence, consistent with previous reports [12, 14, 23]. To determine q_c at equilibrium, we extrapolate its value at zero incident fluence through a linear fit (Fig. 6.7b), which agrees with the value obtained in high resolution X-ray measurements at a similar temperature.[2]

Upon close scrutiny of the transient CDW wavevector, $\tilde{q}_a$, it does not resemble values seen in other rare-earth tritellurides that exhibit an equilibrium a-axis order (Fig. 6.7c). In particular, the equilibrium q_a measured in other rare-earth tritellurides are noticeably larger than $\tilde{q}_a$ beyond experimental uncertainty. According to the trend of q_a with rare-earth mass as discussed in Sect. 2.3.3, $LaTe_3$ would exhibit the largest q_a. Instead, $\tilde{q}_a$ is closer in value to the markedly smaller wavevector of the c-axis CDW, q_c. Thus, the observed $\tilde{q}_a$ of the transient CDW highlights that it is not a trivial extension to an equilibrium a-axis CDW.

We can gain some insight into the origin of the anomalous wavevector from previous inelastic X-ray scattering measurements and density functional theory calculations on $DyTe_3$, which is in the same CDW family [24]. In $DyTe_3$, when the c-axis CDW develops at 308 K, strong CDW fluctuations are also seen along the a-direction in the form of phonon softening, namely, a marked decrease in the phonon frequency. As shown in Fig. 6.7c, these fluctuations occur at a wavevector $q_{a,\mathrm{soft}}$, which is comparable in magnitude to that of the c-axis CDW, q_c. However, when the a-axis CDW eventually forms at 68 K, it does so at a larger wavevector, q_a (so that $q_{a,\mathrm{soft}} \approx q_c \ll q_a$). The reason for these relationships among the wavevectors is the following: At high temperature, the Fermi surface has negligible a/c anisotropy ($q_{a,\mathrm{soft}} \approx q_c$); however, when the a-axis CDW forms at low temperature, it does so after the c-axis CDW has already opened a gap at portions of the Fermi surface, which changes the nesting conditions ($q_c \ll q_a$) [17]. Returning to $LaTe_3$, we observe $\tilde{q}_a \approx q_c$ (see Sect. 6.5.1 for an explanation of the slight discrepancy). This relationship suggests that the transient a-axis CDW looks more akin to one that would have formed at high temperature had the c-axis CDW not prevented it from doing so.

[2] The value of q_c obtained from our electron diffraction is in nearly perfect agreement with the value reported by N. Ru [13], but is slightly smaller than that from Malliakas and coworkers [12] (Fig. 2.9c). This difference may arise from different crystal growth methods; the $LaTe_3$ crystals used in this dissertation are grown by the same procedure as in N. Ru's studies [13, 27].

6.4 Mechanisms of the Light-Induced CDW

In this section, we discuss several scenarios that could lead to the existence of the out-of-equilibrium density wave. We start with a few candidates that have some serious drawbacks and then present the most probable mechanism compatible with our observations.

6.4.1 Candidate Scenarios

Light-Induced Strain

As introduced in Sect. 2.3, the CDW properties in the $R\text{Te}_3$ family are significantly modified by the chemical pressure exerted by the rare-earth ion, leading to systematic trends in T_{c1} and T_{c2} (Fig. 2.8b). Across the series, the lattice constants, a and c, also change systematically (more than 2% from La to Tm) due to the changing size of the rare-earth ions (Fig. 2.9b). Therefore, it is conceivable that photoexcitation could transiently strain the crystal, which would cause the lattice constants to change and induce the a-axis CDW. Furthermore, as demonstrated in a recent experiment where uniaxial strain can switch the directionality of the CDW [28], it may be the case that such strain can lead to a reduction in the a/c lattice anisotropy, which could also result in a preference of the a-axis CDW over the c-axis order.

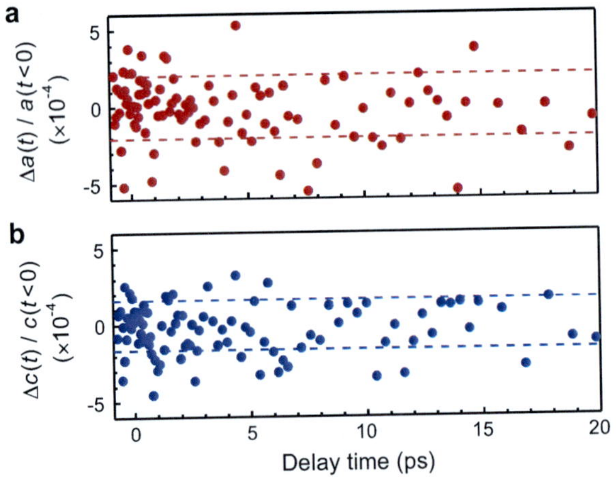

Fig. 6.8 Absence of photoinduced change in the lattice constants. Time evolution of normalized change in lattice constants along the a-axis (**a**) and c-axis (**b**) at 5.4 mJ/cm^2 incident fluence. The lattice constant in each direction is computed by determining the average distance between 47 pairs of Bragg peaks, whose locations are obtained by fitting the peak profile to a Gaussian distribution. Dashed lines demarcate one standard deviation for data points across all time delays, indicating no observable time-dependent change in either lattice constant beyond 0.02%

To test this scenario, we examined the change in the lattice parameters as a function of time after photoexcitation by tracking the positions of lattice Bragg peaks. In Fig. 6.8, we plot the in-plane lattice parameters as a function of pump-probe delay time. These plots demonstrate no observable change in the lattice parameters beyond the experimental uncertainty level of 0.02%. For reference, the *a* and *c* lattice constants of $LaTe_3$ at room temperature differ by 0.35% [13]. Furthermore, the *a* or *c* parameter of $LaTe_3$ is more than 2% larger compared to that of $TbTe_3$, which is the lightest member in the family with $T_{c2} > 0$ at ambient pressure [13]. Therefore, given the negligible change in the lattice parameters after photoexcitation, the scenario of the nonequilibrium CDW induced by strain must be ruled out.

Kibble–Zurek-Like Domains

The Kibble–Zurek mechanism describes a phase transition involving a thermal quench, which gives rise to topological defects, possibly in the form of domain walls [29, 30] (Fig. 4.1a). Considering this scheme, it is plausible that upon photoexcitation, the equilibrium CDW completely vanishes, and during relaxation, some regions of the sample relax into an *a*-axis CDW while others relax into a *c*-axis CDW in a probabilistic fashion, resulting in the formation of domains. Following this initial relaxation process, the *c*-axis domains would then engulf the *a*-axis domains over the observed relaxation timescale. It should be noted that the transient lattice temperature stays below T_{c1} at all time delays, so the CDW melting is nonthermal in nature [19].

While this scenario is worth considering, there is a clear observation that is antithetical to this picture. Even for small fluences, the *a*-axis CDW grows in intensity, as evidenced in Fig. 6.6a. This observation suggests that it is not necessary to fully melt the *c*-axis order to create the *a*-axis CDW. Therefore, one should consider a local melting scenario instead, where photons may suppress the *c*-axis order in "patches" and within those patches, *a*-axis order may grow. This local picture reminds us of the photoinduced topological defects in the *c*-axis CDW, as studied in Chap. 4. We will discuss the applicability of this mechanism in the following section.

6.4.2 Transient CDW Seeded by Topological Defects

To explain all of these observations within a consistent framework, we propose a picture where the nonequilibrium CDW arises due to the generation of topological defect/anti-defect pairs in the *c*-axis CDW through local absorption of high-energy photons (Fig. 6.9b) [19, 31]. The presence of these defects was recently evidenced and extensively characterized in $LaTe_3$ upon photoexcitation [19] and visualized by scanning tunneling microscopy of palladium-intercalated $ErTe_3$ [32]. In spatial regions where the dominant *c*-axis order is suppressed, for instance near topological

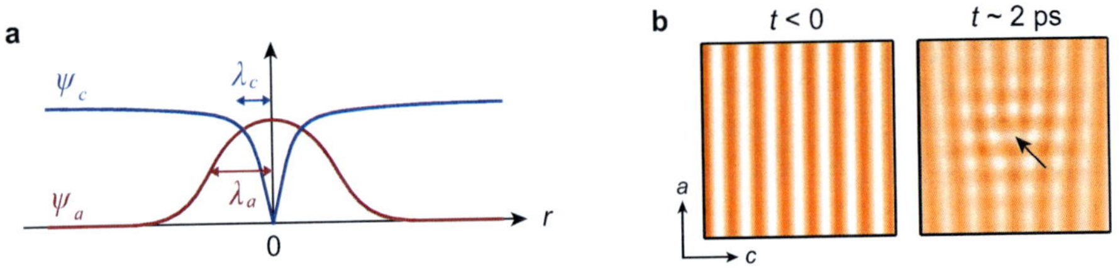

Fig. 6.9 Transient CDW seeded by topological defects. (**a**) Schematic of CDW order parameter amplitudes, ψ_c and ψ_a, near a topological defect at $r = 0$ in the equilibrium c-axis CDW. Characteristic length scales of suppression in ψ_c and enhancement in ψ_a are labeled by λ_c and λ_a, respectively. (**b**) Schematic of charge density waves in real space before (*left*) and approximately 2 ps after (*right*) photoexcitation. Stripe brightness indicates the strength of the CDW amplitude. A dislocation (black arrow) is used as an example of a topological defect in the c-axis CDW after photoexcitation

defects, the normal-state Fermi surface is restored. The normal-state Fermi surface is unstable to the formation of a CDW along both a- and c-axis [24]. However, at a defect core in $LaTe_3$, the instability along the c-axis is topologically inhibited, which allows the sub-dominant a-axis phase to form instead (Fig. 6.9a) [33–35].

The benefit of this picture is that it can explain several observations that are difficult to capture in other theoretical scenarios discussed above. First, the transient CDW forms despite only a partial suppression of the c-axis CDW, as shown in Fig. 6.6. In equilibrium, we know that any finite c-axis CDW amplitude necessarily forbids the presence of an a-axis CDW in $LaTe_3$. The presence of topological defects, however, explains this apparent puzzle considering that it allows for the local suppression of the c-axis CDW. This local constraint also accounts for the anomalous wavevector, $\tilde{q}_a$, since the transient CDW nucleates in the absence of the c-axis order. Furthermore, the coincidence of relaxation timescales is naturally explained in this scenario: As the defects annihilate, the transient CDW can no longer be sustained and the equilibrium c-axis CDW necessarily recovers [19].

This picture based on photoinduced defects is further supported by an equilibrium analogue of another charge-ordered compound, $Bi_{0.35}Sr_{0.18}Ca_{0.47}MnO_3$ [36]. Below the charge-ordering temperature, two transverse lattice displacements occur along the two orthogonal axes, a and c, with wavevectors $q_a \approx 1/3a^*$ and $q_c \approx 1/3c^*$. From atomic-resolution scanning transmission electron microscopy (STEM), shown in Fig. 6.10a, the charge-ordered phase of $Bi_{0.35}Sr_{0.18}Ca_{0.47}MnO_3$ is characterized by local regions of stripes where only one of the two superlattice displacements dominate. However, in regions where this dominant order is disrupted by a defect, such as a shear or an edge dislocation, the orthogonal charge order emerges (Fig. 6.10b, c). This example again emphasizes the instrumental role of defects in determining the local order, especially when several nearly degenerate phases are present.

To place the proposed mechanism on a firmer theoretical footing, we have performed a Ginzburg–Landau analysis in two-dimensional space involving two complex order parameters, ψ_c and ψ_a, which denote the equilibrium and transient

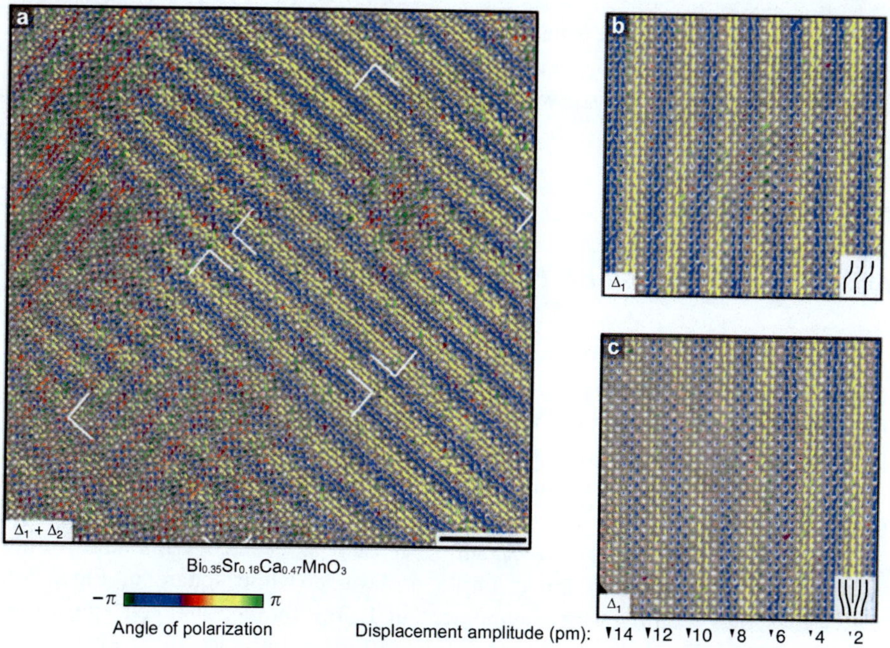

Fig. 6.10 Phase competition near defects in a charge-ordered manganite. (**a**) Atomic-resolution imaging of $Bi_{0.35}Sr_{0.18}Ca_{0.47}MnO_3$ in its charge-ordered state, using high-angle annular dark-field scanning transmission electron microscopy with a 300-keV beam. The image is projected along the *b*-axis and overlaid with picometer scale periodic lattice displacements represented by the triangles, whose color indicates the angle of the polarization vector. In the two square regions, the dominant charge order (Δ_1) appears as blue/yellow stripes while an orthogonal charge order (Δ_2) appears as red/green. The scale bar corresponds to 4 nm. (**b**,**c**) Enlarged images of the two ROIs in **a**, with only the dominant charge order (Δ_1) plotted after Fourier filtering out the orthogonal order. They feature two types of defects in the dominant order: shear (**b**) and edge dislocation (**c**). Schematics of the defects are shown in the bottom right corner. Area of the triangles corresponds to the amplitude of the atomic displacement, which shows a reduction near the defect center. (Adapted from Ref. [36])

CDW orders, respectively. As we show in Sect. 6.5.2, in the presence of phase competition, the minimum-energy solution near a defect core in ψ_c yields a nonzero ψ_a. In addition, we find that the characteristic length scale of the transient CDW, λ_a, can extend well beyond the confines of the defect core, λ_c (Fig. 6.9a). In particular, the ratio of λ_a/λ_c can become large if the normal-state anisotropy between *a*- and *c*-axis is small, which makes the observation of the transient CDW possible even though the defects may be local.

This work provides an illustration of the kind of phenomena that can be observed in far-from-equilibrium systems where phase competition plays a significant role in determining material properties. We expect the mechanism of seeing competing states near topological defects to be general, and that other ordered states of matter will exhibit a similar phenomenology under the influence of photoexcitation. Not only does this result provide a path forward to discovering other states of matter in the presence of phase competition, it also paves the way for the manipulation and control of other ordered phases with light.

6.5 Supplementary

6.5.1 *Temperature- and Time-Dependent CDW Wavevector*

As illustrated in Fig. 2.9c, the wavevector of the equilibrium *c*-axis CDW varies with temperature. Hence, to draw the trend of q_c across the RTe_3 series in Fig. 6.7c, for fair comparison, we use the values of q_c reported by Malliakas and coworkers [12] that are closest to T_{c1}. We note that the exact temperature is not essential to the observed trend because a similar one can be drawn at other temperatures as well (Fig. 2.9c).

In Sect. 6.3.2, we noted that $q_{a,\mathrm{soft}} \approx q_c$ and $\widetilde{q}_a \approx q_c$. While the former relationship looks exact in the example of $DyTe_3$ (white square in Fig. 6.7c), the latter shows slightly worse agreement in $LaTe_3$ (orange and red squares in Fig. 6.7c). This is because the q_c value quoted for $DyTe_3$ is at 300 K, close to its $T_{c1} = 306(3)$ K [14]. On the other hand, q_c for $LaTe_3$ is quoted at 500 K, still considerably lower than its projected $T_{c1} = 670$ K [37]. From the temperature trend in Fig. 2.9c, we speculate that the relationship $\widetilde{q}_a \approx q_c(T = T_{c1})$ is also exact. Upon a closer examination, photoexcitation also results in a slight increase in the magnitude of q_c, as shown in Fig. 6.11. This increase is attributed to a photoinduced reduction of the hopping integral ($t_\perp$ in Fig. 2.10a) [38, 39], and it leads to a better agreement between q_c and $\widetilde{q}_a$ in the transient state.

Compared to q_c, the value of q_a in equilibrium appears to have a much smaller temperature dependence. For example, q_a varies by less than 0.07% from $0.84T_{c2}$ to $0.18T_{c2}$ in $TbTe_3$, about 10 times smaller than the variation in q_c across the same temperature range in terms of T_{c1} [23]. Similarly, the magnitude of q_a for $ErTe_3$ [$T_{c2} = 159(5)$ K] varies within experimental uncertainty from $0.312(5)a^*$ at 155 K

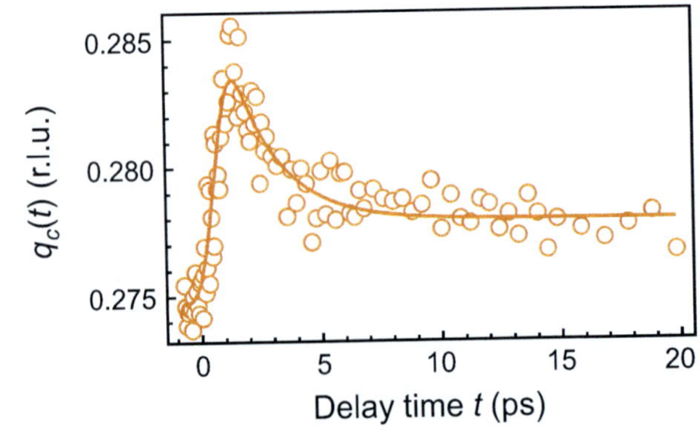

Fig. 6.11 Photoinduced change of the equilibrium *c*-axis CDW wavevector. The incident pump laser fluence was 4.0 mJ/cm^2 while the sample was kept at 307 K. Solid curve is a guide to the eye, obtained by fitting to a single-exponential relaxation described by Eq. (3.14). As the CDW is transiently suppressed, the wavevector increases in amplitude

to $0.313(5)a^*$ at 10 K [25]. Recently, X-ray measurements on $DyTe_3$ ($T_{c2} = 68$ K) suggested a sizable change in q_a from $0.310a^*$ (50 K) to $0.325a^*$ (10 K) [24]. However, the large step size ($0.01a^*$) used for the X-ray scan may necessitate future studies for a more accurate evaluation of the change.

6.5.2 Ginzburg–Landau Formalism of Two Competing Orders

Ginzburg–Landau Free Energy

To describe the properties of the competition between the two CDW order parameters, ψ_a and ψ_c, we introduce the following Ginzburg–Landau free energy density in two spatial dimensions:

$$\mathcal{W} = \left(r_c|\psi_c|^2 + \frac{\beta_c|\psi_c|^4}{2} + \kappa_c|\nabla_{\mathbf{r}}\psi_c|^2 \right) + \left(r_a|\psi_a|^2 + \frac{\beta_a|\psi_a|^4}{2} + \kappa_a|\nabla_{\mathbf{r}}\psi_a|^2 \right) + \eta|\psi_c|^2|\psi_a|^2. \tag{6.1}$$

Here $r_{i=c,a}$, β_i, κ_i, and η are the model parameters. The last term in Eq. (6.1) describes the competition between the two phases. We choose this term to be quadratic in both orders for two reasons. First, such a term is allowed by symmetry, and second, it is the lowest order term that would give rise to the observed competition. Therefore, this is the simplest model capturing all the essential physics. The mean-field phase diagram of the above model has two regimes [40]: for $\beta_a\beta_c > \eta^2$ (tetracritical regime), the two phases can coexist ($\psi_a \neq 0$, $\psi_c \neq 0$); for $\beta_a\beta_c < \eta^2$ (bicritical regime), only one phase may develop ($\psi_c \neq 0$, $\psi_a = 0$). These two regimes are depicted in Fig. 6.12. Given the evolution of the CDWs with rare-earth mass in RTe$_3$, we assume that the bicritical regime applies for $LaTe_3$, as only one CDW exists in equilibrium. On the other hand, the tetracritical case may describe the tritellurides where both a- and c-axis CDWs exist at finite temperature.

An alternative point of view for the bicritical regime in $LaTe_3$ is as follows: Due to the small a/c-anisotropy, we expect that the unrenormalized transition temperature of the sub-dominant order, T_{c2}^*, is close to the observed transition temperature of the dominant order, T_{c1}, with $T_{c1} \gtrsim T_{c2}^*$. Here, $r_a = A_a(T - T_{c2}^*)$ and $r_c = A_c(T - T_{c1}^*)$, where A_a and A_c are positive constants. At the simple mean-field level, $T_{c1}^* = T_{c1}$: since the dominant order sets in first, its transition temperature is not renormalized by the sub-dominant order. On the other hand, the renormalized transition temperature for the sub-dominant order, T_{c2}, which would be observed in experiments, can be suppressed to a negative value, since $T_{c2} = T_{c2}^* - \eta|\psi_c(T_{c2})|^2/A_a$. This is likely the case for $LaTe_3$, where the a-axis CDW does not appear in equilibrium.

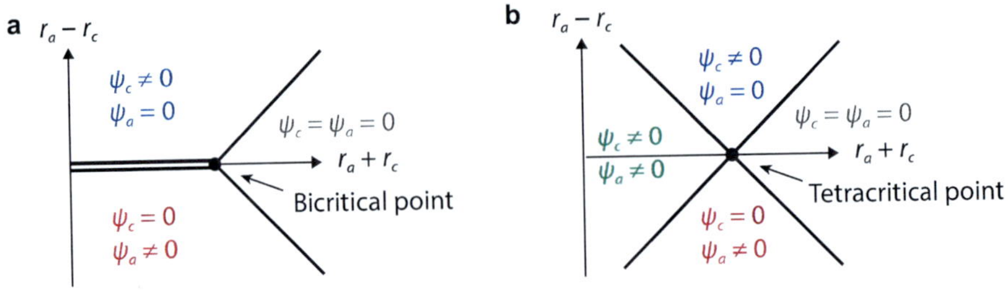

Fig. 6.12 Mean-field phase diagram of two competing orders. Two types of phase diagrams are possible for the free energy described by Eq. (6.1), exhibiting a bicritical point (**a**) when $\beta_a\beta_c < \eta^2$, or a tetracritical point (**b**) when $\beta_a\beta_c > \eta^2$. The order parameters are assumed to be spatially uniform ($\kappa_c = \kappa_a = 0$). A double-line represents a first-order transition while a single-line represents a second-order transition. (Adapted with permission from Ref. [40], Cambridge University Press)

We anticipate that thermal fluctuations of both order parameters, which are neglected in the mean-field treatment, may have a strong impact on the equilibrium phase diagram. In this regard, the effect of the sub-dominant CDW fluctuations on the ground state properties is expected to be pronounced, especially in the temperature range from T_{c2} to T_{c2}^*. Furthermore, we expect that these fluctuations may also renormalize the actual transition temperature of the dominant CDW, T_{c1}.

Vortex Solution

In Sect. 6.4.2, we presented a consistent interpretation of the experimental results: Photoexcitation creates topological defects in the dominant order parameter ψ_c, and at the cores of these defects, the competing phase ψ_a can develop. Below we study the properties of these two competing orders near a topological defect in ψ_c.

Minimizing the total free energy, $\int d^2\mathbf{r}\,\mathcal{W}(\mathbf{r})$, one obtains the following equations:

$$-\kappa_c\nabla_{\mathbf{r}}^2\psi_c + r_c\psi_c + \beta_c|\psi_c|^2\psi_c + \eta|\psi_a|^2\psi_c = 0, \quad (6.2)$$

$$-\kappa_a\nabla_{\mathbf{r}}^2\psi_a + r_a\psi_a + \beta_a|\psi_a|^2\psi_a + \eta|\psi_c|^2\psi_a = 0. \quad (6.3)$$

Below we assume that $\kappa_a = \kappa_c = \kappa$, $\beta_a = \beta_b = \beta$, $r_c < r_a < 0$, and $\eta^2 \geq \beta^2$. The latter two conditions imply the bicritical regime, where only ψ_c develops. The first two conditions are justified by the small a/c-anisotropy; the two orders are only differentiated by r_a and r_c. We emphasize that our subsequent analysis can be easily generalized to the tetracritical regime, where the main conclusions do not change.

Assuming the photoinduced topological defect in the c-axis CDW takes the form of a vortex (i.e., a CDW dislocation), similar to the observation in refs. [32, 36], we seek a vortex solution in ψ_c and solve for ψ_a in a self-consistent way. In the cylindrical coordinates (r, ϕ) where the vortex is located at $r = 0$, the solution has the form: $\psi_c(r, \phi) = \psi_c^\infty f(r)e^{im\phi}$, $\psi_a(r, \phi) = \psi_a^\infty g(r)$, where $\psi_{i=c,a}^\infty = \sqrt{-r_i/\beta}$ and $m = \pm 1, \pm 2, \ldots$ is the vorticity of the vortex. $f(r)$ and $g(r)$ are smooth functions of r, representing normalized order parameters. Then, Eqs. (6.2)–(6.3) take the following form:

$$\xi_c^2(f'' + \frac{1}{r}f') = \frac{\xi_c^2 m^2}{r^2} f - f + f^3 + \alpha_c f g^2, \tag{6.4}$$

$$\xi_a^2(g'' + \frac{1}{r}g') = -g + g^3 + \alpha_a f^2 g, \tag{6.5}$$

where $\xi_{i=c,a}^2 = -\frac{\kappa}{r_i}$, $\alpha_a = \frac{\eta}{\beta}\frac{r_c}{r_a} > 1$, and $\alpha_c = \frac{\eta}{\beta}\frac{r_a}{r_c} > 0$. From these equations, we can calculate the asymptotic behavior of the order parameters:

$$\begin{aligned} f_{r\to\infty} &= 1 - C_1 \exp(-\sqrt{2}r/\xi_c), & f_{r\to 0} &= C_3 r^{|m|}, \\ g_{r\to\infty} &= C_2 \exp(-(\sqrt{\alpha_a - 1})r/\xi_a), & g_{r\to 0} &= C_4 + C_5 r^2. \end{aligned} \tag{6.6}$$

Here C_i are constants that may be obtained by numerically solving Eqs. (6.4)–(6.5). From these analyses we learn that

$$\frac{\lambda_a}{\lambda_c} = \frac{\xi_a}{\xi_c}\sqrt{\frac{2}{\alpha_a - 1}} = \sqrt{\frac{\frac{2r_c}{r_a}}{\frac{\eta}{\beta}\frac{r_c}{r_a} - 1}}, \tag{6.7}$$

where $\lambda_a = \xi_a/\sqrt{\alpha_a - 1}$ is the spatial extent of the sub-dominant phase ψ_a, and $\lambda_c = \xi_c/\sqrt{2}$ is the characteristic length scale of the defect core in ψ_c (Fig. 6.9a). Notably, when the a-axis and c-axis become more isotropic ($r_c/r_a \to 1+$), the ratio λ_a/λ_c becomes larger. In the case of $R\text{Te}_3$ where a/c-anisotropy is small, the relatively large spatial extent of ψ_a would make the transient CDW more easily detected in a diffraction experiment.

References

1. Tokura, Y.: Critical features of colossal magnetoresistive manganites. Rep. Prog. Phys. **69**, 797–851 (2006)
2. Norman, M.R.: The challenge of unconventional superconductivity. Science **332**, 196–200 (2011)
3. Abbamonte, P., Rusydi, A., Smadici, S., Gu, G.D., Sawatzky, G.A., et al.: Spatially modulated 'Mottness' in $La_{2-x}Ba_xCuO_4$. Nat. Phys. **1**, 155–158 (2005)
4. Tranquada, J.M.: Spins, stripes, and superconductivity in hole-doped cuprates. AIP Conf. Proc. **1550**, 114–187 (2013)

5. Leggett, A.J.: A theoretical description of the new phases of liquid ^{3}He. Rev. Mod. Phys. **47**, 331–414 (1975)
6. Nova, T.F., Disa, A.S., Fechner, M., Cavalleri, A.: Metastable ferroelectricity in optically strained $SrTiO_3$. Science **364**, 1075–1079 (2019)
7. Li, X., Qiu, T., Zhang, J., Baldini, E., Lu, J., et al.: Terahertz field-induced ferroelectricity in quantum paraelectric $SrTiO_3$. Science **364**, 1079–1082 (2019)
8. Fausti, D., Tobey, R.I., Dean, N., Kaiser, S., Dienst, A., et al.: Light-induced superconductivity in a stripe-ordered cuprate. Science **331**, 189–191 (2011)
9. Matsubara, M., Okimoto, Y., Ogasawara, T., Tomioka, Y., Okamoto, H., et al.: Ultrafast photoinduced insulator-ferromagnet transition in the perovskite manganite $Gd_{0.55}Sr_{0.45}MnO_3$. Phys. Rev. Lett. **99**, 207401 (2007)
10. Nasu, K. (ed.): Photoinduced Phase Transitions. World Scientific, Singapore (2004)
11. Sie, E.J., Nyby, C.M., Pemmaraju, C.D., Park, S.J., Shen, X., et al.: An ultrafast symmetry switch in a Weyl semimetal. Nature **565**, 61–66 (2019)
12. Malliakas, C.D., Kanatzidis, M.G.: Divergence in the behavior of the charge density wave in $RETe_3$ (RE = Rare-Earth element) with temperature and RE element. J. Am. Chem. Soc. **128**, 12612–12613 (2006)
13. Ru, N.: Charge Density Wave Formation in Rare-earth Tellurides. Ph.D. thesis, Stanford University, Stanford (2008)
14. Ru, N., Condron, C.L., Margulis, G.Y., Shin, K.Y., Laverock, J., et al.: Effect of chemical pressure on the charge density wave transition in rare-earth tritellurides RTe_3. Phys. Rev. B **77**, 035114 (2008)
15. DiMasi, E., Aronson, M.C., Mansfield, J.F., Foran, B., Lee, S.: Chemical pressure and charge-density waves in rare-earth tritellurides. Phys. Rev. B **52**, 14516–14525 (1995)
16. Straquadine, J.A.W., Weber, F., Rosenkranz, S., Said, A.H., Fisher, I.R.: Suppression of charge density wave order by disorder in Pd-intercalated $ErTe_3$. Phys. Rev. B **99**, 235138 (2019)
17. Moore, R.G., Brouet, V., He, R., Lu, D.H., Ru, N., et al.: Fermi surface evolution across multiple charge density wave transitions in $ErTe_3$. Phys. Rev. B **81**, 073102 (2010)
18. Lee, C., Rohwer, T., Sie, E.J., Zong, A., Baldini, E., et al.: High resolution time- and angle-resolved photoemission spectroscopy with 11 eV laser pulses. Rev. Sci. Instrum. **91**, 043102 (2020)
19. Zong, A., Kogar, A., Bie, Y.-Q., Rohwer, T., Lee, C., et al.: Evidence for topological defects in a photoinduced phase transition. Nat. Phys. **15**, 27–31 (2019)
20. Hellmann, S., Rohwer, T., Kalläne, M., Hanff, K., Sohrt, C., et al.: Time-domain classification of charge-density-wave insulators. Nat. Commun. **3**, 1069 (2012)
21. Schmitt, F., Kirchmann, P.S., Bovensiepen, U., Moore, R.G., Rettig, L., et al.: Transient electronic structure and melting of a charge density wave in $TbTe_3$. Science **321**, 1649–1652 (2008)
22. Weathersby, S.P., Brown, G., Centurion, M., Chase, T.F., Coffee, R., et al.: Mega-electron-volt ultrafast electron diffraction at SLAC National Accelerator Laboratory. Rev. Sci. Instrum. **86**, 073702 (2015)
23. Banerjee, A., Feng, Y., Silevitch, D.M., Wang, J., Lang, J.C., et al.: Charge transfer and multiple density waves in the rare earth tellurides. Phys. Rev. B **87**, 155131 (2013)
24. Maschek, M., Zocco, D.A., Rosenkranz, S., Heid, R., Said, A.H., et al.: Competing soft phonon modes at the charge-density-wave transitions in $DyTe_3$. Phys. Rev. B **98**, 094304 (2018)
25. Ru, N., Condron, C.L., Margulis, G.Y., Shin, K.Y., Laverock, J., et al.: Erratum: Effect of chemical pressure on the charge density wave transition in rare-earth tritellurides RTe_3 [Phys. Rev. B 77, 035114 (2008)]. Phys. Rev. B **77**, 249908(E) (2008)
26. Moore, R.G., Lee, W.S., Kirchman, P.S., Chuang, Y.D., Kemper, A.F., et al.: Ultrafast resonant soft X-ray diffraction dynamics of the charge density wave in $TbTe_3$. Phys. Rev. B **93**, 024304 (2016)
27. Ru, N., Fisher, I.R.: Thermodynamic and transport properties of YTe_3, $LaTe_3$, $CeTe_3$. Phys. Rev. B **73**, 033101 (2006)

28. Straquadine, J.A.W., Ikeda, M.S., Fisher, I.R.: Evidence for realignment of the charge density wave state in $ErTe_3$ and $TmTe_3$ under uniaxial stress via elastocaloric and elastoresistivity measurements. arXiv 2005.10461 (2020)
29. Kibble, T.W.B.: Topology of cosmic domains and strings. J. Phys. A **9**, 1387–1398 (1976)
30. Zurek, W.H.: Cosmological experiments in condensed matter systems. Phys. Rep. **276**, 177–221 (1996)
31. Vogelgesang, S., Storeck, G., Horstmann, J.G., Diekmann, T., Sivis, M., et al.: Phase ordering of charge density waves traced by ultrafast low-energy electron diffraction. Nat. Phys. **14**, 184–190 (2018)
32. Fang, A., Straquadine, J.A.W., Fisher, I.R., Kivelson, S.A., Kapitulnik, A.: Disorder-induced suppression of charge density wave order: STM study of Pd-intercalated $ErTe_3$. Phys. Rev. B **100**, 235446 (2019)
33. Arovas, D.P., Berlinsky, A.J., Kallin, C., Zhang, S.-C.: Superconducting vortex with antiferromagnetic core. Phys. Rev. Lett. **79**, 2871–2874 (1997)
34. Lake, B., Aeppli, G., Clausen, K.N., McMorrow, D.F., Lefmann, K., et al.: Spins in the vortices of a high-temperature superconductor. Science **291**, 1759–1762 (2001)
35. Hoffman, J.E., Hudson, E.W., Lang, K.M., Madhavan, V., Eisaki, H., et al.: A four unit cell periodic pattern of quasi-particle states surrounding vortex cores in $Bi_2Sr_2CaCu_2O_{8+\delta}$. Science **295**, 466–469 (2002)
36. Savitzky, B.H., El Baggari, I., Admasu, A.S., Kim, J., Cheong, S.-W., et al.: Bending and breaking of stripes in a charge ordered manganite. Nat. Commun. **8**, 1883 (2017)
37. Hu, B.F., Cheng, B., Yuan, R.H., Dong, T., Wang, N.L.: Coexistence and competition of multiple charge-density-wave orders in rare-earth tritellurides. Phys. Rev. B **90**, 085105 (2014)
38. Rettig, L., Cortés, R., Chu, J.-H., Fisher, I.R., Schmitt, F., et al.: Persistent order due to transiently enhanced nesting in an electronically excited charge density wave. Nat. Commun. **7**, 10459 (2016)
39. Zhou, F., Williams, J., Sun, S., Malliakas, C.D., Kanatzidis, M.G., et al.: Nonequilibrium dynamics of spontaneous symmetry breaking into a hidden state of charge-density wave. Nat. Commun. **12**, 566 (2021)
40. Chaikin, P.M., Lubensky, T.C.: Mean-field theory. In: Principles of Condensed Matter Physics, pp. 144–212. Cambridge University Press, Cambridge (1995)

Chapter 7
Phase Competition Out of Equilibrium

Abstract Engineering novel states of matter by an ultrafast laser pulse is at the forefront of materials research. One promising approach relies on systems that host competing ground states, where light suppresses the dominant phase in equilibrium whilestrengthening a subdominant order. However, the interplay among competing orders remains difficult to study due to a lack of experimental probes that can measure intertwined dynamics all at once. Here, we examine $DyTe_3$ and $LaTe_3$: Both possess two competing charge modulations, out of which only the dominant one appears in equilibrium at high temperature. Using a pump-pump-probe scheme with ultrafast electron diffraction, we find that both CDWs emerge simultaneously after photoexcitation. This joint growth lasts for a few picoseconds and features fluctuating patches of both density waves—in striking resemblance to the critical state near the equilibrium transition temperature. Our investigation unveils a co-developing relation between an otherwise competing pair of CDWs. It further establishes photoinduced fluctuations as a generic pathway for realizing transient states in solids.

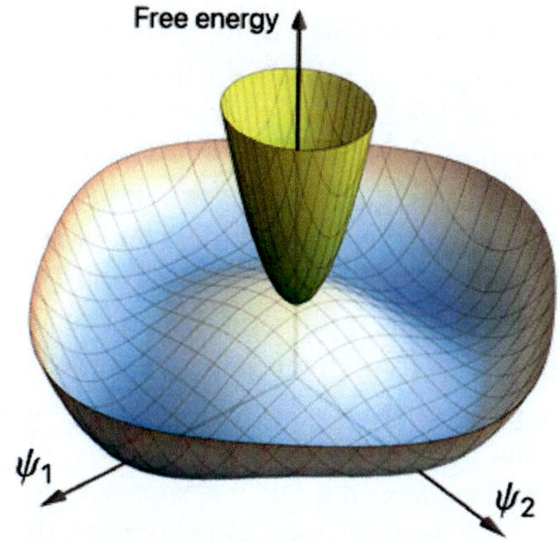

Landau free energy of two competing order parameters, ψ_1 and ψ_2

A. Zong, *Emergent States in Photoinduced Charge-Density-Wave Transitions*, Springer Theses, https://doi.org/10.1007/978-3-030-81751-0_7

7.1 Light Control of Matter and Phase Competition

Competing phases of matter harbor the opportunity to selectively enhance one state over another as we tune, for instance, magnetic field, chemical doping, and uniaxial strain [1]. With recent advances in producing high-flux ultrashort laser pulses, photoexcitation emerges as another powerful knob that has enabled the exciting discovery of subdominant orders or "hidden" states when a dominant phase transiently vanishes [2–10]. This mechanism has been proposed for light-induced superconductivity at the expense of CDWs in cuprates [2] and a light-induced ferromagnetic metal out of an antiferromagnetic insulator in strained manganites [4, 5]. However, besides heuristic arguments based on the competing nature of two or more phases, insights into how a subdominant order emerges are lacking. In particular, the relationship between competing orders is unclear after photoexcitation. For example, while equilibrium charge stripes give way to photoinduced superconductivity in $La_{1.8-x}Eu_{0.2}Sr_xCuO_4$ and $La_{2-x}Ba_xCuO_4$ [11, 12], at the same doping level in $YBa_2Cu_3O_{6+\delta}$, photoexcitation instead suppresses superconductivity and enhances the charge order [3]. It remains unknown whether a generic mechanism exists to account for the wide-ranging observations or whether each light-induced order depends on system-specific details.

Understanding these photoinduced states calls for a close examination of the interplay between competing orders during their nonequilibrium evolution. Typically, one has to rely on different techniques for different orders, which inevitably introduces the challenge of varying experimental conditions that often renders data interpretation difficult. Therefore, it is paramount to use a time-resolved probe that can concurrently inspect all participating order parameters. To this end, the rare-earth tritelluride family $R\mathrm{Te}_3$ offers an excellent platform to study competing phases within a single probe. All members host two competing incommensurate CDWs, which originate from the nearly square-shaped Te planes (Fig. 2.8). The dominant CDW has a modulation along the c-axis while the subdominant one along the a-axis, which gets suppressed in equilibrium but appears after photoexcitation [9, 10]. Both orders can be captured in the same snapshot using ultrafast electron diffraction, providing a unique opportunity to unravel the relationship between competing orders out of equilibrium.

In this chapter, we focus on $LaTe_3$ ($T_c \approx 670\,\mathrm{K}$) and $DyTe_3$ ($T_c = 306(3)\,\mathrm{K}$[1]) [13, 14]. They share nearly identical properties except for their different transition temperatures, giving us access to both the critical regime near T_c and a state with only the c-axis long-range order (LRO). We find that despite the phase competition, the nonequilibrium evolution is characterized by a stage where both CDWs grow in a symmetric manner, as evidenced by their response to a second photoexcitation event. The timespan of this growth stage, equivalent to the lifetime of the light-induced a-axis order, increases with the laser pulse fluence. As we demonstrate in

[1] In this chapter, we are only concerned with the high-temperature CDW transition, so we denote T_{c1} by T_c for brevity.

a non-perturbative Ginzburg–Landau calculation of the quench dynamics, this co-development reflects proliferating fluctuations in the respective CDWs, similar to the fluctuating state near T_c. These findings suggest that photoinduced fluctuations make it possible for competing orders to grow simultaneously, which may account for a large class of light-induced orders in other systems where phase competition is pronounced.

7.2 Competing CDWs In and Out of Equilibrium

As we discovered in Chap. 6, shining an intense laser pulse on $LaTe_3$ is capable of suppressing its equilibrium c-axis CDW while inducing a transient a-axis CDW. This light-induced density wave is evident in the electron diffraction pattern in the $(H, 0, L)$ plane (Fig. 7.1a, b), taken before and 1.6 ps after the incidence of an 80-fs, 800-nm laser pulse. The diffraction intensity of the dominant CDW peaks diminishes (blue arrows) while the subdominant CDW peaks appear (red arrows). As explained in Sect. 7.5.2, the transverse polarization of both CDWs leads to brighter $(H \pm q_a, 0, L)$ satellites along the c-axis than along the a-axis; the opposite is observed for the $(H, 0, L \pm q_c)$ peaks, where $q_a \approx q_c \approx 2/7$. In this transient state, the CDW satellites along both axes share a similar intensity and wavevector (Fig. 7.1b), hinting at a restored symmetry between the two orthogonal CDWs despite the asymmetry of the crystal lattice.

Remarkably, the time-resolved diffraction patterns in $LaTe_3$ were reproduced by static images of $DyTe_3$, taken below and near its T_c (Fig. 7.1c, d). At 100 K, pairs of CDW satellite peaks are found along the c-axis (blue arrows in Fig. 7.1c), but no satellite peaks are observed along the orthogonal a-axis.[2] When the sample is heated to T_c, the c-axis peaks significantly weaken but remain visible (Fig. 7.1d). At the same time, diffuse spots arise along the a-axis, which are localized at a wavevector $q_a \approx q_c$. Similar to $LaTe_3$ in a photoexcited state, the diffuse peaks in $DyTe_3$ at T_c have approximately equal wavevector and intensity along the a and c directions.

The striking similarity between Fig. 7.1b and 7.1d allows us to interpret the photoexcited state using an equilibrium picture close to T_c. In the momentum space, the diffuse satellite peaks are indicative of two Kohn anomalies in the phonon dispersion along the a- and c-axis (Fig. 7.1e). In the real space, this critical regime is characterized by short-range CDW patches in both directions (Fig. 7.1f), where the correlation length is inversely proportional to the momentum width of the Kohn anomaly (see Sect. 7.5.1). From inelastic X-ray measurements [16], the phonon energies at q_a and q_c are approximately 1 to 2 meV, corresponding to a fluctuating timescale of 2 to 4 ps for these CDW patches. The same timescale is observed as the lifetime of the light-induced CDW, whose intensity evolution is plotted in Fig. 7.2a. This correspondence suggests that the light-induced CDW is indistinguishable from

[2]The image was taken above $T_{c2} = 68$ K for $DyTe_3$ [16].

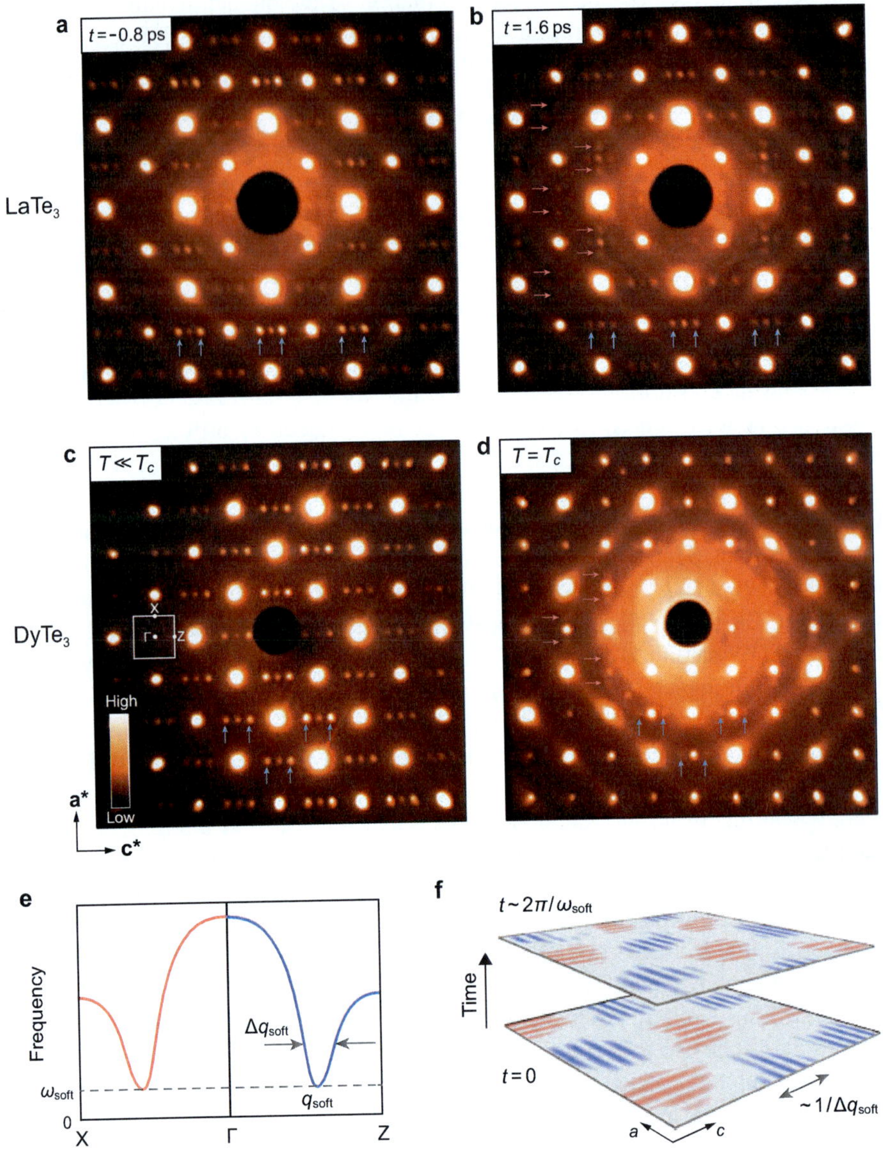

Fig. 7.1 **Competing density wave fluctuations in rare-earth tritellurides.** (**a**,**b**) Time-resolved electron diffractions of $LaTe_3$ before (**a**) and 1.6 ps after (**b**) photoexcitation by an 80-fs, 800-nm laser pulse with an incident fluence of 2.1 mJ/cm^2, measured at 307 K. Blue and red arrows indicate the equilibrium CDW peaks along the c-axis and the light-induced CDW peaks along the a-axis, respectively. (**c**,**d**) Equilibrium diffractions of $DyTe_3$ ($T_c = 306(3)$ K [15]) taken at 100 K (**c**) and 307 K (**d**). (**e**) Schematic phonon dispersion right above T_c along Γ-X and Γ-Z, featuring two Kohn anomalies at q_{soft}. (**f**) Schematic of short-range, fluctuating CDWs right above T_c

a soft phonon, confirming the intimate link between the photoexcited state and the critical regime near T_c.

Despite the similar phenomenology from the diffraction snapshots, it should be emphasized that the transient lattice temperature of $LaTe_3$ stays far below its estimated T_c of 670 K for the pump fluence used in the experiment [13]. Therefore, the similarity between the photoexcited state and the critical state near T_c does not result from a trivial scenario of pulse-induced lattice heating. In the following, we obtain additional insights of both states by studying their ultrafast dynamics.

7.3 Co-Development of Competing Orders

7.3.1 *Two-Part Growth of the Light-Induced CDW*

Leveraging the capability of UED to observe the dynamics of both CDW orders, we examine the intensity evolution of each satellite peak in Fig. 7.2, paying close attention to the initial change. Within 350 fs after photoexcitation, a timescale not limited by the instrument resolution, the equilibrium c-axis order is maximally suppressed [17–19]. The growth of the subdominant a-axis order is markedly slower, peaking at a pump-probe delay of 2 ps. In addition, its intensity evolution exhibits a kink at $t_{\text{kink}} = 350$ fs (dashed line), corresponding to the time delay when the c-axis intensity reaches the minimum. This correspondence is not coincidental and hints at the competing relation between the two CDW instabilities: The growth

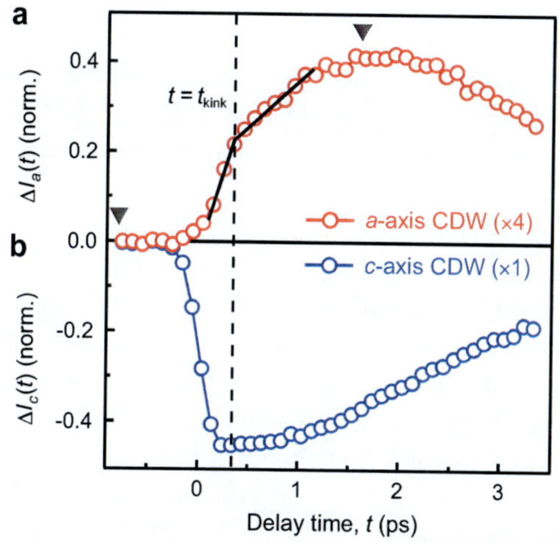

Fig. 7.2 Emergence of the light-induced CDW. Changes in the integrated intensities of the equilibrium c-axis CDW peak (I_c, **b**) and the transient a-axis CDW peak (I_a, **a**). Peaks in multiple Brillouin zones are averaged for improved signal-to-noise ratio (see Sec. 7.5.2). Both traces are normalized by the average value of I_c before photoexcitation. Triangles indicate the time points for the snapshots in Fig. 7.1a, b. Dashed line marks the kink in the evolution of I_a and solid lines are guides to the eye. Data was taken at 307 K with an incident pump laser fluence of 2.1 mJ/cm^2

of the subdominant order slows down as the dominant order starts to recover. The slowing-down can be interpreted as a competition for the finite number of carriers that scatter the soft phonons. As the c-axis order begins to recover, a subset of carriers starts to scatter in the $\mathbf{q}_c$ direction, so the a-axis fluctuations do not grow as quickly as before.

7.3.2 Symmetric CDW Growths

The time evolution revealed in Fig. 7.2 suggests that after t_{kink}, the system spends a significant time in a "confused" state where both the dominant and the subdominant orders develop simultaneously. This behavior is in contrast to one's expectation of competing orders in equilibrium, where the growth of one phase is at the cost of another [20, 21]. To gain some insights into this coexisting phase, we apply a second laser pulse to $LaTe_3$ right after t_{kink} and watch the response in both orders. As a reference experiment, we also photoexcite $DyTe_3$ at its T_c, where fluctuations of both CDWs abound.

We start by examining the photoinduced response in $DyTe_3$, shown in Fig. 7.3. The diffuse satellite spots display an initial dip in intensity followed by a fast recovery, a trend perfectly mirrored in both axes (Fig. 7.3c, d). This dynamics is in stark contrast to diffuse scattering intensities at other momenta away from Bragg or CDW peaks, where only a single-exponential rise is observed (Fig. 7.3b). The dip can be understood in two equivalent ways. From the phonon perspective, it represents a stiffening of the soft mode in the Kohn anomaly [22]. As electronic states are excited to high energy, there is a transient reduction of dielectric screening of the lattice ions, leading to an increase in the renormalized phonon frequency and hence a decrease in diffuse intensity (see Sect. 7.5.1). An alternative viewpoint is based on the short duration of the electron pulse ($\sim$150 fs), so the low-energy phonon (1 to 2 meV) appears "frozen" in each diffraction frame. In this picture, the dip indicates a smaller atomic displacement in the fluctuating patches, averaged over all electron pulses and probed volumes. The second perspective naturally connects to the photoinduced melting of a long-range CDW, in contrast to the suppression of short-range fluctuations presented here. Both processes occur over a timescale of $\sim$0.4 ps and there is no fundamental distinction between them on the sub-picosecond level. In Fig. 7.3c, d, the intensities are seen to rise after the dip, indicating an increased phonon population from laser-induced heating. After subtracting this thermal diffuse contribution, the dip is only partially recovered (Fig. 7.3e), suggesting an elevated lattice temperature above T_c, where the Kohn anomaly becomes less pronounced.

Next, we study $LaTe_3$, focusing on the response of co-developing CDWs to a second laser pulse (Fig. 7.4). The two pump pulses share the same incident fluence. To assess the absorbed fluence, we note that the maximum value attained in thermal diffuse scattering doubles after the second pulse (Fig. 7.4b). This observation affirms that energy absorption is minimally affected by the presence of excited carriers.

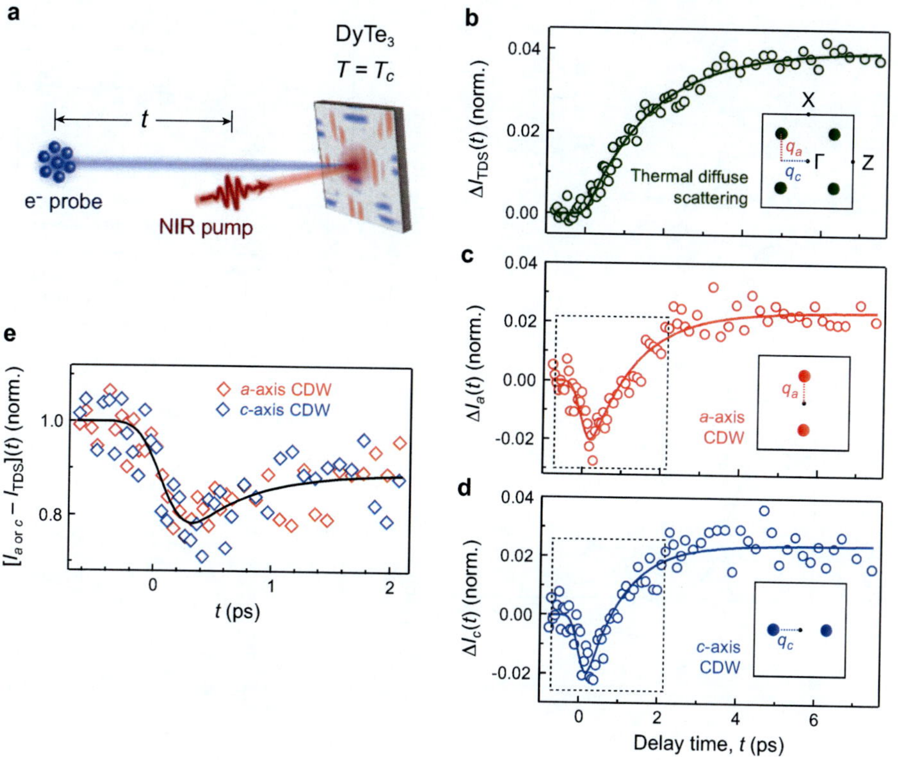

Fig. 7.3 Response of fluctuating orders to photoexcitation. (**a**) Schematic of the setup for $DyTe_3$ kept close to its T_c, with an incident fluence of 3.3 mJ/cm^2. (**b–d**) Changes in the integrated intensities for thermal diffuse scattering (I_{TDS}), a-axis diffuse CDW peak (I_a), and c-axis diffuse CDW peak (I_c). Integration areas are marked by solid circles in the insets. Traces are normalized by the average value of I_c before photoexcitation. (**e**) Zoomed-in view of dashed rectangles in **c,d** after subtracting the thermal diffuse background (I_{TDS}). Traces are normalized by average values at $t < 0$. Solid curves in **b–d** are fits to Eq. (3.14). The fitted curve in **e** uses the averaged data between a- and c-axis

We now move on to analyze the CDW peaks (Fig. 7.4c, d). Unlike their distinct behavior after the first pulse, the intensity evolutions of both CDWs share nearly identical trends after the second pulse. For a direct comparison between the two orders, we zoom in to their dynamics right after the second pulse and plot them together in Fig. 7.4e, where contributions from thermal diffuse scatterings have also been removed. Similar to short-range CDWs in $DyTe_3$ near its T_c, the co-developing orders in $LaTe_3$ feature a transient reduction in the fluctuation amplitude, followed by a recovery that lasts for more than 2 ps. Unlike $DyTe_3$, the satellite intensities in $LaTe_3$ are fully recovered compared to their values just before the second pulse, suggesting the nonthermal nature of these CDW fluctuations.

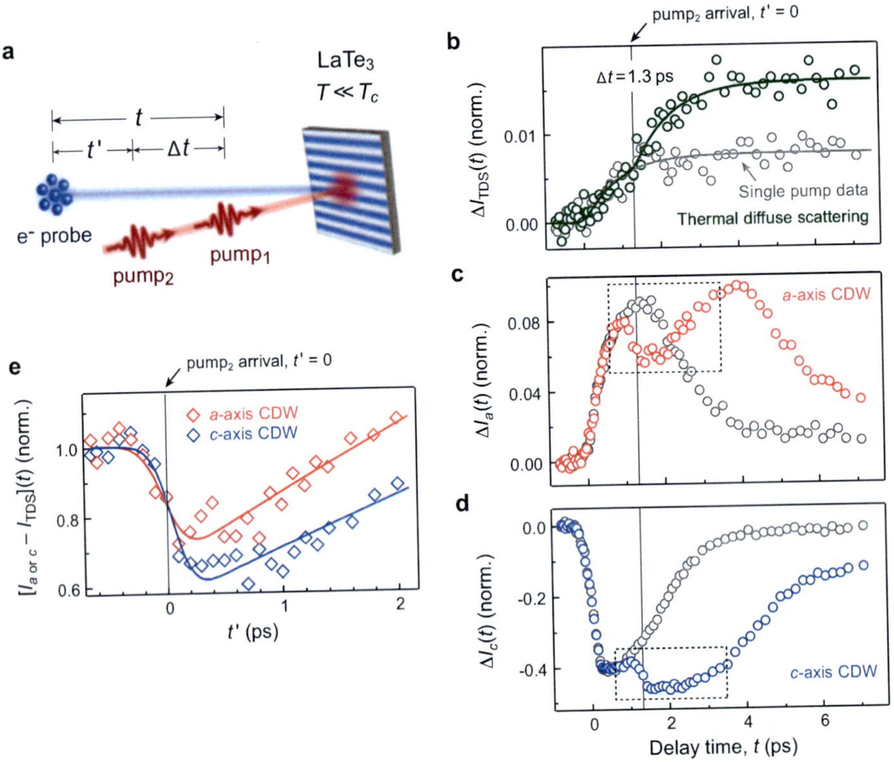

Fig. 7.4 Co-development of competing orders out of equilibrium. (**a**) Schematic of the two-pump measurement for $LaTe_3$. Sample was kept at 307 K and the incident fluence of each pump was 1.0 mJ/cm^2. (**b–d**) Changes in the integrated intensities of thermal diffuse scattering (I_{TDS}), light-induced a-axis peak (I_a), and equilibrium c-axis peak (I_c), with the same integration areas and normalization procedures used in Fig. 7.3. Vertical lines indicate the arrival time of pump$_2$ at $\Delta t = 1.3$ ps. For reference, dynamics in the absence of pump$_2$ is shown in gray. (**e**) Zoomed-in view of dashed rectangles in **c**,**d** after subtracting the thermal diffuse background (I_{TDS}). Traces are plotted as a function of the relative delay between the probe pulse and pump$_2$ (t'). Intensities are normalized by the average value in the interval $t' \in [-0.7\,\text{ps}, -0.2\,\text{ps}]$. A slightly larger reduction along the c-axis is attributed to a mismatch between pumped and probed volumes, leading to additional melting of residual long-range CDW by the second pulse

7.3.3 *Lifetime of Light-Induced Orders*

Similar to the single-pump measurement (Fig. 7.2), we observe a stage of co-developing orders after the second pulse in $LaTe_3$ (Fig. 7.4e). The timespan of this concurrent growth increases with the pump fluence: As more energy is injected into the system, both CDW peaks experience a slowing-down in their recovery (Fig. 7.5). An important implication of this observation is that the lifetime of the light-induced order would appear to increase with deposited energy [9]. Therefore, a long-lasting photoinduced order may not imply a protracted absence of the competing order. Instead, fluctuations of both orders can develop concurrently, albeit at a slower rate.

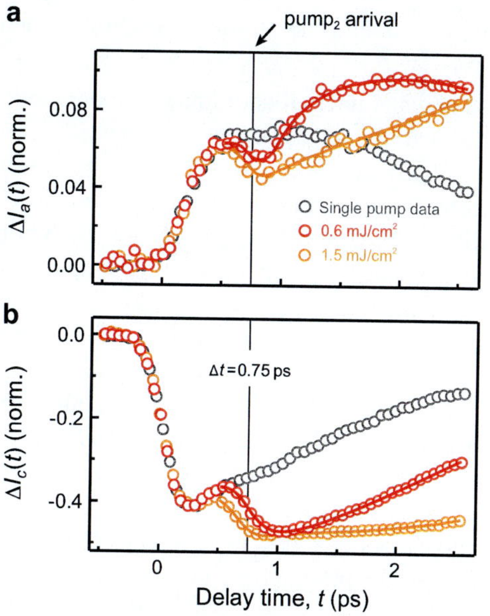

Fig. 7.5 Fluence-dependent growth of fluctuating domains. Changes in the integrated intensities for the a-axis (**a**) and the c-axis (**b**) CDW peaks following photoexcitation by two pulses. For reference, dynamics induced by a single pulse is shown in gray. The arrival time of the second pulse is marked by the vertical line. The incident fluence of pump$_1$ was 1.1 mJ/cm^2; varying fluences of pump$_2$ were labeled. Traces are normalized by the average value of I_c before photoexcitation. Solid curves are guides to the eye

7.4 Photoinduced Fluctuations of Competing Orders

The similarities between the excited state in $LaTe_3$ and the critical state in $DyTe_3$—both in their diffraction snapshots (Fig. 7.1a–d) and in their photoinduced dynamics (Figs. 7.3e and 7.4e)—suggest that the co-development of competing orders is made possible by the fluctuating nature of CDWs in the absence of any long-range order. While the fluctuations near T_c are thermal and follow the scaling relations prescribed by the theory of renormalization group, the light-induced fluctuations may not conform to a thermodynamic distribution. To gain some insights into the nonequilibrium evolution of CDW fluctuations, we compute their dynamics using a time-dependent interacting $O(N)$ model. The simplest Ginzburg–Landau free energy functional that contains all necessary ingredients for the two competing CDWs has the following form:

$$\mathcal{F}[\boldsymbol{\psi}_c, \boldsymbol{\psi}_a] = \mathcal{F}_c[\boldsymbol{\psi}_c] + \mathcal{F}_a[\boldsymbol{\psi}_a] + \mathcal{F}_{\text{int}}[\boldsymbol{\psi}_c, \boldsymbol{\psi}_a], \text{ where} \tag{7.1}$$

$$\mathcal{F}_{i=a,c}[\boldsymbol{\psi}_i] = \int d^d\mathbf{x} \left[\frac{r_i}{2}\boldsymbol{\psi}_i^2 + \frac{K_i}{2}(\nabla\boldsymbol{\psi}_i)^2 + u_i(\boldsymbol{\psi}_i^2)^2\right], \tag{7.2}$$

$$\mathcal{F}_{\text{int}}[\boldsymbol{\psi}_c, \boldsymbol{\psi}_a] = 2\eta \int d^d\mathbf{x}\, \boldsymbol{\psi}_c^2\boldsymbol{\psi}_a^2, \text{ where } \eta > 0. \tag{7.3}$$

Here, r_i, u_i, K_i, and η are model parameters. $\boldsymbol{\psi}_{i=a,c}(\mathbf{x})$ is an N-component vector of real numbers, which is used to denote the order parameter for the symmetry-breaking transition [23, 24]

$$\boldsymbol{\psi}_{i=a,c}(\mathbf{x}) = \left[\psi_{i,1}(\mathbf{x}), \ldots, \psi_{i,N}(\mathbf{x})\right]. \tag{7.4}$$

If $N = 2$, one can alternatively think of $\boldsymbol{\psi}_i(\mathbf{x})$ as a complex number. The gradient term $(\nabla \boldsymbol{\psi}_i)^2$ means $\sum_{j=1}^{d} \sum_{\alpha=1}^{N} \left(\partial_j \psi_{i,\alpha}\right)^2$, namely, the scalar product over both spatial and internal coordinates. In our case, $N = 2$ for incommensurate CDWs and the spatial dimension is $d = 3$.

The mean-field phase diagram of the model in Eq. (7.1) agrees with the phenomenology of rare-earth tritellurides. Specifically, this model has two qualitatively different regimes [25] (Fig. 6.12): (i) bicritical regime ($u_a u_c < \eta^2$) corresponds to a situation when only the dominant order, ψ_c, can develop in equilibrium; (ii) tetracritical regime ($u_a u_c > \eta^2$) represents the case when the two order parameters can coexist. We focus on the former situation for $LaTe_3$ and further impose $u_a = u_c = u$ and $K_a = K_c = K$ because the anisotropy between the orthogonal CDWs is small. The equilibrium asymmetry between the two CDWs lies in a slightly smaller value of r_c compared to r_a, so the CDW always occurs along the c-axis.

Photoexcitation is modeled by an impulsive change to the phenomenological parameters r_a and r_c in Eq. (7.2). If we define $r \equiv (r_c + r_a)/2$ and $g \equiv (r_c - r_a)/2$, the quench protocol is

$$r(t) = r_0 + \Theta(t - t_1)e^{-(t-t_1)/\tau_{\mathrm{QP}}} F_1 + \Theta(t - t_2)e^{-(t-t_2)/\tau_{\mathrm{QP}}} F_2, \tag{7.5}$$

where r_0 is the equilibrium value; $t_{1,2}$ and $F_{1,2}$ denote the arrival times and fluences of the two excitation pulses. g is assumed to be constant throughout the temporal evolution. τ_{QP} represents the typical relaxation timescale of excited quasiparticles, which are responsible for the modification of the free energy potential. A representative $r(t)$ is shown in Fig. 7.6a.

7.4.1 Dynamics of Fluctuation Amplitudes

Without loss of generality, we assume that spontaneous symmetry breaking occurs along the first component of the order parameter, $\psi_{i,1}$. In the temporal evolution after photoexcitation, the presence of LRO is hence represented by a nonzero expectation value

$$\psi_i(t) \equiv \langle \psi_{i,1}(t) \rangle. \tag{7.6}$$

Irrespective of whether LRO exists, the order parameter can also exhibit transverse and longitudinal fluctuations, represented by the transient equal-time two-point correlation functions [26],

$$D^{\parallel}_{\mathbf{k},i}(t) \equiv \langle \psi_{i,1}(-\mathbf{k};\, t)\, \psi_{i,1}(\mathbf{k};\, t) \rangle_c \quad \text{(Higgs modes)}, \tag{7.7}$$

$$D^{\perp}_{\mathbf{k},i}(t) \equiv \langle \psi_{i,\alpha\neq 1}(-\mathbf{k};\, t)\, \psi_{i,\alpha\neq 1}(\mathbf{k};\, t) \rangle_c \quad \text{(Goldstone modes)}. \tag{7.8}$$

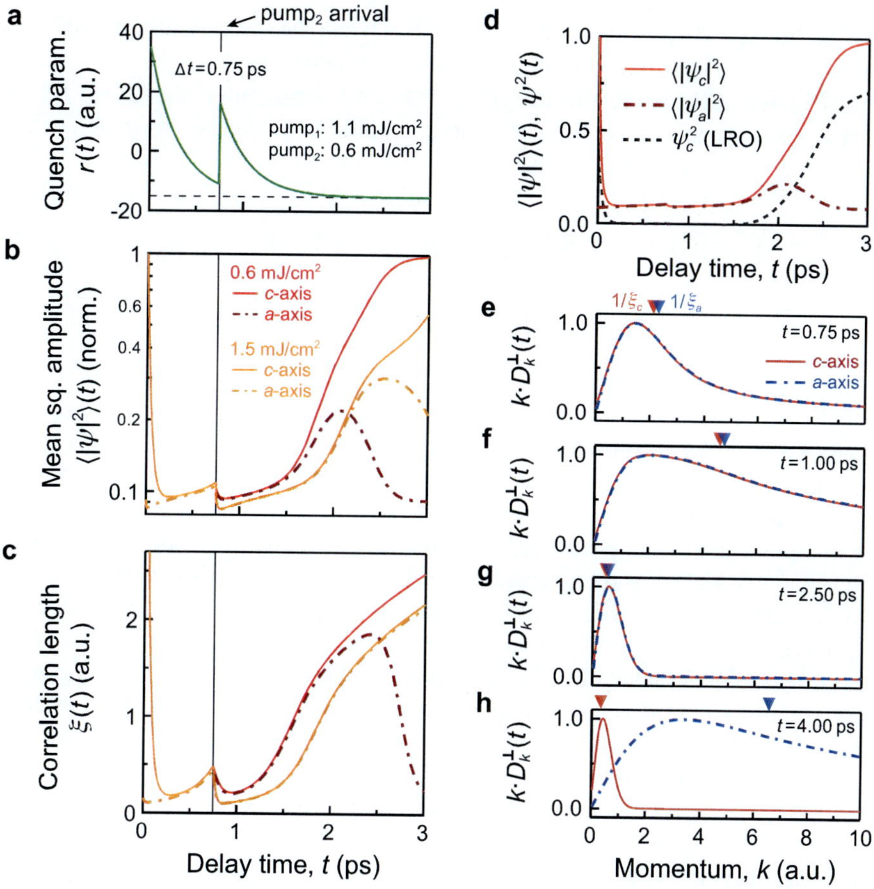

Fig. 7.6 Simulated dynamics of competing CDW orders by time-dependent Ginzburg–Landau formalism. (**a**) Quench protocol for the double-pump experiment, where the simulated fluence ratio of the first and the second pulse follows the experimental value of 1.1 and 0.6 mJ/cm^2. Horizontal dashed line represents the equilibrium value of $r \equiv (r_a + r_c)/2$. τ_{QP} is taken to be 0.3 ps in Eq. (7.5), obtained from transient reflectivity measurements [27]. (**b,c**) Simulated mean square order parameter amplitude $\langle|\psi_{a,c}|^2\rangle(t)$ and correlation length $\xi_{a,c}(t)$ for two different pump$_2$ fluences, matching those in Fig. 7.5. Values of $\langle|\psi_{a,c}|^2\rangle(t)$ are normalized by $\langle|\psi_c|^2\rangle(0)$. (**d**) Evolution of the long-range order (LRO) component $\psi_c^2(t)$ (gray dashed curve) of the mean square order parameter amplitude $\langle|\psi_c|^2\rangle(t)$. Red curves are the same as those in **b**. All curves are normalized by $\langle|\psi_c|^2\rangle(0)$. There is no LRO for the a-axis CDW: $\psi_a^2(t) = 0$ for all t. (**e–h**) Momentum distributions of transverse fluctuations at indicated time delays; each curve is individually normalized by its maximum value. $D_{k,i}^{\perp}(t)$ is the transverse correlation function defined in Eq. (7.8). The extracted inverse correlation lengths, $1/\xi_{a,c}$, are indicated by the triangles. The simulation parameters used are the same as the red curves in **c**

The evolutions of $\psi_i(t)$, $D^{\parallel}_{\mathbf{k},i}(t)$, and $D^{\perp}_{\mathbf{k},i}(t)$ can be numerically solved for following their respective equations of motion, detailed in Sect. 7.5.4.

In electron diffraction, the integrated intensity surrounding the CDW peaks measures the mean square order parameter amplitude at each pump-probe delay t, averaged over the probed sample volume. We denote this observable by

$$\langle|\psi_i|^2\rangle(t) \equiv \int d^3\mathbf{x}\left[\langle\psi^2_{i,1}(\mathbf{x},t)\rangle + \langle\psi^2_{i,2}(\mathbf{x},t)\rangle\right] \tag{7.9}$$

$$= \underbrace{\psi_i^2(t)}_{\text{LRO}} + \underbrace{\int \frac{d^3\mathbf{k}}{(2\pi)^3}\left[D^{\parallel}_{\mathbf{k},i}(t) + D^{\perp}_{\mathbf{k},i}(t)\right]}_{\text{Fluctuations}}, \tag{7.10}$$

where the second line follows from the Fourier transform. As labeled in Eq. (7.10), the integrated intensity contains contributions from both the LRO and the fluctuations. The evolution of $\langle|\psi_a|^2\rangle(t)$ and $\langle|\psi_c|^2\rangle(t)$ are plotted in Fig. 7.6b. After the first photoexcitation event, $\langle|\psi_c|^2\rangle(t)$ decreases significantly due to the loss of LRO $\psi_c^2(t)$, which is shown separately in Fig. 7.6d (gray dashed curve). As the LRO is suppressed, short-range fluctuations of both CDWs start to develop in an indistinguishable manner. The effect of the second pump is to transiently suppress the co-developing fluctuations, which swiftly rebound and grow over the next picosecond. As the fluence of the second pump increases, the joint growth of both $\langle|\psi_a|^2\rangle(t)$ and $\langle|\psi_c|^2\rangle(t)$ slows down, in agreement with experimental observations in Fig. 7.5. The mean square amplitudes of the two CDWs start to diverge once the LRO $\psi_c^2(t)$ starts to recover from zero (Fig. 7.6d). This behavior suggests that the out-of-equilibrium coexistence of competing orders is only possible when they remain as fluctuations without any LRO.

7.4.2 *Evolution of Correlation Lengths*

We can further infer the correlation length of fluctuating CDW patches from the transverse correlation function $D^{\perp}_{\mathbf{k},i}(t)$. In equilibrium, the corresponding transverse correlation length increases as one approaches T_c from above and remains divergent in the symmetry-broken phase, as expected in a second-order phase transition [26]. In Fig. 7.6e–h, $kD^{\perp}_{k,i}(t)$ are shown for several representative time delays, corresponding to the simulated dynamics in Fig. 7.6b. Right after the second pulse, the distribution shifts to higher momenta (Fig. 7.6e, f), indicating a reduction of correlation in the real space. As both CDWs develop in amplitude, their correlation lengths also grow in a symmetric manner (Fig. 7.6g). Finally, as LRO along the c-axis develops, the correlation functions of the two CDWs depart (Fig. 7.6h) and the a-axis order loses its phase coherence.

From the correlation function $D^{\perp}_{\mathbf{k},i}(t)$, we can extract a correlation length ξ_i, whose reciprocal is indicated by the triangles in Fig. 7.6e–h. In the simulation, $1/\xi_i$

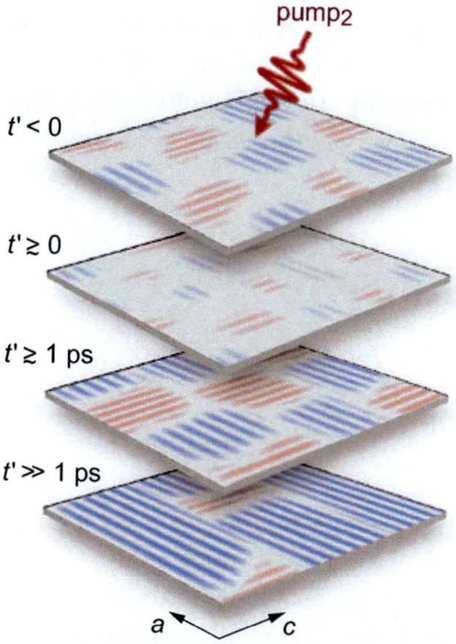

Fig. 7.7 Schematic evolution of competing fluctuations upon photoexcitation. The intensity of the stripes indicates local CDW amplitude. The competing CDWs share the same growth phase for approximately 1 ps or longer (depending on the fluence). During this period, both local amplitude and correlation length increase

is taken as the centroid of $D^{\perp}_{\mathbf{k},i}$ up to a UV cutoff, though the exact definition is not important. Figure 7.6c shows the evolutions of $\xi_{a,c}(t)$ for two pump fluences of the second pulse. In general, photoexcitation decreases the correlation length; the amount of suppression positively scales with the fluence. This result agrees with the experimental observation that satellite peaks broaden after photoexcitation [10, 27], suggesting the transient formation of topological defects in the superlattice. At higher fluence, more defects are created, leading to a slower recovery of the correlation lengths that follow a scaling law at long time delays [26, 27].

Taken together, our experimental and theoretical results are summarized by the schematic in Fig. 7.7. After the first pulse produces a state filled with fluctuating, short-range CDWs ($t' < 0$), the second pulse further decreases the patch size and the average atomic displacement at the CDW wavevectors ($t' \gtrsim 0$). For the next few picoseconds, both correlation length and amplitude develop without breaking the global symmetry between the a- and c-axis ($t' \gtrsim 1$ ps). Finally, the system relaxes to a quasi-equilibrium state where LRO of the dominant CDW takes over ($t' \gg 1$ ps). While the competing CDWs are depicted as spatially separated, a checkerboard pattern could possibly develop, whose presence can be confirmed by diffuse peaks at wavevectors $\mathbf{q}_a \pm \mathbf{q}_c$. All of our data return a null result though future experiments are needed as background scatterings may have masked the tiny signal.

This work offers a clear picture of phase competition when different orders are captured within the same measurement. We have demonstrated that two incompatible CDWs in equilibrium can develop simultaneously in a nonequilibrium regime, where photoinduced fluctuations play a key role. The proliferating fluctuations in an excited state [26, 28] motivate a natural connection to a critical point, be it thermal or quantum, where hidden states are known to abound [16, 29]. Hence, we envision that the joint growth of competing orders be general whenever long-range orders are suppressed by photoexcitation. Similar to an equilibrium critical point, this out-of-equilibrium phenomenology should hold regardless of microscopic details, providing a guiding principle in our search for other light-induced orders.

7.5 Supplementary

7.5.1 *Diffraction Intensity Near a Kohn Anomaly*

In Fig. 7.1f, the characteristic length scale of a fluctuating CDW domain is labeled as $1/\Delta q_{\mathrm{soft}}$, where Δq_{soft} is the width of the Kohn anomaly in the phonon dispersion. In this section, we formalize the relation between the momentum width of the Kohn anomaly and the corresponding width of the diffuse scattering peak, the latter of which measures the reciprocal of the domain size.

Under the kinematic approximation, the diffraction intensity at wavevector $\mathbf{k}$ is the sum of contributions from elastic scattering and inelastic scatterings with $n \geq 1$ phonons:

$$I(\mathbf{k}) = I_0(\mathbf{k}) + \sum_{n\geq 1} I_n(\mathbf{k}). \tag{7.11}$$

As the scattering cross sections with two or more phonons are relatively small, we are mostly concerned with I_0 and I_1, given by [30]

$$I_0(\mathbf{k}) = I_c \left| \sum_{\alpha} f_\alpha(\mathbf{k}) e^{-W_\alpha(\mathbf{k})} e^{-i\mathbf{k}\cdot\mathbf{r}_\alpha} \right|^2, \tag{7.12}$$

$$I_1(\mathbf{k}) = I_c \sum_{j} \frac{n_{j,\mathbf{q}} + 1/2}{\omega_{j,\mathbf{q}}} \left| F_{1j}(\mathbf{k}) \right|^2, \tag{7.13}$$

where I_c is some constant of proportionality, $\mathbf{k}$ is the total scattering wavevector, $\mathbf{q} \equiv \mathbf{k} - \mathbf{G}$ is the reduced crystal momentum, defined with respect to the closest reciprocal lattice vector $\mathbf{G}$ for a given $\mathbf{k}$. $\mathbf{r}_\alpha$ is the position of atom α in the unit cell, $W_\alpha(\mathbf{k})$ is the Debye–Waller factor, and $f_\alpha(\mathbf{k})$ is the atomic form factor. Index j runs over the phonon branches and $n_{j,\mathbf{q}}$, $\omega_{j,\mathbf{q}}$ correspond to the population and frequency for the phonon at reduced momentum $\mathbf{q}$ in branch j. $\left|F_{1j}(\mathbf{k})\right|^2$ is the one-phonon structure factor, given by

$$\left|F_{1j}(\mathbf{k})\right|^2 = \left| \sum_{\alpha} e^{-W_\alpha(\mathbf{k})} \frac{f_\alpha(\mathbf{k})}{\sqrt{m_\alpha}} \left(\mathbf{k} \cdot \mathbf{e}_{j,\alpha,\mathbf{q}}\right) \right|^2, \tag{7.14}$$

where m_α is the mass of atom α and $\mathbf{e}_{j,\alpha,\mathbf{q}}$ is the displacement polarization vector for atom α under the motion of phonon in branch j with momentum $\mathbf{q}$. As we explain in

Sect. 7.5.2, the transverse polarization of the CDWs in $R\mathrm{Te}_3$ leads to a momentum-dependent satellite intensity in the diffraction pattern, which is accounted for by the $\mathbf{k} \cdot \mathbf{e}_{j,\alpha,\mathbf{q}}$ term. The Debye–Waller factor $W_\alpha(\mathbf{k})$ describes the intensity reduction in both I_0 and I_1, which results from the excitation of phonons in all branches and momenta. It reads

$$W_\alpha(\mathbf{k}) = \frac{1}{4m_\alpha} \sum_{j,\mathbf{q}} \left|u_{j,\mathbf{q}}\right|^2 \left|\mathbf{k} \cdot \mathbf{e}_{j,\alpha,\mathbf{q}}\right|^2. \tag{7.15}$$

Here, $u_{j,\mathbf{q}}$ is the vibration amplitude for a particular phonon mode,

$$\left|u_{j,\mathbf{q}}\right|^2 = \frac{2\hbar\left(n_{j,\mathbf{q}} + 1/2\right)}{N_c\,\omega_{j,\mathbf{q}}}, \tag{7.16}$$

where N_c is the number of unit cells.

From Eq. (7.13), the presence of a Kohn anomaly—a dip in the frequency $\omega_{j,\mathbf{q}}$ at a particular momentum $\mathbf{q}_{\mathrm{soft}}$ for phonon branch j—would lead to a locally enhanced one-phonon scattering. The enhancement is primarily due to the $n_{j,\mathbf{q}}$ term in the numerator and the $\omega_{j,\mathbf{q}}$ term in the denominator. If we choose a scattering vector $\mathbf{k}$ that is aligned with the polarization vector $\mathbf{e}_{j,\alpha,\mathbf{q}}$ for the soft mode in question, we expect the one-phonon structure factor $\left|F_{1j}(\mathbf{k})\right|^2$ to be a slow-varying function of $\mathbf{q}$ within a Brillouin zone, so it is less affected by the Kohn anomaly. To see the relationship between $\omega_{j,\mathbf{q}}$ and $I_1(\mathbf{k})$, a schematic of a Kohn anomaly is sketched in Fig. 7.8a (blue curve), using experimentally determined parameters for DyTe_3 near its T_c [16]. The corresponding I_1 from this soft phonon branch is shown in the red curve, where the contribution from $|F_{1j}|^2$ is neglected. Here, we use the Bose–Einstein distribution for the phonon population $n_{j,\mathbf{q}} = 1/\left(e^{\omega_{j,\mathbf{q}}/k_B T} - 1\right)$, where T is set to 307 K, at the T_c of DyTe_3. As expected, a diffuse peak is developed near q_{soft}, whose width matches that of the Kohn anomaly up to a small factor. In Fig. 7.8b, the width of the diffuse peak is plotted for a range of widths for the Kohn anomaly, showing a quasi-linear relationship. Therefore, it is indeed justified to use the width of the Kohn anomaly to estimate the width of the diffuse peak up to a small constant factor, which in turn indicates the spatial extent of the fluctuating domains (Fig. 7.1e, f).

7.5.2 *Transverse Atomic Displacement and Bragg Peak Dynamics*

From the diffraction patterns in Fig. 7.1a–d, the c-axis CDW peaks at $(H, 0, L \pm q_c)$ are most prominent when $|H| > |L|$. On the other hand, the a-axis satellites at $(H \pm q_a, 0, L)$ are brightest for $|H| < |L|$. Here, we adopt the convention of $q_a \approx q_c \approx 2/7$, so $H + L$ is an odd integer. This momentum dependence suggests the transverse

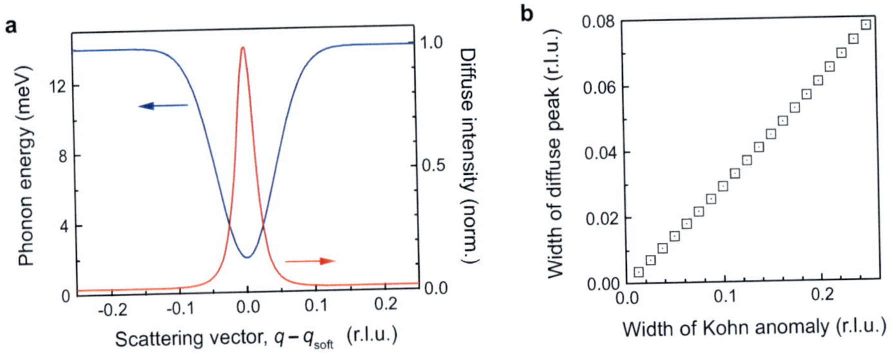

Fig. 7.8 Widths of Kohn anomaly and thermal diffuse peak. (**a**) Sketch of a Kohn anomaly in the phonon dispersion (blue curve, left axis), adopting experimentally observed parameters for $DyTe_3$ near its T_c [16]. The calculated one-phonon diffuse intensity at $T = 307$ K for this dispersion is superimposed (red curve, right axis). Intensity is normalized by its maximum value. (**b**) FWHMs of the diffuse peak as a function of FWHMs of the Kohn anomaly, showing a quasi-linear relation. The FWHM of the diffuse peak is obtained by fitting it to a Lorentzian profile

polarization of the atomic displacement underlying the periodic lattice distortion [16, 31]. As we recall from Eq. (2.21) in Sect. 2.2.1, the intensity of the CDW peak contains a factor of $\mathbf{u} \cdot \mathbf{k}$, where $\mathbf{u}$ is the atomic displacement and $\mathbf{k}$ is the scattering vector. For a transverse mode, $\mathbf{u} \perp \mathbf{q}$, where $\mathbf{q}$ is the CDW ordering vector. Hence, $\mathbf{u} \parallel \mathbf{a}^*$ for the c-axis CDW and $\mathbf{u} \parallel \mathbf{c}^*$ for the a-axis CDW. The maximum intensity occurs when $\mathbf{k} \parallel \mathbf{u}$, leading to the momentum pattern observed. For this reason, we divide the diffraction pattern into four quadrants, shown in Fig. 7.9a. When plotting the intensity evolutions of CDW peaks—$I_a(t)$ and $I_c(t)$—we only average over peaks in the corresponding quadrants to maximize the signal. The integration areas for $I_a(t)$ and $I_c(t)$ are marked by red and blue circles in Fig. 7.9a, respectively.

Another consequence of the transverse polarization is encoded in the intensity evolution of the Bragg peak, shown in Fig. 7.9c. The yellow markers correspond to quadrants I & III while the purple markers correspond to quadrants II & IV, as indicated by the color-coded ROIs in Fig. 7.9a. While the yellow curve shows an initial increase in intensity, the purple curve only exhibits an intensity decay over ~2 ps. The decay arises from the photoinduced Debye–Waller factor that redistributes the spectral weight from the Bragg peak to elsewhere in the momentum space as phonon branches are populated. The initial rise can only be accounted for by the transient suppression of the equilibrium c-axis CDW. As described by Eq. (2.23), this intensity gain of the Bragg peak also contains the factor $\mathbf{u} \cdot \mathbf{k}$, so the effect is the most pronounced when $\mathbf{u} \parallel \mathbf{k}$. For the c-axis CDW, $\mathbf{u} \parallel \mathbf{a}^*$, so the intensity gain is mostly manifested in Bragg peaks from quadrants I & III.

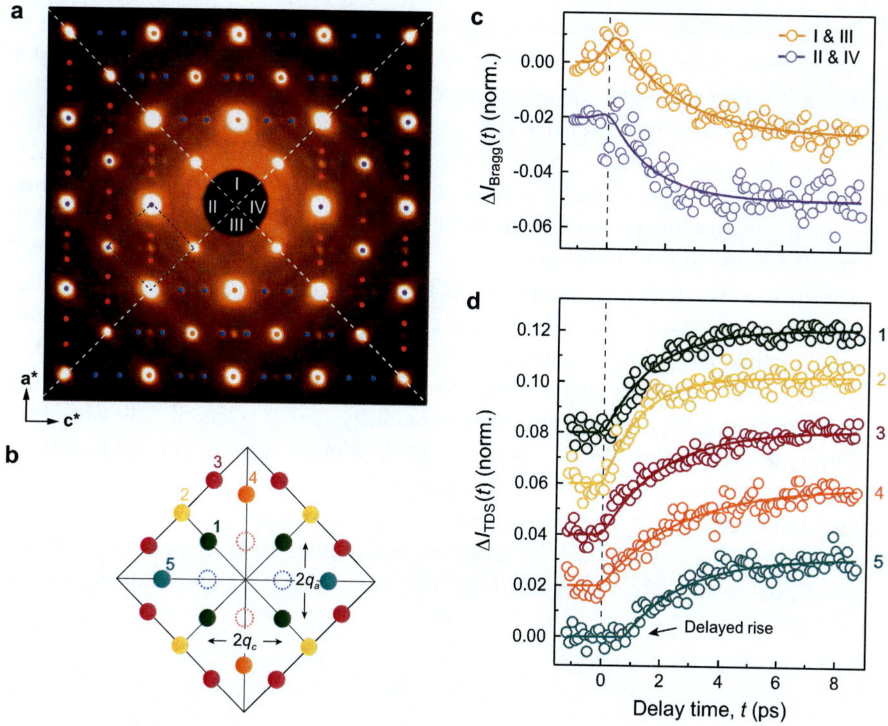

Fig. 7.9 Evolution of lattice Bragg peaks and diffuse scatterings. (**a**) The same diffraction pattern of $LaTe_3$ as in Fig. 7.1b, overlaid with circular ROIs that indicate the integration regions for CDW peak intensities I_c (blue) and I_a (red). ROIs are divided into quadrants I–IV (white dashed lines) due to the transverse nature of the lattice distortion. Yellow and purple ROIs on Bragg peaks correspond to integration areas for **c**. (**b**) Zoomed-in view of the dashed diamond in **a** centered around the $(H\ 0\ L)$ order where $H + L$ is an odd integer. Color-coded circles are integration areas for diffuse scatterings (I_{TDS}) plotted in **d**, where intensities in areas of the same color across multiple diffraction orders are averaged. Locations of CDW peaks are indicated by the dashed circles. (**c**) Changes in the integrated intensity of Bragg peaks in quadrant I & III (yellow) and quadrant II & IV (purple). Curves are fits to Eq. (3.14). (**d**) Evolutions of the change in the diffuse scattering intensity at locations 1 to 5, labeled in **b**. Curves are fits to a single-exponential rise. Intensities in **c**,**d** are normalized to their averaged value at $t < 0$. Curves are vertically offset by 0.02 for clarity. The incident laser fluence for **a**,**c**,**d** was 2.1 mJ/cm^2

7.5.3 Momentum-Dependent Diffuse Scattering Dynamics

In Figs. 7.3b and 7.4b, we plotted the temporal evolutions of diffuse scattering at $\mathbf{q}_a \pm \mathbf{q}_c$, showing a single-exponential rise after photoexcitation. In this section, we examine the momentum dependence of the diffuse scattering dynamics. In Fig. 7.9b, we identified five representative momentum positions together with their symmetry-equivalent points within the diamond-shaped ROI. The intensity evolution of thermal diffuse scattering from each location is plotted in Fig. 7.9d. The short-time dynamics of all curves is well described by a single-exponential rise, indicating the transient excitation of incoherent phonons. In particular, for locations 1 to 4, the intensity rise starts immediately after photoexcitation and all curves plateau at a quasi-equilibrium value of a 4% increase. For comparison, under the same excitation condition, the maximum intensity rise at $\pm\mathbf{q}_a$ is approximately 10%, plotted in Fig. 7.2a. The time constants for the intensity rise vary among different momenta, ranging from 1 to 2.5 ps. These rise times encode information about the momentum-dependent electron–phonon and phonon–phonon coupling strengths [32], which are beyond the scope of this discussion.

The diffuse scattering dynamics at location 5 is markedly different from the others. First, there is a 0.8 ps delay of the rise, indicated by the arrow in Fig. 7.9d. Second, the intensity only increases by 3% when it plateaus, distinctly smaller than the others. These features suggest that there is another dynamics apart from the increase in the phonon population. One possibility is that location 5 lies in close proximity to high-order *c*-axis CDW peaks [33]. Though they are not resolved from the background in the static diffraction pattern, their presence is evidenced in the differential intensity plots before and after the pulse incidence—one example was shown in Fig. 6.5a in Chap. 6. The photoinduced melting of these high-order peaks will both delay and offset the rise of diffuse scattering, explaining the anomaly at location 5.

7.5.4 Time-Dependent O(N) Model with Competing Orders

In this section, we explain the equations of motion for the long-range order $\psi_i(t)$ and correlation functions $D^{\parallel}_{\mathbf{k},i}(t)$, $D^{\perp}_{\mathbf{k},i}(t)$ corresponding to the free energy functional $\mathcal{F}[\boldsymbol{\psi}_c, \boldsymbol{\psi}_a]$ in Eq. (7.1). To describe the photoinduced evolution, we assume overdamped order parameter dynamics (model-A [34]) and follow the formalism in Ref. [26]. We neglected the coherent oscillatory dynamics associated with the phononic degree of freedom. The overdamped dynamics is expected to be reliable in describing long time dynamics after a strong laser pulse, which results in a proliferation of low-energy, low-momenta order parameter collective modes. Adding the phonon contribution will better capture the short-time dynamics, especially the initial response time [19]. The equations of motion are a simple generalization from the single-CDW result [26]:

$$\frac{d\psi_i(t)}{dt} = -\Gamma\, r_{\mathrm{eff},i}\, \psi_i, \tag{7.17}$$

$$\frac{dD^{\perp}_{\mathbf{k},i}(t)}{dt} = 2T\Gamma - 2\Gamma(K\mathbf{k}^2 + r_{\mathrm{eff},i})D^{\perp}_{\mathbf{k},i}, \tag{7.18}$$

$$\frac{dD^{\parallel}_{\mathbf{k},i}(t)}{dt} = 2T\Gamma - 2\Gamma(K\mathbf{k}^2 + r_{\mathrm{eff},i} + 8u\psi_i^2)D^{\parallel}_{\mathbf{k},i}, \tag{7.19}$$

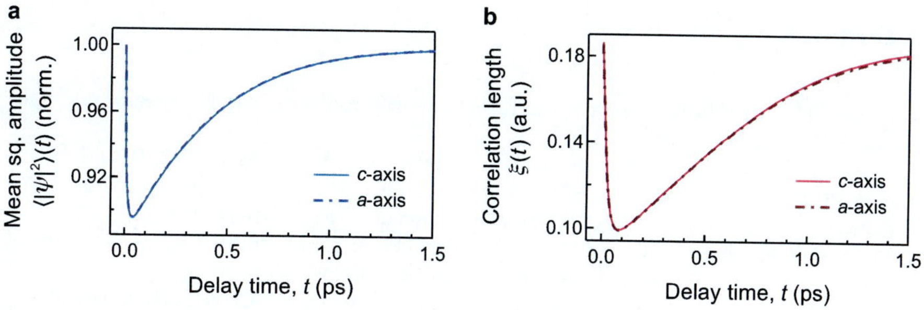

Fig. 7.10 Simulated dynamics of fluctuating CDWs above T_c. Temporal evolutions of $\langle|\psi_i|^2\rangle(t)$ (**a**) and $\xi_i(t)$ (**b**) when the initial state is above T_c. $\langle|\psi_i|^2\rangle(t)$ is normalized by its value at $t = 0$. The CDW instabilities along the a- and c- axis show nearly identical dynamics

where T is the temperature of a phononic bath that provides thermalization of the excited system, and is assumed to be unchanged after photoexcitation. Γ is a measure of the relaxation rate of the long-range order, assumed to be the same for the two CDWs. In principle, one may impose different Γ values for different order parameters [28], but the photoinduced symmetry between the two CDWs in $LaTe_3$ justifies our choice. $r_{\text{eff},i}$ is a self-consistent "mass" term, defined as

$$r_{\text{eff},a}(t) = 4u\Big[\psi_a^2 + \int \frac{d^3\mathbf{k}}{(2\pi)^3}(D^{\parallel}_{\mathbf{k},a} + (N-1)D^{\perp}_{\mathbf{k},a})\Big] + 4\eta\Big[\psi_c^2 + \int \frac{d^3\mathbf{k}}{(2\pi)^3}(D^{\parallel}_{\mathbf{k},c} + (N-1)D^{\perp}_{\mathbf{k},c})\Big] + r_a(t), \tag{7.20}$$

and a symmetric expression holds for $r_{\text{eff},c}$.

We make a few remarks on Eqs. (7.17)–(7.19). In equilibrium, there is no LRO for the a-axis CDW ($\psi_a = 0$), so it remains zero throughout the time evolution as dictated by Eq. (7.17). It also implies that the dynamics of $D^{\perp}_{\mathbf{k},a}(t)$ and $D^{\parallel}_{\mathbf{k},a}(t)$ are identical because in the initial condition, there is no distinction between transverse and longitudinal fluctuations in the absence of symmetry breaking along the a-axis.

In Fig. 7.6, we studied the effect of photoexcitation on $LaTe_3$ in its broken-symmetry phase. The above formalism is also capable of capturing the dynamics when the initial state lacks any LRO, such as in $DyTe_3$ near its T_c. Figure 7.10 shows the evolutions of mean square order parameter amplitude and correlation length for the fluctuating CDWs above T_c, mimicking the dynamics measured experimentally in Fig. 7.3e. In particular, there is no observable distinction between the two competing orders because LRO never sets in during the nonequilibrium evolution.

References

1. Basov, D.N., Averitt, R.D., Hsieh, D.: Towards properties on demand in quantum materials. Nat. Mater. **16**, 1077–1088 (2017)
2. Kaiser, S.: Light-induced superconductivity in high-T_c cuprates. Phys. Scr. **92**, 103001 (2017)
3. Wandel, S., Boschini, F., Neto, E.H.D.S., Shen, L., Na, M.X., et al.: Light-enhanced charge density wave coherence in a high-temperature superconductor. arXiv 2003.04224 (2020)
4. McLeod, A.S., Zhang, J., Gu, M.Q., Jin, F., Zhang, G., et al.: Multi-messenger nanoprobes of hidden magnetism in a strained manganite. Nat. Mater. **19**, 397–404 (2020)
5. Zhang, J., Tan, X., Liu, M., Teitelbaum, S.W., Post, K.W., et al.: Cooperative photoinduced metastable phase control in strained manganite films. Nat. Mater. **15**, 956–960 (2016)
6. Stoica, V.A., Laanait, N., Dai, C., Hong, Z., Yuan, Y., et al.: Optical creation of a supercrystal with three-dimensional nanoscale periodicity. Nat. Mater. **18**, 377–383 (2019)
7. Stojchevska, L., Vaskivskyi, I., Mertelj, T., Kusar, P., Svetin, D., et al.: Ultrafast switching to a stable hidden quantum state in an electronic crystal. Science **344**, 177–180 (2014)
8. Gerasimenko, Y.A., Vaskivskyi, I., Litskevich, M., Ravnik, J., Vodeb, J., et al.: Quantum jamming transition to a correlated electron glass in $1T$-TaS_2. Nat. Mater. **18**, 1078–1083 (2019)
9. Kogar, A., Zong, A., Dolgirev, P.E., Shen, X., Straquadine, J., et al.: Light-induced charge density wave in $LaTe_3$. Nat. Phys. **16**, 159–163 (2020)
10. Zhou, F., Williams, J., Sun, S., Malliakas, C.D., Kanatzidis, M.G., et al.: Nonequilibrium dynamics of spontaneous symmetry breaking into a hidden state of charge-density wave. Nat. Commun. **12**, 566 (2021)
11. Fausti, D., Tobey, R.I., Dean, N., Kaiser, S., Dienst, A., et al.: Light-induced superconductivity in a stripe-ordered cuprate. Science **331**, 189–191 (2011)
12. Nicoletti, D., Casandruc, E., Laplace, Y., Khanna, V., Hunt, C.R., et al.: Optically induced superconductivity in striped $La_{2-x}Ba_xCuO_4$ by polarization-selective excitation in the near infrared. Phys. Rev. B **90**, 100503 (2014)
13. Hu, B.F., Cheng, B., Yuan, R.H., Dong, T., Wang, N.L.: Coexistence and competition of multiple charge-density-wave orders in rare-earth tritellurides. Phys. Rev. B **90**, 085105 (2014)
14. Ru, N., Condron, C.L., Margulis, G.Y., Shin, K.Y., Laverock, J., et al.: Effect of chemical pressure on the charge density wave transition in rare-earth tritellurides RTe_3. Phys. Rev. B **77**, 035114 (2008)
15. Ru, N.: Charge Density Wave Formation in Rare-earth Tellurides. Ph.D. thesis, Stanford University, Stanford (2008)
16. Maschek, M., Zocco, D.A., Rosenkranz, S., Heid, R., Said, A.H., et al.: Competing soft phonon modes at the charge-density-wave transitions in $DyTe_3$. Phys. Rev. B **98**, 094304 (2018)
17. Schmitt, F., Kirchmann, P.S., Bovensiepen, U., Moore, R.G., Rettig, L., et al.: Transient electronic structure and melting of a charge density wave in $TbTe_3$. Science **321**, 1649–1652 (2008)
18. Trigo, M., Giraldo-Gallo, P., Kozina, M.E., Henighan, T., Jiang, M.P., et al.: Coherent order parameter dynamics in $SmTe_3$. Phys. Rev. B **99**, 104111 (2019)
19. Zong, A., Dolgirev, P.E., Kogar, A., Ergeçen, E., Yilmaz, M.B., et al.: Dynamical slowing-down in an ultrafast photoinduced phase transition. Phys. Rev. Lett. **123**, 097601 (2019)
20. Dagotto, E.: Complexity in strongly correlated electronic systems. Science **309**, 257–262 (2005)
21. Frano, A., Blanco-Canosa, S., Keimer, B., Birgeneau, R.J.: Charge ordering in superconducting copper oxides. J. Phys. Condens. Matter **32**, 374005 (2020)
22. Otto, M.R., Pöhls, J.-H., René de Cotret, L.P., Stern, M.J., Sutton, M., et al.: Mechanisms of electron-phonon coupling unraveled in momentum and time: The case of soft phonons in $TiSe_2$. Sci. Adv. **7**, eabf2810 (2021)
23. Mazenko, G.F., Zannetti, M.: Instability, spinodal decomposition, and nucleation in a system with continuous symmetry. Phys. Rev. B **32**, 4565–4575 (1985)
24. Bray, A.J.: Theory of phase-ordering kinetics. Adv. Phys. **43**, 357–459 (1994)

25. Chaikin, P.M., Lubensky, T.C.: Mean-field theory. In: Principles of Condensed Matter Physics, pp. 144–212. Cambridge University Press, Cambridge (1995)
26. Dolgirev, P.E., Michael, M.H., Zong, A., Gedik, N., Demler, E.: Self-similar dynamics of order parameter fluctuations in pump-probe experiments. Phys. Rev. B **101**, 174306 (2020)
27. Zong, A., Kogar, A., Bie, Y.-Q., Rohwer, T., Lee, C. et al.: Evidence for topological defects in a photoinduced phase transition. Nat. Phys. **15**, 27–31 (2019)
28. Sun, Z., Millis, A.J.: Transient trapping into metastable states in systems with competing orders. Phys. Rev. X **10**, 021028 (2020)
29. Scalapino, D.J.: A common thread: The pairing interaction for unconventional superconductors. Rev. Mod. Phys. **84**, 1383–1417 (2012)
30. René de Cotret, L.P., Pöhls, J.-H., Stern, M.J., Otto, M.R., Sutton, M., et al.: Time- and momentum-resolved phonon population dynamics with ultrafast electron diffuse scattering. Phys. Rev. B **100**, 214115 (2019)
31. Maschek, M., Rosenkranz, S., Heid, R., Said, A.H., Giraldo-Gallo, P., et al.: Wave-vector-dependent electron-phonon coupling and the charge-density-wave transition in $TbTe_3$. Phys. Rev. B **91**, 235146 (2015)
32. Stern, M.J., René de Cotret, L.P., Otto, M.R., Chatelain, R.P., Boisvert, J.-P., et al.: Mapping momentum-dependent electron-phonon coupling and nonequilibrium phonon dynamics with ultrafast electron diffuse scattering. Phys. Rev. B **97**, 165416 (2018)
33. Straquadine, J.A.W., Weber, F., Rosenkranz, S., Said, A.H., Fisher, I.R.: Suppression of charge density wave order by disorder in Pd-intercalated $ErTe_3$. Phys. Rev. B **99**, 235138 (2019)
34. Hohenberg, P.C., Halperin, B.I.: Theory of dynamic critical phenomena. Rev. Mod. Phys. **49**, 435–479 (1977)

Chapter 8
Ultrafast Manipulation of Mirror Domains in 1T-TaS_2

Abstract Domain walls (DWs) are singularities in an ordered medium that often host exotic phenomena such as charge ordering, insulator–metal transition, or superconductivity. The ability to locally write and erase DWs is highly desirable, as it allows one to design material functionality by patterning DWs in specific configurations. In this chapter, we demonstrate such capability at room temperature in a charge density wave (CDW), a macroscopic condensate of electrons and phonons, in ultrathin 1T-TaS_2. A single femtosecond light pulse is shown to locally inject or remove mirror DWs in the CDW condensate, with probabilities tunable by pulse energy and temperature. Two mirror-opposite CDWs can be distinguished by their nonlinear optical response, and the mirror boundary is atomically sharp, featuring periodically arranged dislocations where the CDW amplitude is locally suppressed. Using time-resolved electron diffraction, we are able to simultaneously track anti-synchronized CDW amplitude oscillations from both the lattice and the CDW, where photo-injected DWs lead to a red-shifted frequency. The demonstration of reversible DW manipulation may pave new ways for engineering correlated materials with light.

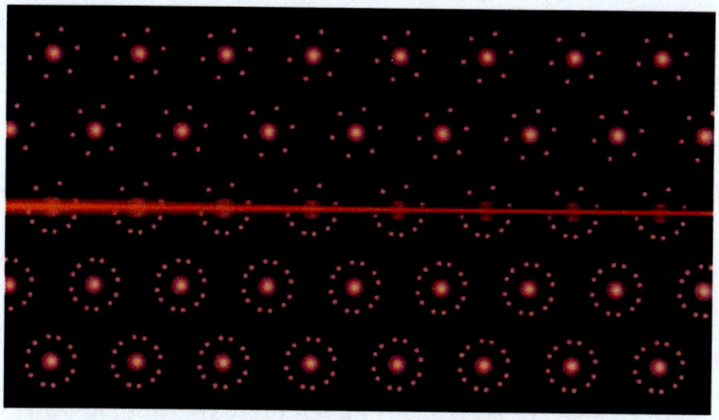

Artist's impression of laser writing or erasure of planar chiral domains in 1T-TaS_2

A. Zong, *Emergent States in Photoinduced Charge-Density-Wave Transitions*, Springer Theses, https://doi.org/10.1007/978-3-030-81751-0_8

8.1 Emergent Phenomena at Domain Walls

Domains are ubiquitous in systems with long-range order and are identified by different configurations of the same order parameter. Intense studies of domain formation have not only underpinned important technologies such as nonvolatile memory [1], but also led to surprising discoveries at DWs (Fig. 8.1): metallicity in a ferroelectric or antiferromagnetic insulator [2, 3], superconductivity in a ferroelastic oxide [4], charge density waves (CDWs) in a 2D semiconductor [5], or Wigner crystallization in a Mott insulator [6]. These unique properties at DWs arise because the order parameter is disrupted locally, leading to a renormalization of single-particle spectra and quasiparticle interactions.

A particularly interesting ground state in which DWs occur is the CDW phase, where electrons and phonons cooperatively condense to form a superlattice below a transition temperature. Recent studies have unveiled novel electrical and optical properties at CDW DWs [7–12], which not only hold potential for CDW-based devices [7, 8, 13] but also yield insight into the fundamental question of how CDWs interact with other phases such as the Mott insulating state [9, 10] and superconductivity [11, 14].

Despite these recent advances, there has been limited progress in manipulating CDW DWs, with no reports of stable yet rewritable DWs at room temperature. To create (or remove) a CDW DW, the challenge is to provide sufficiently high energy to lodge (or dislodge) long-lived defects in the superlattice, without causing irreversible damage to the underlying crystal. Here, we used a single femtosecond pulse with a tailored fluence to locally inject or erase mirror DWs in a CDW, which were directly monitored by transmission electron diffraction. The femtosecond pulse can transiently perturb the CDW and introduce topological defects [15–22], discontinuities that cannot be removed by smoothing the order parameter. These defects then follow a nonequilibrium pathway of relaxation into various DW structures. By carefully tuning the pulse fluence, we were able to eliminate these CDW DWs using a subsequent light pulse. The processes of DW injection and removal are entirely reversible.

In this chapter, we study the octahedrally coordinated polytype of a layered dichalcogenide, 1T-TaS_2. It has two degenerate CDW configurations, α and β, which are related by mirror reflection. As introduced in Sect. 2.4, when 1T-TaS_2 transitions from its incommensurate CDW (IC) phase to the nearly commensurate CDW (NC) phase, a mirror symmetry is spontaneously broken. Its superlattice peaks acquire either a counterclockwise (α) or a clockwise (β) rotation, corresponding to two mirror-symmetric orientations of the hexagram super-cells (Fig. 2.12). A cool-down into the commensurate CDW (C) phase does not alter the orientation of the hexagram, locking the planar chiral structure of the CDW. Despite the weakly first-order nature of the IC-to-NC transition, single crystals of 1T-TaS_2 are observed to adopt only one planar chirality for a macroscopically large sample (see Sect. 2.4.2).

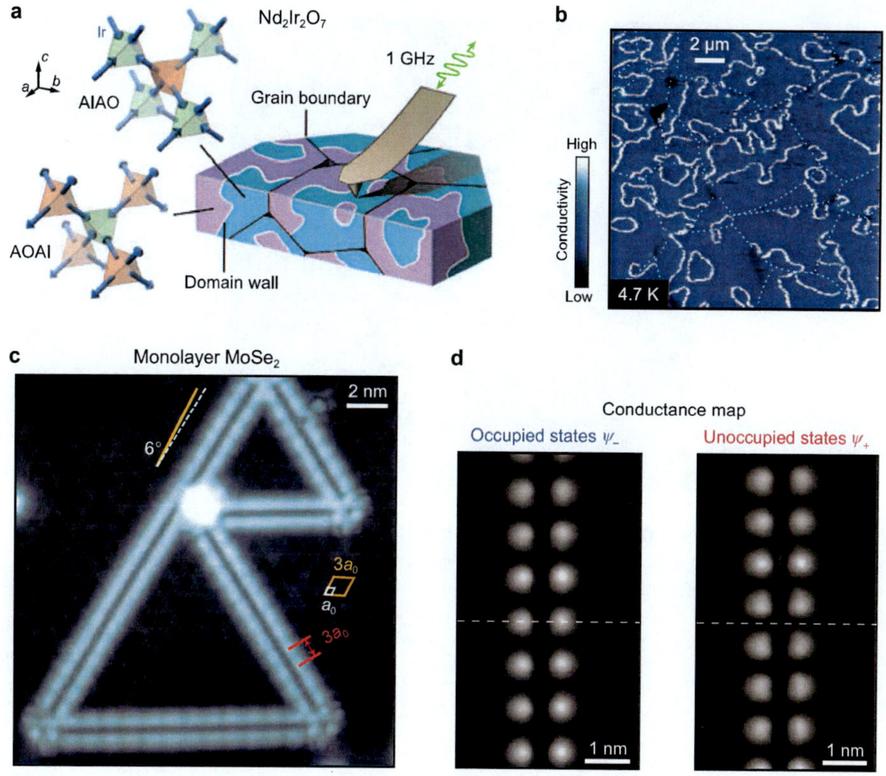

Fig. 8.1 Emergent phenomena at domain walls. (**a**) Spin configuration of the all-in-all-out (AIAO) and all-out-all-in (AOAI) orders in $Nd_2Ir_2O_7$. Their conductivity at the GHz range can be probed by microwave impedance microscopy (MIM). (**b**) An MIM image of a polished Nd_2IrO_7 polycrystal surface at 4.7 K after cooling from 40 K without any applied magnetic field. Dotted lines are grain boundaries and dark spots are voids between grains. White curves are AIAO/AOAI magnetic domain walls, which are conducting against an insulating bulk. (**c**) STM image of a semiconducting monolayer $MoSe_2$ grown on bilayer graphene by molecular beam epitaxy (MBE). It shows a nearly commensurate 3×3 moiré pattern marked by the yellow diamond, with a single unit cell of $MoSe_2$ indicated by the white diamond. Bright lines are mirror twin boundaries (MTBs), whose orientation (dashed white line) is 6° misaligned from the moiré lattice (solid yellow line). There is a modulation of the electronic states near the MTBs with a period of $3a_0$. (**d**) Constant-height differential conductance map along a MTB below (ψ_-) and above (ψ_+) its 73 meV energy gap, showing a 1D density wave. Brighter regions represent a larger density of state. The density modulation is 180° out of phase between the occupied and unoccupied states, highlighted by the horizontal dashed line. (Adapted with permission from Ref. [3], AAAS (**a**,**b**); Ref. [5], Springer Nature Ltd (**c**,**d**))

8.2 Single-Pulse Manipulation of CDW Mirror Domain Walls

To locally inject α/β DWs in a single-domain $1T$-TaS_2, we focused an 80-fs, 800-nm (1.55-eV) pulse onto a $150 \times 150\ \mu m^2$ region of interest (ROI) in a 50-nm-thick sample and probed the ROI by MeV transmission electron diffraction.

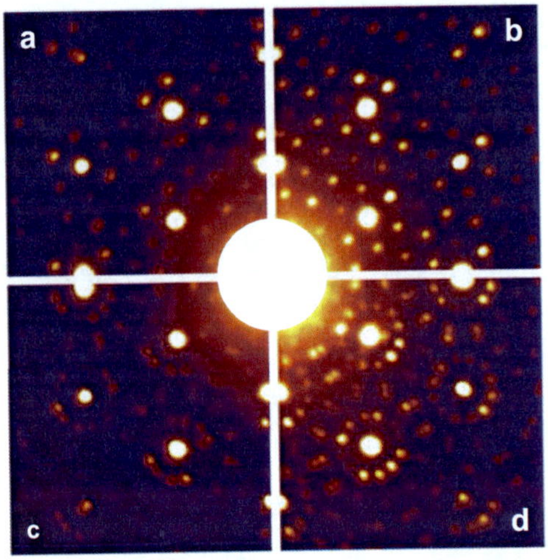

Fig. 8.2 MeV electron diffraction of switchable CDW mirror domains by a single femtosecond pulse. (**a**,**c**) α and α/β states of the NC phase at 300 K, respectively, which can be repeatedly switched from one to the other after applying a single 800-nm (1.55-eV), 80-fs pulse at 7 mJ/cm^2 incident fluence. (**b**,**d**) The corresponding α and α/β states in the C phase at 40 K, which can no longer be switched into the other by a single pulse up to the highest fluence attempted (11 mJ/cm^2)

Figure 8.2a, c shows the diffraction patterns before and after the exposure of a single pulse with an incident fluence of 7 mJ/cm^2 at room temperature. Two sets of satellite peaks appear in Fig. 8.2c, signaling the presence of α/β DWs. The locality of DW injection was verified by probing a neighboring region unexposed to the pulse, where only α satellites were visible as in Fig. 8.2a. These α/β DWs are stable up to $T_{\text{IC-NC}} = 354$ K and stay unchanged in atmosphere, as confirmed by a room temperature test for one week. There was no irreversible damage due to the laser pulse; a temperature cycle to the IC phase at 370 K (Fig. 8.3d) and back to room temperature removed all α/β DWs and restored a uniform α domain as before.

Remarkably, applying the same laser pulse to the α/β state (Fig. 8.2c) erased the β domains in the ROI, returning it to the original state (Fig. 8.2a). Such reversible switching between α-only and α/β states was repeated by more than 1,000 pulses on four different samples. The result was also reproduced in a separate keV UED setup, which uses a different wavelength and pulse duration (1038 nm, 190 fs), shown in Fig. 8.3. As we discuss below, the switch is non-deterministic and its probability can be tuned by temperature and pulse fluence. Once switched, the sample remained in the α/β state upon cooling from the NC to the C phase (Fig. 8.2c, d). The transition temperature $T_{\text{NC-C}}$ is the same in both α and α/β states, again suggesting no pulse-inflicted sample degradation. A single pulse, up to the highest fluence attempted (11 mJ/cm^2), was unable to create or annihilate α/β DWs in the C phase.

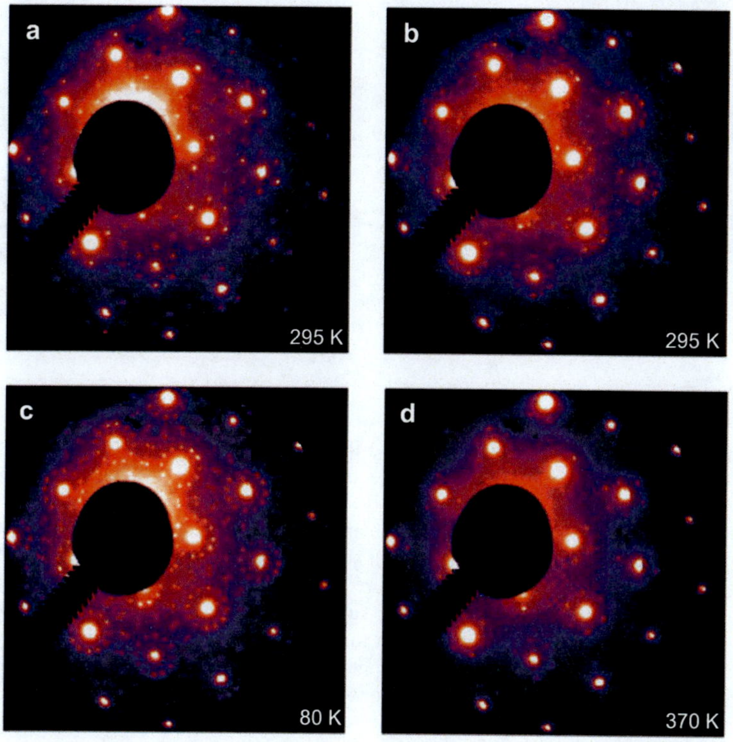

Fig. 8.3 Diffractions of switchable domains measured in the keV UED setup. (**a**,**b**) Diffractions of α and α/β states, respectively, in the NC phase at 295 K. A single 1038-nm (1.19-eV), 190-fs pulse of 7 mJ/cm^2 incident fluence was able to reversibly switch between patterns in **a** and **b**. (**c**) Diffraction of an α/β state in the C phase by cooling the sample in **b** to 80 K. (**d**) Diffraction in the IC phase after heating the sample in **b** to 370 K at a rate of ~0.5 K/minute. Upon cooling to 295 K at the same rate, the pattern in **a** was reproduced, where mirror DWs were removed through the thermal cycle

8.2.1 Morphology of Photoinduced Domain Walls

The diffraction patterns in Figs. 8.2c, d and 8.3b, c do not discriminate between inter-layer and intra-layer DWs because of the large electron beam spot (90 μm to 300 μm). To distinguish between the two possibilities, we performed selected area diffraction (SAD) on the same sample ROI as in Fig. 8.2 using a 1.1-μm-diameter beam of a transmission electron microscope (TEM). We observed intra-layer DWs (Fig. 8.4), with separate α (red circle) and β (blue circle) domains, divided by α/β DWs (yellow circle). Overlaid color masks denote the approximate domain locations. The yellow region suggests the presence of submicron domains, and we cannot rule out the possibility of inter-layer DWs in the area at this spatial resolution. The unmasked corner (Fig. 8.4a, top right) shows the bright-field electron micrograph in grayscale, featuring bend contours due to the underlying strain [23].

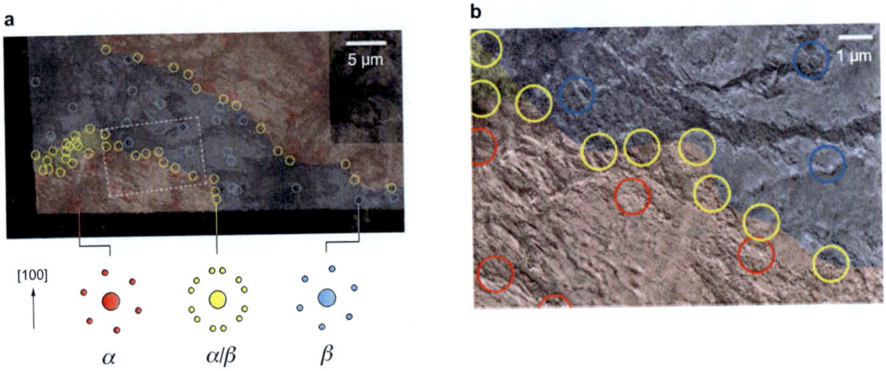

Fig. 8.4 Transmission electron micrograph of light-induced mirror domains. (**a**) Room temperature, bright-field TEM image of the pulse-induced α/β state, taken with 120 keV electron beam energy (Tecnai G2 Spirit TWIN, FEI). Selected area diffractions (SADs) were performed at each circle, whose diameter indicates a 1.1 μm beam size. Schematic SAD patterns are shown at the bottom. Overlaid red and blue masks indicate approximate locations of α and β domains. The yellow mask (*bottom left*) suggests a region with submicron domains. The top right corner is not masked to expose the grayscale micrograph where bend contours are clearly visible. Black edges at the bottom and left are copper frames of the supporting TEM grid. (**b**) Enlarged view of the dashed rectangle in **a**, taken at 8.75 times higher magnification. No correspondence was observed between the locations of α/β DWs and bend contours, suggesting that pulse-induced α/β DWs are not caused by macroscopic lattice deformations

We note the absence of correlation between locations of bend contours and α/β DWs, even under high magnification (Fig. 8.4b). This suggests that pulse-induced α/β DWs are unrelated to any macroscopic lattice deformation.

To obtain detailed morphology of the photoinduced mirror boundary, atomic resolution scanning transmission electron microscopy (STEM) was performed (Fig. 8.5a). Bright spots indicate the Ta atom locations at room temperature, and the Fourier transformed image reveals two sets of satellite peaks due to the two CDW mirror domains present in the field of view (red and blue triangles in Fig. 8.5b). To visualize the domain wall, a subset of the satellite peaks is selected for one particular CDW wavevector (arrows in Fig. 8.5b), and the corresponding spatial map is reconstructed by the inverse Fourier transform (Fig. 8.5c). The domain boundary is atomically sharp, and it is akin to a low-angle grain boundary between the two mirror-symmetric orientations. This is in contrast to pulse-induced domain walls in the commensurate CDW phase of $1T$-TaS_2 [9, 10], which are characterized by phase slips between two hexagram superlattices (Fig. 2.14c) instead of orientational misalignment.

To quantitatively characterize the CDW states near the mirror domain wall, maps of the CDW orientation (Fig. 8.5d) and modulation amplitude (Fig. 8.5e) are computed. Both images feature quasi-periodic dislocations along the boundary, which are approximately 5 nm apart. At the dislocation point, the CDW amplitude is suppressed. At other parts along the boundary, the amplitude remains finite and the corresponding CDW orientation smoothly evolves from one mirror configuration to the other. This is another distinct feature from previously reported phase slip DWs

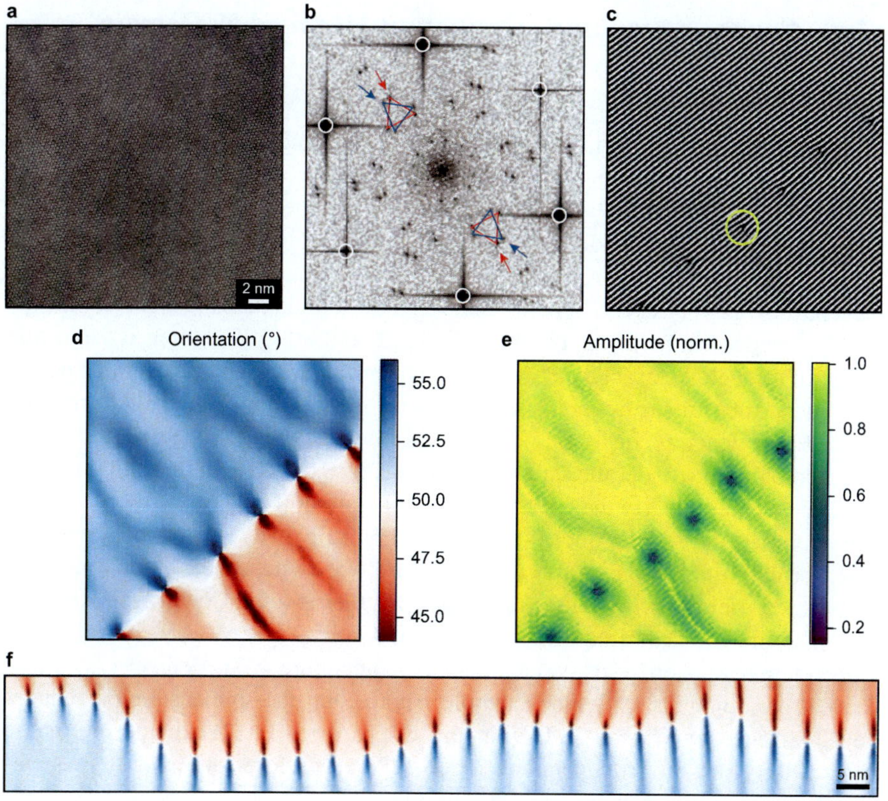

Fig. 8.5 Atomically resolved mirror domain walls by scanning transmission electron microscopy. (**a**) High-angle annular dark field STEM image of the photoinduced α/β state at room temperature. Bright spots are Ta atoms. The field of view contains two mirror CDW domains (see **c**–**e**). (**b**) Fourier transform of the map in **a**, showing hexagonal lattice Bragg spots (white circles) and satellite peaks due to the NCCDW. Two CDW orientations are observed; examples of their respective second-order satellite peaks are marked by red and blue triangles [24]. (**c**) Fourier-filtered real-space image using the four peaks in **b** indicated by the arrows. Two CDW mirror domains meet in a low-angle grain boundary with periodically arranged dislocations, an example of which is marked by the circle. (**d**,**e**) Local orientation and normalized modulation amplitude of the stripes in **c**. The amplitude collapses at dislocations but remains finite in between. (**f**) Local orientations of a larger field of view, showing a wavy domain wall

at low temperature where the CDW is expected to be suppressed throughout the domain boundary. As the CDW is intimately tied to the Mott insulating state of $1T$-TaS_2, the mirror boundary induced in our case may provide an interesting example where conducting hotspots emerge periodically at these dislocation points, potentially forming another 1D density wave on top of the commensurate superlattice, namely, a CDW within a CDW. A DW orientation map with a larger field of view is shown in Fig. 8.5f, confirming the quasi-periodicity of the dislocations in a wavy boundary.

8.2.2 Coherent Amplitude Oscillation Mediated by Domain Walls

To investigate how these mirror DWs modify the underlying CDW order, we characterized its amplitude mode (AM) frequency in an α/β state. This mode arises from a broken translational symmetry during the CDW formation [25]. It manifests as a breathing mode of the CDW hexagrams (Fig. 8.6d). In a time-resolved diffraction experiment, the AM causes an oscillatory transfer of intensities between the central Bragg peaks and superlattice peaks of both domains [26], as confirmed in our measurement performed at 40 K (Figs. 8.6 and 8.11). The intensity sum rule is obeyed through the exact π phase shift between the Bragg and superlattice peaks. We obtained the mode frequency after Fourier transforming the oscillatory parts (Fig. 8.6c, e), which is consistent with the AM frequency measured separately in the single-domain state under the same conditions (Fig. 8.6e, vertical line). In this case, α/β DWs did not modify the spectroscopic signature of the CDW.

By contrast, the AM significantly softened (Fig. 8.6g) in a repeated measurement: after warming the sample to room temperature, we switched the sample into the α state and then back into the α/β state; we then cooled down again to 40 K to re-measure the AM frequency. In the CDW phase, a soft AM is associated with a suppressed order parameter due to temperature [27], defects [7, 12], or external perturbations [19]. For example, significant discommensuration reduces the AM frequency from 2.25 THz in the C phase to 2.1 THz in the NC phase at the phase transition temperature $T_{\text{C-NC}}$ [28]. Therefore, different frequencies present in Fig. 8.6e, g are suggestive of two distinct concentrations and distributions of pulse-induced DWs (Fig. 8.6f, h). For Fig. 8.6e, very few α/β DWs are present, similar to regions masked in red and blue in Fig. 8.4. In Fig. 8.6g, DWs are dense, similar to the yellow-masked region in Fig. 8.4. As we show in Sect. 8.5.2, this proposed DW configuration is further evidenced by a short CDW correlation length. The significant presence of a DW network hence leads to a softer AM. Additional high resolution, large-area microscopy measurement is required to confirm the exact distribution of DWs. Nonetheless, we stress that a single pulse can change the DW distribution, a key feature that enables domains to be reoriented.

8.3 Mechanism of Photo-Manipulation of Domain Walls

To understand how a single pulse creates or annihilates α/β DWs, we examine the sequence of events upon photoexcitation. With sufficient pulse energy, the initial carrier excitation melts the NC phase within 1 ps [18]. In the meantime, IC order nucleates in the discommensurate regions [18] and domains continue to grow over a nanosecond timescale [16, 17]. We verified the existence of this nonequilibrium IC phase by taking a snapshot of the diffraction pattern at a pump–probe delay of

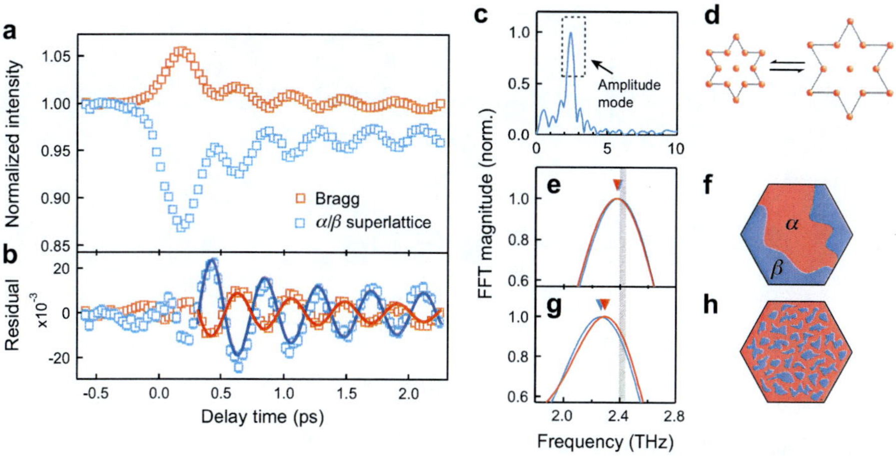

Fig. 8.6 Coherent excitation of the amplitude mode in the commensurate CDW phase. (**a**) Integrated intensities of Bragg and superlattice peaks of both α and β domains in the photoinduced α/β state as a function of pump–probe delay time. Intensities are normalized by values before pump incidence. See Sect. 8.5.2 for contributions from each mirror domain. (**b**) Coherent, oscillatory part of the intensity after subtracting a fitted single-exponential, incoherent part from **a**. Solid curves are fits to an exponentially decaying cosine function. (**c**) Fourier transformed spectrum of the superlattice intensity in **b**, featuring the prominent AM peak. The FFT was computed with zero padding and normalized to the maximum value. Uncertainties reflect fitting errors to an exponentially decaying cosine. (**d**) Schematic of the AM, corresponding to a breathing mode of hexagrams. (**e**) Zoomed-in view of the dashed rectangle in **c**. Color codes are the same as used in **a**. Peak positions are marked by inverted triangles. The gray vertical line indicates the AM frequency of a single-domain sample measured separately under the same condition. Widths of the triangles and the vertical line indicate frequency uncertainties. (**g**) The same as in **e**, but after switching the α/β state to a single domain and back to an α/β state at room temperature. The AM markedly softened. (**f**,**h**) Schematic DW distributions corresponding to **e** and **g**, respectively. DWs are sparse in **f** but dense in **h**. All data were taken at a base temperature of 40 K and an incident fluence of 1 mJ/cm^2

1 ns when the IC phase has fully developed (Fig. 8.7, inset). Since the broken mirror symmetry is restored in the IC phase, we postulate that the transient transition to the IC phase is a necessary intermediate step for creating or removing DWs. The energy required for this ultrafast NC-to-IC transition thus represents the minimal energy barrier in any domain reorientation.

In the following, we describe the mechanism that allows the system to relax from the transient IC phase into either a pure α or an α/β NC phase. We will first explain why a simple thermal quench scenario is not adequate and then suggest a defect-based picture that takes into account the domain switching probabilities.

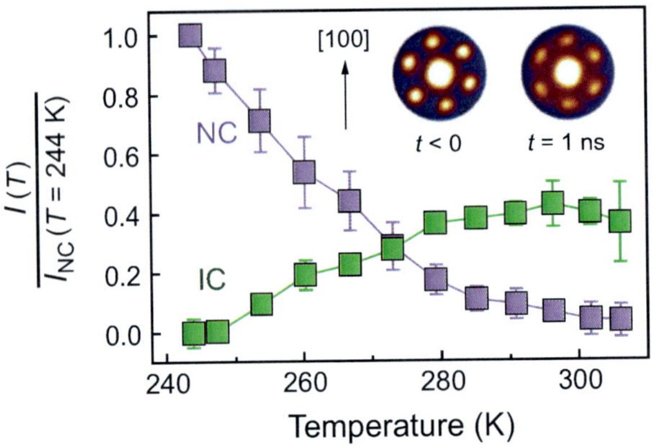

Fig. 8.7 Transient transition to the incommensurate CDW phase. Normalized integrated intensities of NC and IC CDW peaks at various temperatures, taken at a pump–probe delay of 1 ns and an incident fluence of 5 mJ/cm^2. *Inset*, room temperature diffractions taken before photoexcitation ($t < 0$) and at 1 ns delay, showing the nonequilibrium NC-to-IC transition. Data were averaged over (3 0 0) and its five symmetry equivalent peaks

8.3.1 *Fast Thermal Quench and Domain Switching*

Whenever a broken-symmetry phase has multiple degenerate configurations, it is possible for different domains to coexist. In the case of ferromagnets and ferroelectrics, domains arise due to the presence of a self-induced demagnetization and depolarization field, respectively [2]. If these fields do not exist, such as in the present case of a CDW compound, domains can also form due to a fast thermal quench. In this scenario, the phase transition is traversed within a timescale shorter than the time required for different parts of the sample to relax into the same ground state. As the lattice temperature transiently increases after laser incidence, one may be tempted to interpret the domain switching phenomena as a result of a fast thermal quench. However, the experimental observation is not consistent with a simple thermal quench scenario and we note two reasons below.

First, the 1*T* polytype of TaS_2 is prepared by a rapid thermal quench from 920°C (1193 K) to room temperature because the 2*H* polytype is the stable one at room temperature [29]. For comparison, we estimate the sample temperature after a laser pulse. Using the data presented in Fig. 8.2, an ambient temperature of 300 K and an incident fluence of 7 mJ/cm^2 correspond to a maximum transient lattice temperature of 1110 K [30]. Comparing thermal quench during crystal growth and sample cooling after photoexcitation, the changes in temperature are similar. However, only one domain orientation remains in thermally quenched single crystals (see Sect. 2.4.2). On the other hand, domain switching is observed to occur after photoexcitation in our experiments. Hence, a rapidly decreasing sample temperature does not suffice to account for the injection and removal of mirror DWs.

Second, at a higher laser pulse fluence, the IC-to-NC relaxation time is empirically found to be longer [31]; such slower relaxation back to the low-temperature CDW phase is a generic feature across many other CDW systems [21, 32–35]. In a thermal quench scenario, a longer relaxation time disfavors the formation of α/β domains. However, as we will see in the next section, this fluence dependence is contradictory to the experimental observation: the probability of injecting (or removing) domain walls increases (or decreases) as the fluence increases.

The above discussion suggests that a fast thermal quench is not adequate to explain the observed domain switching; some additional ingredient is required. As we discuss below, one possibility is the presence of photoinduced defects in the transient IC phase. Similar to our studies of the rare-earth tritellurides in Chap. 4, these defects are a recurring theme of ultrafast phase transitions far from equilibrium.

8.3.2 Probability of Creating or Annihilating Domain Walls

To get some hints about the relaxation pathway from the transient IC phase, we carried out systematic experiments to characterize the probabilities of mirror DW creation (P_+) and annihilation (P_-). At each temperature and fluence, a diffraction pattern was captured after a single pulse was incident on the sample, for a total of 200 pulses. The pulses were spaced three seconds apart to allow the full relaxation of the sample. Integrated intensities of a pair of α/β superlattice peaks were computed in each diffraction pattern. For reference, without any pulse exposure, Fig. 8.8a shows the image-to-image change in integrated intensity in an α-only state under the same CCD camera setting, indicating the level of statistical fluctuation. Intensities were normalized by I_0, the mean integrated intensity of superlattice peaks in the α-only state.

An example of pulse-to-pulse intensity change in α and β peaks is shown in Fig. 8.8e, taken at 300 K with an incident fluence of 6.4 mJ/cm^2. It is worth noting that if the sample is in an α/β (or α) state after one pulse, it is much more likely that the previous state is also an α/β (or α) state than the opposite case. It shows one example of the history-dependent nature of the resulting state after the single-pulse photoexcitation. In the corresponding scatter plot (Fig. 8.8c), we applied k-means clustering [36] to assign each diffraction pattern into one of the two categories: α state (light color) and α/β state (dark color). Intensity spread within the cluster originates from statistical fluctuation (Fig. 8.8a) and partial creation or annihilation of mirror DWs. The same clustering analysis was applied to other temperatures and fluences. A minimum threshold fluence is marked by the appearance of an additional cluster (Fig. 8.8b–d).

Given the two categories of states, the switching probabilities, $P_\pm$, can be calculated as

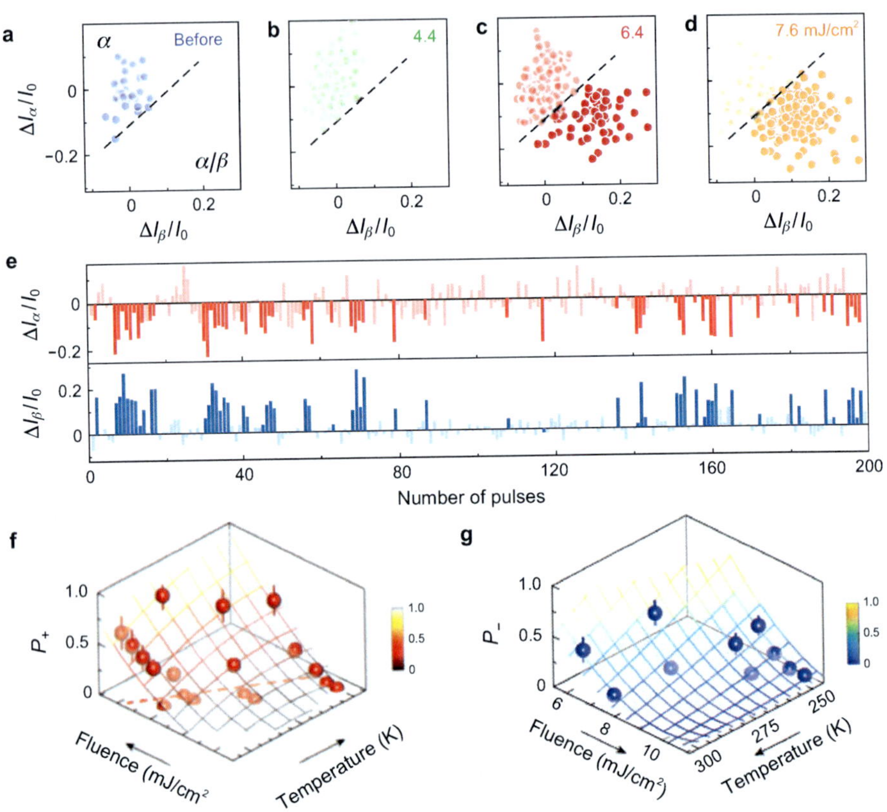

Fig. 8.8 Probabilities of mirror domain switching. (**a–d**) Distributions of image-to-image change in integrated intensity of a pair of α and β superlattice peaks at 300 K. Values were normalized by I_0, the mean integrated intensity of superlattice peaks in the α-only state. Panel **a** contains data points from 30 diffraction images before exposing the sample to laser pulses, indicating the extent of statistical fluctuation. Each of the 200 points in **b–d** was taken after exposing the sample to a single laser pulse of respective incident fluence; pulses were spaced three seconds apart. α (light color) and α/β (dark color) states are determined by k-means clustering. Dashed lines indicate perpendicular bisectors of the two cluster centroids. Cluster center of the α state sets the zeros in both axes. (**e**) Normalized changes in intensities of α and β superlattice peaks shown in **c**, plotted against the number of laser pulses incident on the sample. Darker and lighter colors denote α/β and α states, respectively. (**f**) Fluence and temperature dependence of P_+, the probability of injecting DWs by switching from an α-only to an α/β state with a single pulse. Arrows indicate the direction of increasing temperature or fluence. The mesh is a guide to the eye, obtained by fitting the raw data (spheres) to a second-order two-dimensional polynomial. Dashed line marks the minimum threshold fluence for a nonzero P_+. (**g**) The same as **f**, but for P_-, the probability of removing DWs with a single pulse

$$P_+ = \frac{n_{\alpha\to\alpha/\beta}}{n_{\alpha\to\alpha} + n_{\alpha\to\alpha/\beta}}, \tag{8.1}$$

$$P_- = \frac{n_{\alpha/\beta\to\alpha}}{n_{\alpha/\beta\to\alpha/\beta} + n_{\alpha/\beta\to\alpha}}, \tag{8.2}$$

where $n_{i \to j}$ denotes the number of events in which switching from state i to state j occurs. Error bars in $P_\pm$ were calculated using the Wilson interval [37], reflecting a 95% confidence level for the probability estimates.

In Fig. 8.8f, g, we plot the empirically determined $P_\pm$ as a function of both fluence and temperature and superimpose a guide to the eye (mesh) by fitting the raw data (spheres) with a second-order two-dimensional polynomial. For each temperature, P_+ remains zero below a threshold fluence (dashed line) because the pulse lacks the energy to bring the sample to the mirror-symmetric IC phase. The threshold fluence increases, while temperature decreases, as the transient IC order is suppressed in favor of a residual NC order (Fig. 8.7). P_- is not available at low fluences as no α/β state was created, but we expect a vanishing P_- below a minimum fluence threshold similar to that of P_+ due to the energy barrier for the nonequilibrium NC-to-IC transition.

8.3.3 A Defect-Mediated Nonthermal Pathway

Beyond the minimum threshold, the probability P_+ increases with an increasing fluence (Fig. 8.8f). By contrast, the probability of removing α/β DWs, P_-, decreases with fluence (Fig. 8.8g). This growing tendency to create DWs at high fluence can be understood if we consider the following microscopic picture of DW creation and annihilation. Starting from a single-domain NC phase (Fig. 8.9, left), a strong femtosecond pulse instantaneously suppresses the NC order while seeding many nucleation centers for the IC phase ($t \sim 1$ ps), where the local potential barriers of an NC-to-IC transition are among the lowest [18]. Patches of the IC order have varied phases in the order parameter, and the energy deposited creates strong phase fluctuations that inhibit a complete coalescence of IC domains. Hence, a high concentration of topological defects remains when the IC order starts to relax ($t \sim$ 1–100 ns). These CDW defects introduce local strains that break the α/β degeneracy at different locations, as sketched in the Landau-type potentials in Fig. 8.9. Hence, as the system relaxes, both orientations develop in the NC phase (Fig. 8.9, right).

The erasure of DWs can be similarly instigated by another femtosecond pulse (Fig. 8.9, bottom half), but with a reduced energy to limit the number of IC domains and pulse-induced defects. However, a minimum threshold fluence is still required to fully melt the NC order and reorient α and β domains ($\sim$5 mJ/cm^2 at 300 K) (Fig. 8.7). In this case, the much-reduced defect concentration at the start of the IC-to-NC relaxation ($t \sim$ 1–100 ns) favors a single NC domain, similar to a thermodynamic phase transition. As we show in Sect. 8.5.3, this proposed mechanism for domain switching is further supported by a simple model, where we relate the defect concentration in the transient IC phase to the size of reoriented domains. The simulated change in defect concentration after each pulse well captures the fluence dependences of both P_+ and P_- (Fig. 8.12).

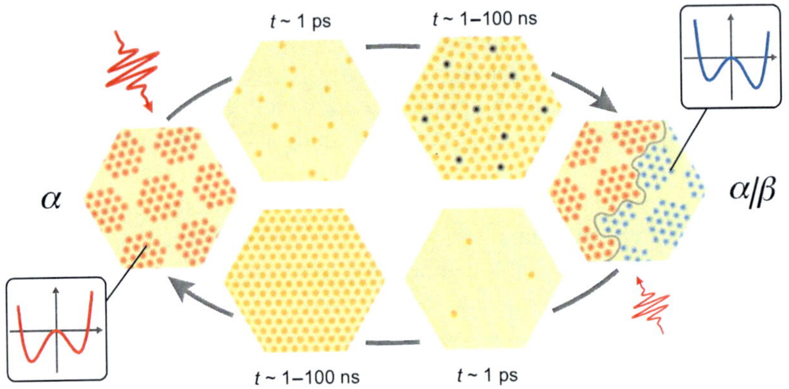

Fig. 8.9 Proposed mechanism for photoinduced domain wall manipulation. A real-space schematic of creating (*upper half*) and annihilating (*lower half*) mirror DWs. Periodic patterns represent superlattices in the NC (red, α; blue, β) and the IC phase (yellow). Black dots mark the positions of unannihilated topological defects in the transient IC phase before it relaxes to the equilibrium NC phase. Landau-type potentials indicate the breaking of local α/β degeneracy in different parts of the sample (see Eq. (8.11))

In this picture, the key to manipulating mirror DWs lies in the control of the defect concentration in the transient IC phase. The mechanism differs from a conventional thermal quench scenario, where mirror domains may arise as sample quickly cools after pulse-induced heating. As the defect concentration depends on whether α/β DWs are present before the pulse arrival, the domain switchings depend on the sample history, an important characteristic for potential device applications. It is crucial to note that these are CDWs and not lattice defects, so their concentration can be effectively adjusted by optical pulses at various temperatures and fluences. In $1T$-TaS_2, these pulse-induced defects are useful for domain reorientation as they are long-lived ($\geq$1 ns) and tunable by laser pulses, thanks to a rugged potential landscape with multiple local minima [38], where barriers are comparable to the pulse energy applied.

8.4 All-Optical Read and Write of CDW Domains

The ability to switch between states with distinct orientations holds strong potential in data storage applications such as CDW-based memory [39]. Here, α and β domains act as "on" and "off" states, which are extremely stable at room temperature. While we could optically write and erase these DWs, we have so far resorted to electron diffraction or microscopy to distinguish the two mirror-opposite regions. For real-life applications, it would be much more practical to have an all-optical read and write capability. One possibility has been demonstrated by rotational anisotropy second harmonic generation (RA-SHG), performed for

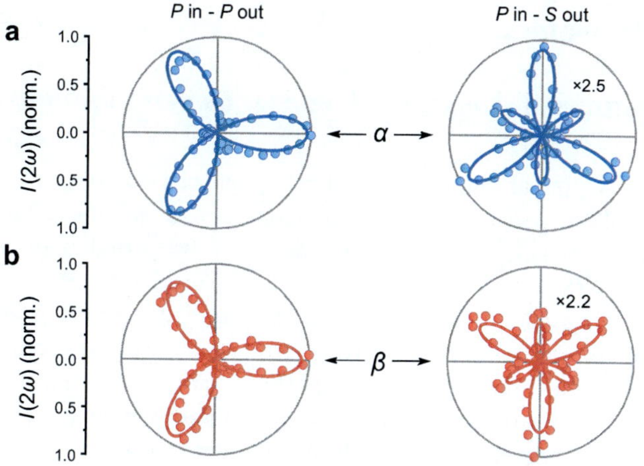

Fig. 8.10 Nonlinear optical signature of CDW mirror domains. Bulk $1T$-TaS_2 crystals were characterized by rotational anisotropy second harmonic generation (RA-SHG). The plot shows the second harmonic (400 nm) intensity as a function of azimuthal angle (ϕ in Fig. 1.2c) from mirror-opposite domains in the NC phase, taken at 340 K. The 800-nm fundamental beam is incident at a 10° angle with respect to the (001) sample normal. The solid curve in **a** is a fit to the data using the surface point group C_3. The fit in **b** was generated by performing a mirror operation to the numerical susceptibility tensor obtained from **a**. Data is normalized to the maximum value in the P_{in}-P_{out} polarization channel for each planar chirality

different incident and outgoing light polarizations (Fig. 8.10). For the surface point group C_3 in the NC state, a differentiating observable between α and β domains is the relative phase of the second harmonic pattern between the P_{in}-P_{out} channel and the P_{in}-S_{out} channel [40].

Beyond the potential of creating a novel memory device, the photo-manipulation of DWs may hold other application opportunities. At these mirror DWs, the CDW amplitude is quasi-periodically suppressed, as revealed by the atomically resolved STEM map (Fig. 8.5e). In the Mott insulating C phase, we therefore expect these DWs to host conducting channels or hotspots in an insulating bulk [3, 9, 10], a promising property for nanoelectronics [2]. To achieve custom functionality, one can pattern the light pulse, as done in holographic lithography, to create a specific network of DWs. More broadly, our results show that photoexcitation can be used as a reversible pathway between different states, providing a platform to realize unique properties in quantum matter through DW engineering.

8.5 Supplementary

8.5.1 Signature of Amplitude Mode vs. Generic Phonons

In general, many optical phonons near the Γ point can be coherently excited by a femtosecond laser pulse [41]. These phonons in turn modulate the diffraction intensity $I(\mathbf{k}) \propto |F(\mathbf{k})|^2$ at momentum transfer $\mathbf{k}$ by periodically changing the structure factor $F(\mathbf{k})$ due to the movement of ions in a unit cell. To see why the observed intensity oscillations in Figs. 8.6 and 8.11 arise from the amplitude mode instead of another coherent phonon, we note that the observed oscillation always starts with the suppression of superlattice peaks, regardless of the diffraction order. By contrast, for a generic phonon oscillation, whether the superlattice peaks are initially suppressed or enhanced depends on the diffraction order.

To understand this difference between the amplitude mode and the generic phonons, we explicitly evaluate the diffraction structure factor $F(\mathbf{k})$. Generalizing the result in Ref. [26] to the case of a polyatomic crystal with multiple CDW wavevectors, such as in $1T$-TaS_2, we first write the density function of ions $\rho(\mathbf{r})$ in the presence of superlattice modulations specified by wavevectors $\mathbf{q}_{i=1,2,\ldots}$

$$\rho(\mathbf{r}) = \sum_{\mathbf{R},\alpha,i} \delta\left[\mathbf{r} - (\mathbf{R} + \mathbf{r}_\alpha) - \mathbf{u}_{\alpha,i} \sin\left(\mathbf{q}_i \cdot (\mathbf{R} + \mathbf{r}_\alpha) + \phi_{\alpha,i}\right)\right], \tag{8.3}$$

where $\delta[\,\cdot\,]$ is the Dirac δ-function, $\mathbf{R}$ specifies unit cell locations, and $\mathbf{r}_\alpha$ are equilibrium ionic positions within the unit cell in the absence of a charge density wave. $\mathbf{u}_{\alpha,i}$ and $\phi_{\alpha,i}$ are the displacement vector and phase for superlattice modulation $\mathbf{q}_i$, which are generally different for different ions α. Harmonics of the superlattice modulation are not explicitly included but can be easily incorporated into $\mathbf{q}_i$. Taking the Fourier transform of $\rho(\mathbf{r})$ and including the atomic scattering factor $f_\alpha(\mathbf{k})$, the structure factor is

$$F_n(\mathbf{k}) = \sum_{\alpha,i} \delta(\mathbf{k} + n\mathbf{q}_i - \mathbf{G}) f_\alpha(\mathbf{k}) J_n(\mathbf{k} \cdot \mathbf{u}_{\alpha,i}) e^{i\mathbf{G}\cdot\mathbf{r}_\alpha + in\phi_{\alpha,i}}, \tag{8.4}$$

where $\mathbf{G}$ is the reciprocal lattice vector that indicates the Bragg order, J_n is the Bessel function of the first kind, and $n = 0, \pm 1, \pm 2, \ldots$ corresponds to Bragg peak, first-order superlattice peak, second-order superlattice peak, and so on.

We first consider the effect of a generic optical phonon oscillation on superlattice peaks. For simplicity, we only consider the first order ($n = \pm 1$). A femtosecond pulse sets the atomic positions $\mathbf{r}_\alpha$ to oscillate in time, leading to the modulation of $|F_{\pm 1}(\mathbf{k})|^2$. However, for the same initial change of $\mathbf{r}_\alpha$, the superlattice diffraction intensity may either increase or decrease, depending on the Bragg order $\mathbf{G}$ in the phase factor $e^{i\mathbf{G}\cdot\mathbf{r}_\alpha}$ in Eq. (8.4).

Next, we consider the amplitude mode, in which the amplitude of $\mathbf{u}_{\alpha,i}$ oscillates around its equilibrium value. In a typical diffraction experiment, $\mathbf{k} \cdot \mathbf{u}_{\alpha,i} \ll 1$, so to the lowest order, $J_{\pm 1}(\mathbf{k} \cdot \mathbf{u}_{\alpha,i}) \approx \pm \mathbf{k} \cdot \mathbf{u}_{\alpha,i}/2$. When the femtosecond pulse initially suppresses the CDW, there is a reduction of CDW amplitude $|\Delta \mathbf{u}_{\alpha,i}|$ with the same percentage decrease $|\Delta \mathbf{u}_{\alpha,i}|/|\mathbf{u}_{\alpha,i}|$ for all ions α. Regardless of the Bragg order **G**, these reduced magnitudes in $\mathbf{u}_{\alpha,i}$ necessarily lead to smaller $\left|J_{\pm 1}(\mathbf{k} \cdot \mathbf{u}_{\alpha,i})\right|$, which always causes an initial decrease in the superlattice diffraction intensity.

The above analysis thus enables us to assign the observed oscillation specifically to the amplitude mode. This is further supported by the observed 180° phase relation between intensities of Bragg and superlattice peaks (Figs. 8.6b and 8.11a, b), indicating the coherent intensity transfer as the CDW amplitude oscillates. The observed frequency of 2.42(3) THz (Fig. 8.6e, gray line) is consistent with values reported by other techniques such as optical pump–probe and Raman spectroscopies [28, 42], confirming our assignment.

8.5.2 Coherent Amplitude Mode and Domain Wall Configuration

Owing to the high electron energy in the MeV UED setup, high-order diffractions can be captured in the same active area of a CCD. This allows us to reduce experimental statistical fluctuations by considering a large number of peaks in the oscillation analysis. In Fig. 8.6, the intensities were averaged over 57 Bragg peaks and 274 α/β superlattice peaks. We also confirmed that the oscillation waveform and frequency are independent of the choice of peaks. In the α/β state, while Bragg peak intensity contains contributions from both domains, it is possible to separately analyze the superlattice peaks in each domain. We plot the oscillatory components in intensities of α, β peaks in Fig. 8.11a, b, which correspond to two distinct DW configurations proposed in Fig. 8.6f, h. The respective Fourier transform spectra of each domain are shown in Fig. 8.11c, d, where the peak positions are consistent with an exponentially decaying sinusoidal fit.

In the first configuration (Figs. 8.6e and 8.11a, c), amplitude mode oscillations in both domains have identical phase and frequency, which are the same as the case of a single-domain sample (Fig. 8.6e, gray line). This observation is consistent with the proposed large domains of either orientation (Fig. 8.6f), where the femtosecond pulse simultaneously excites the same amplitude mode in both α and β domains. In the second configuration (Figs. 8.6g and 8.11b, d), where α/β DWs were re-introduced into an α-only domain by a femtosecond pulse at room temperature, the amplitude mode significantly softens in the α orientation (2.08 ± 0.04 THz), while it remains unchanged in the β orientation (2.40 ± 0.04 THz). The exact reason for this distinction between α and β domains is subject to future investigations, but the asymmetry is not unexpected given the probabilistic nature of DW distribution and the global strain field that favors a single α domain under thermal equilibrium. The oscillation frequency in the combined superlattice peaks is the intensity-weighted average of individual frequencies from each domain (Fig. 8.11d), and the 180° phase

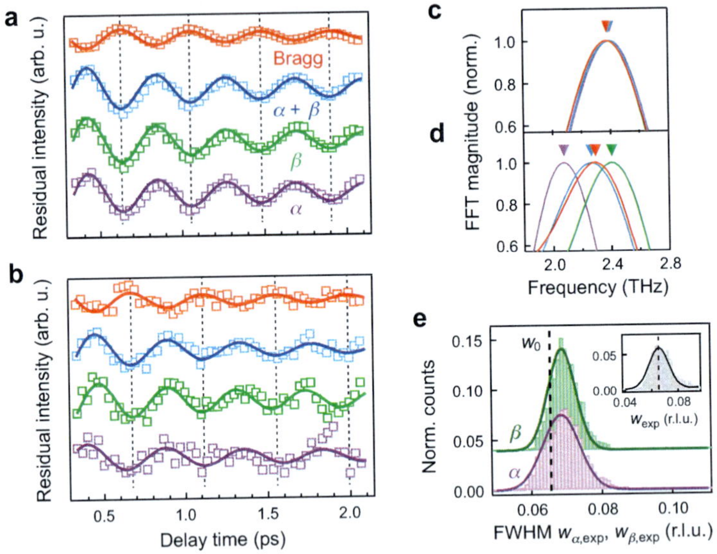

Fig. 8.11 Coherent amplitude mode oscillation and superlattice peak width. (**a**,**b**) Oscillatory parts of the diffraction intensity in the commensurate CDW phase at 40 K, corresponding to two distinct DW configurations in Fig. 8.6f, h, respectively. The same analysis was performed as in Fig. 8.6b, but separate contributions from α and β superlattices were considered here. Traces are vertically displaced for clarity. Dashed lines mark the Bragg intensity maxima. The smaller statistical noise in **a** than in **b** reflects a more optimized electron beam during the measurement. In both cases, oscillation frequencies were consistently determined by either exponentially decaying sinusoidal fits (solid curves) or Fourier transforms. (**c**,**d**) Fourier transformed spectra from **a**,**b**, respectively, where we zoom into the AM peak. (**e**) Histograms of experimentally fitted FWHM of α and β superlattice peaks, $w_{\alpha,\text{exp}}$, and $w_{\beta,\text{exp}}$, in reciprocal lattice units (r.l.u.). A total of 3,066 peaks were used from the time-resolved diffraction images corresponding to **b**. Curves are Gaussian fits to the histograms. The dashed line indicates the instrumental resolution, w_0, determined from the narrowest Bragg peaks, whose width is assumed to be resolution-limited. Right-shifted histograms from w_0 indicate short correlation lengths in both superlattices. *Inset*, the same histogram plot for a control experiment where only the α domain is present. In this case, the widths of both Bragg and superlattice peaks are limited by instrumental resolution

relation between the Bragg and superlattice peaks is preserved (Fig. 8.11b). As explained in Sect. 8.2.2, the softened amplitude mode suggests a significant presence of DWs within the α domain, similar to the discommensurate network in the NC phase where neighboring hexagrams suffer a phase slip without any relative rotation.

By analyzing the peak widths, we find a significant presence of α/β DWs as well, as proposed in the schematic in Fig. 8.6h. In particular, broad superlattice peaks in α and β orientations reveal a short CDW correlation length in both domains. A much softened amplitude mode in the α superlattice peaks further indicates the existence of discommensuration-like DWs within the α orientation. These observations hence suggest the following picture of DW formation after a strong pulse arrives at the initial α-only state. As the system relaxes from the transient IC to the equilibrium NC phase (Fig. 8.9), some remnant topological defects induce the local formation of β domains, while others lead to phase slips in the original α domain, in addition to the discommensuration DWs in a thermodynamic NC phase. These intra-α phase slips account for the observed soft amplitude mode, while dense α/β DWs account for the small size of β domains. Though additional studies are required to confirm this mechanism, it captures all current experimental observations within a consistent framework.

In the following calculation, we estimate the superlattice correlation lengths in both α and β orientations. To minimize statistical errors from choosing any particular peak, the widths of 21 pairs of α/β superlattice peaks were computed in 73 time-resolved diffraction patterns, which were averaged from 3,285 raw images. For each peak, a 13×13 pixel^2 ROI was chosen, and FWHMs along the horizontal (w_x) and vertical directions (w_y) were extracted from Gaussian fits to integrated intensities in each direction. The experimentally determined width w_{exp} was taken as the geometric mean of w_x and w_y [16]. We plot the histograms of w_{exp} for both α and β superlattice peaks (Fig. 8.11e), together with Gaussian fits to determine the average widths $w_{\alpha,\text{exp}}$ and $w_{\beta,\text{exp}}$. The vertical line indicates the instrumental resolution, w_0, calculated from the narrowest Bragg peaks assuming their width is resolution-limited. We note that both $w_{\alpha,\text{exp}}$ and $w_{\beta,\text{exp}}$ are significantly larger than the resolution limit at w_0, indicating broadened superlattice peaks due to the short correlation lengths. For comparison, we performed the same analysis in a control dataset where only a single α domain is present (Fig. 8.11e, inset). In this case, the widths of both superlattice and Bragg peaks are resolution-limited.

To estimate the superlattice correlation lengths $\xi_{i=\alpha,\beta} = 1/w_i$, the instrumental width w_0 is excluded [17],

$$w_i = \sqrt{w_{i,\text{exp}}^2 - w_0^2}, \quad \text{where } i = \alpha, \beta, \tag{8.5}$$

and we obtained $\xi_\alpha \approx \xi_\beta \approx 16\,\text{nm}$, about 13.5 superlattice unit cells. This indicates that the sizes of individual α and β domains are both small.

For α domains, the small size is consistent with the significant presence of discommensuration that dramatically softens the amplitude mode. For β domains, an unchanged amplitude mode frequency implies a dense network of α/β DWs that limit its size, as sketched in Fig. 8.6h. The large number of α/β DWs necessarily intermix micro-domains of both orientations, which can lead to interactions of amplitude modes between α and β domains. Such interactions may account for a shifted starting phase of oscillation in either domain (green and purple traces in Fig. 8.11b). This is in contrast to the same starting phase when domains are large (Fig. 8.11a), where inter-domain phonon interactions are negligible.

8.5.3 Simulation of Transient Defect Concentration

In the proposed mechanism for injecting or removing α/β DWs (Fig. 8.9), the determining factor is the concentration n_d of topological defects in the transient IC phase as it begins to relax to the equilibrium NC phase. For small n_d, a predominant single orientation is formed, as in thermal equilibrium. For large n_d, domains of either orientation can be locally favored due to defect-induced strains, leading to α/β DWs. In this phenomenological model, we simulate n_d after the arrival of each pulse. From the sequence of n_d, we calculate $\widetilde{P}_+$ and $\widetilde{P}_-$, the probability of switching from low to high n_d and the probability of the inverse process. These probabilities are then compared to experimentally determined probabilities of creating and removing α/β DWs, P_+ and P_-.

Let n_i denote the defect concentration in the transient IC phase after the arrival of the i-th pulse. We express n_{i+1} in terms of n_i as

$$n_{i+1} = n_i - \Delta n_{\text{decrease}} + \Delta n_{\text{increase}}, \tag{8.6}$$

where

$$\Delta n_{\text{decrease}} = \Theta(F - F_{\text{min}}) a_1 n_i, \tag{8.7}$$

$$\Delta n_{\text{increase}} = \Theta(F - F_{\text{min}}) a_2 (F - F_{\text{min}})^{\nu}. \tag{8.8}$$

Here $\Theta(\cdot)$ is the Heaviside step function, F denotes the laser pulse fluence, and F_{min} is the minimum threshold fluence. a_1 and a_2 are the positive constants of proportionality and the exponent ν characterizes a power-law fluence dependence.

These relations can be understood if we consider the underlying physical process, where an incoming pulse with sufficient fluence ($\geq F_{\text{min}}$) can simultaneously anneal defects that give rise to α/β DWs ($\Delta n_{\text{decrease}}$) and introduce new defects into the IC phase ($\Delta n_{\text{increase}}$). In our framework, the threshold fluence F_{min} corresponds to the minimum energy needed for a complete nonequilibrium NC-to-IC phase transition. Above the threshold F_{min}, if all previous defects are annihilated or self-relaxed, we have $a_1 = 1$. In our simulation, $a_1 = 0.43$, implying that some remnant defects are persistent in the NC-to-IC transition; this reproduces the important history-dependent switching property as mentioned in Sect. 8.3.2. For the defect creation process, we assume a power-law fluence dependence. Inter-defect interactions are not included in this simple model, where we consider the dilute limit.

For each fluence F, a sequence $\{n_i\}_{i=1,\ldots,N}$ is generated based on Eqs. (8.6)–(8.8), where $N = 10^6$. An initial value n_0 was randomly selected in the interval $[0, n_{\text{max}}]$, where n_{max} is the maximally allowed defect concentration that corresponds to the maximum size of reoriented domains. The initial seed n_0 does not affect the result due to the long sequence. Each n_i in the sequence is bounded above by n_{max} and Gaussian noise with a standard deviation of $0.1 n_{\text{max}}$ was added to $\Delta n_{\text{decrease}}$ and $\Delta n_{\text{increase}}$. Similar to the calculation of probabilities $P_{\pm}$, we classify each defect concentration n_i into high (H) and low (L) clusters, defined by the cutoff

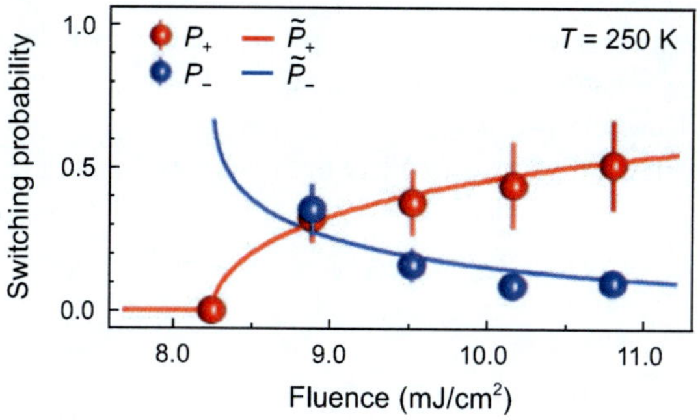

Fig. 8.12 Simulated switching probabilities of transient defect concentration. The simulated probabilities $\widetilde{P}_{\pm}$ are plotted together with experimentally determined domain switching probabilities $P_{\pm}$ at 250 K. At low fluence, P_- and $\widetilde{P}_-$ are unavailable due to a vanishing probability of creating DWs or superlattice defects

at $n_{\max}/2$. Hence, at each fluence, we can compute the corresponding probabilities for switching between high and low defect concentrations:

$$\widetilde{P}_+ = \frac{|\{n_i \mid n_i \in \mathrm{L} \text{ and } n_{i+1} \in \mathrm{H}\}|}{|\{n_i \mid n_i \in \mathrm{L}\}|}, \tag{8.9}$$

$$\widetilde{P}_- = \frac{|\{n_i \mid n_i \in \mathrm{H} \text{ and } n_{i+1} \in \mathrm{L}\}|}{|\{n_i \mid n_i \in \mathrm{H}\}|}, \tag{8.10}$$

where $|\cdot|$ denotes the cardinality of the corresponding set.

In Fig. 8.12, we plot the simulated probabilities $\widetilde{P}_{\pm}$ alongside the experimentally determined probabilities $P_{\pm}$ at 250 K using a threshold fluence $F_{\min} = 8.23\,\mathrm{mJ/cm^2}$, a constant $a_2 = 0.24$, and a power-law exponent $\nu = 0.18$, indicating a good qualitative agreement. In particular, it reproduces the feature where P_+ increases with an increasing fluence, while P_- has the opposite trend. This feature reflects the fluence-dependent probability of defect creation, as captured in Eq. (8.8). Hence, despite the simplicity of this model, the underlying physics for DW manipulation is retained.

8.5.4 *Landau Free Energy for Mirror Symmetry Breaking*

In Fig. 8.9, we used a tilted Landau free energy to illustrate how topological defects break the local α/β degeneracy during the IC-to-NC relaxation, hence giving rise to the formation of α/β DWs. Here, we write down the simplest form of the free energy responsible for breaking the mirror symmetry in the IC–NC phase transition. We use ψ to denote the order parameter, where $\psi < 0$ represents an α state, while $\psi > 0$ represents a β state. For simplicity, we limit ourselves to the discussion of breaking the mirror symmetry only; a complete treatment of the phase transition involving a triple-q charge density modulation and discommensuration has been well formulated elsewhere [43–46].

The local free energy density $f(\psi)$ can be written as

$$f(\psi) = a(T)\psi^2 + \frac{b}{2}\psi^4 + s(\mathbf{r})\psi + \xi^2 |\nabla\psi|^2 , \text{ where} \tag{8.11}$$

$$a(T) = a_0 \frac{T - T_c}{T}, \text{ and} \tag{8.12}$$

$$s(\mathbf{r}) = s_0 + c\left[n_\alpha(\mathbf{r}) - n_\beta(\mathbf{r})\right] . \tag{8.13}$$

The first two terms in Eq. (8.11) give the usual double-well potential. The $|\nabla\psi|^2$ term represents the energy cost of DWs, characterized by an energy scale set by the "stiffness" constant ξ. The α/β degeneracy is lifted by a coupling of the order parameter to the density of topological defects. Similar to an Ising model with random local fields, this coupling can be written as $s(\mathbf{r})\psi$ to the lowest order in ψ. The coefficient $s(\mathbf{r})$ depends on the spatially varying densities of topological defects, $n_\alpha(\mathbf{r})$ and $n_\beta(\mathbf{r})$, which locally favor α and β orientations, respectively. In Eq. (8.13), c is a constant of proportionality and $s_0 > 0$ represents a global strain field that prefers the α orientation under equilibrium conditions. In the NC phase where the mirror symmetry is broken, at location $\mathbf{r}$ where $s(\mathbf{r}) > 0$ (or < 0), α (or β) orientation prevails.

It remains to be explained why topological defects present at the start of the IC-to-NC relaxation may locally favor one orientation over the other. From a microscopic perspective, one possibility is that these superlattice defects act as pinning sites where the Ta hexagram centers are located. Therefore, a certain local arrangement of defects favors a specific tiling pattern of hexagrams in the NC phase, leading to patches of either α or β domains.

References

1. Bader, S.D.: Colloquium: Opportunities in nanomagnetism. Rev. Mod. Phys. **78**, 1–15 (2006)
2. Catalan, G., Seidel, J., Ramesh, R., Scott, J.F.: Domain wall nanoelectronics. Rev. Mod. Phys. **84**, 119–156 (2012)
3. Ma, E.Y., Cui, Y.-T., Ueda, K., Tang, S., Chen, K., et al.: Mobile metallic domain walls in an all-in-all-out magnetic insulator. Science **350**, 538–541 (2015)
4. Aird, A., Salje, E.K.H.: Sheet superconductivity in twin walls: Experimental evidence of WO_{3-x}. J. Phys. Condens. Matter **10**, L377–L380 (1998)
5. Barja, S., Wickenburg, S., Liu, Z.F., Zhang, Y., Ryu, H., et al.: Charge density wave order in 1D mirror twin boundaries of single-layer $MoSe_2$. Nat. Phys. **12**, 751–756 (2016)
6. Aishwarya, A., Howard, S., Padhi, B., Wang, L., Cheong, S.-W. et al.: Visualizing 1D zigzag Wigner crystallization at domain walls in the Mott insulator TaS_2. arXiv 1906.11983 (2019)
7. Stojchevska, L., Vaskivskyi, I., Mertelj, T., Kusar, P., Svetin, D., et al.: Ultrafast switching to a stable hidden quantum state in an electronic crystal. Science **344**, 177–180 (2014)

8. Vaskivskyi, I., Gospodaric, J., Brazovskii, S., Svetin, D., Sutar, P., et al.: Controlling the metal-to-insulator relaxation of the metastable hidden quantum state in $1T$-TaS_2. Sci. Adv. **1**, e1500168 (2015)
9. Ma, L., Ye, C., Yu, Y., Lu, X.F., Niu, X., et al.: A metallic mosaic phase and the origin of Mott-insulating state in $1T$-TaS_2. Nat. Commun. **7**, 10956 (2016)
10. Cho, D., Cheon, S., Kim, K.-S., Lee, S.-H., Cho, Y.-H., et al.: Nanoscale manipulation of the Mott insulating state coupled to charge order in $1T$-TaS_2. Nat. Commun. **7**, 10453 (2016)
11. Yan, S., Iaia, D., Morosan, E., Fradkin, E., Abbamonte, P., et al.: Influence of domain walls in the incommensurate charge density wave state of Cu intercalated $1T$-$TiSe_2$. Phys. Rev. Lett. **118**, 106405 (2017)
12. Ravnik, J., Vaskivskyi, I., Mertelj, T., Mihailovic, D.: Real-time observation of the coherent transition to a metastable emergent state in $1T$-TaS_2. Phys. Rev. B **97**, 075304 (2018)
13. Vaskivskyi, I., Mihailovic, I.A., Brazovskii, S., Gospodaric, J., Mertelj, T., et al.: Fast electronic resistance switching involving hidden charge density wave states. Nat. Commun. **7**, 11442 (2016)
14. Li, L.J., O'Farrell, E.C.T., Loh, K.P., Eda, G., Özyilmaz, B., et al.: Controlling many-body states by the electric-field effect in a two-dimensional material. Nature **529**, 185–189 (2016)
15. Zurek, W.H.: Cosmological experiments in condensed matter systems. Phys. Rep. **276**, 177–221 (1996)
16. Vogelgesang, S., Storeck, G., Horstmann, J.G., Diekmann, T., Sivis, M., et al.: Phase ordering of charge density waves traced by ultrafast low-energy electron diffraction. Nat. Phys. **14**, 184–190 (2018)
17. Laulhé, C., Huber, T., Lantz, G., Ferrer, A., Mariager, S.O., et al.: Ultrafast formation of a charge density wave state in $1T$-TaS_2: Observation at nanometer scales using time-resolved x-ray diffraction. Phys. Rev. Lett. **118**, 247401 (2017)
18. Haupt, K., Eichberger, M., Erasmus, N., Rohwer, A., Demsar, J., et al.: Ultrafast metamorphosis of a complex charge-density wave. Phys. Rev. Lett. **116**, 016402 (2016)
19. Yusupov, R., Mertelj, T., Kabanov, V.V., Brazovskii, S., Kusar, P., et al.: Coherent dynamics of macroscopic electronic order through a symmetry breaking transition. Nat. Phys. **6**, 681–684 (2010)
20. Gerasimenko, Y.A., Karpov, P., Vaskivskyi, I., Brazovskii, S., Mihailovic, D.: Intertwined chiral charge orders and topological stabilization of the light-induced state of a prototypical transition metal dichalcogenide. npj Quantum Mater. **4**, 32 (2019)
21. Zong, A., Kogar, A., Bie, Y.-Q., Rohwer, T., Lee, C., et al.: Evidence for topological defects in a photoinduced phase transition. Nat. Phys. **15**, 27–31 (2019)
22. Trigo, M., Giraldo-Gallo, P., Clark, J.N., Kozina, M.E., Henighan, T., et al.: Ultrafast formation of domain walls of a charge density wave in $SmTe_3$. Phys. Rev. B **103**, 054109 (2021)
23. Williams, D.B., Carter, C.B.: Thickness and bending effects. In: Transmission Electron Microscopy: A Textbook for Materials Science, pp. 407–417. Springer, Boston (2009)
24. Hovden, R., Tsen, A.W., Liu, P., Savitzky, B.H., El Baggari, I., et al.: Atomic lattice disorder in charge-density-wave phases of exfoliated dichalcogenides ($1T$-TaS_2). Proc. Natl. Acad. Sci. USA **113**, 11420–11424 (2016)
25. Pekker, D., Varma, C.M.: Amplitude/Higgs modes in condensed matter physics. Annu. Rev. Condens. Matter Phys. **6**, 269–297 (2015)
26. Overhauser, A.W.: Observability of charge-density waves by neutron diffraction. Phys. Rev. B **3**, 3173–3182 (1971)
27. Grüner, G.: Density Waves in Solids. Addison-Wesley, Boston (1994)
28. Demsar, J., Forró, L., Berger, H., Mihailovic, D.: Femtosecond snapshots of gap-forming charge-density-wave correlations in quasi-two-dimensional dichalcogenides $1T$-TaS_2 and $2H$-$TaSe_2$. Phys. Rev. B **66**, 041101 (2002)
29. Wilson, J.A., Di Salvo, F.J., Mahajan, S.: Charge-density waves and superlattices in the metallic layered transition metal dichalcogenides. Adv. Phys. **24**, 117–201 (1975)

30. Eichberger, M., Schäfer, H., Krumova, M., Beyer, M., Demsar, J., et al.: Snapshots of cooperative atomic motions in the optical suppression of charge density waves. Nature **468**, 799–802 (2010)
31. Lantz, G., Laulhé, C., Ravy, S., Kubli, M., Savoini, M., et al.: Domain size effects on the dynamics of a charge density wave in $1T$-TaS_2. Phys. Rev. B **96**, 224101 (2017)
32. Möhr-Vorobeva, E., Johnson, S.L., Beaud, P., Staub, U., De Souza, R., et al.: Nonthermal melting of a charge density wave in $TiSe_2$. Phys. Rev. Lett. **107**, 036403 (2011)
33. Lee, W.S., Chuang, Y.D., Moore, R.G., Zhu, Y., Patthey, L., et al.: Phase fluctuations and the absence of topological defects in a photo-excited charge-ordered nickelate. Nat. Commun. **3**, 838 (2012)
34. Huber, T., Mariager, S.O., Ferrer, A., Schäfer, H., Johnson, J.A., et al.: Coherent structural dynamics of a prototypical charge-density-wave-to-metal transition. Phys. Rev. Lett. **113**, 026401 (2014)
35. Schmitt, F., Kirchmann, P.S., Bovensiepen, U., Moore, R.G., Chu, J.-H., et al.: Ultrafast electron dynamics in the charge density wave material $TbTe_3$. New J. Phys. **13**, 063022 (2011)
36. Kanungo, T., Mount, D.M., Netanyahu, N.S., Piatko, C.D., Silverman, R., et al.: An efficient k-means clustering algorithm: Analysis and implementation. IEEE Trans. Pattern Anal. Mach. Intell. **24**, 881–892 (2002)
37. Wilson, E.B.: Probable inference, the law of succession, and statistical inference. J. Am. Stat. Assoc. **22**, 209–212 (1927)
38. Han, T.-R.T., Zhou, F., Malliakas, C.D., Duxbury, P.M., Mahanti, S.D., et al.: Exploration of metastability and hidden phases in correlated electron crystals visualized by femtosecond optical doping and electron crystallography. Sci. Adv. **1**, e1400173 (2015)
39. Ogawa, N., Miyano, K.: Charge-density wave as an electro-optical switch and memory. Appl. Phys. Lett. **80**, 3225–3227 (2002)
40. Fichera, B.T., Kogar, A., Ye, L., Gö, Zong, A., et al.: Second harmonic generation as a probe of broken mirror symmetry. Phys. Rev. B **101**, 241106 (2020)
41. Dekorsy, T., Cho, G.C., Kurz, H.: Coherent phonons in condensed media. In: Cardona, M., Güntherodt, G. (eds.) Light Scattering in Solids VIII, pp. 169–209. Springer, Berlin/Heidelberg (2000)
42. Sugai, S., Murase, K., Uchida, S., Tanaka, S.: Comparison of the soft modes in tantalum dichalcogenides. Physica B+C **105**, 405–409 (1981)
43. McMillan, W.L.: Time-dependent Laudau theory of charge-density waves in transition-metal dichalcogenides. Phys. Rev. B **12**, 1197–1199 (1975)
44. McMillan, W.L.: Theory of discommensurations and the commensurate-incommensurate charge-density-wave phase transition. Phys. Rev. B **14**, 1496–1502 (1976)
45. Nakanishi, K., Shiba, H., Takatera, H., Yamada, Y.: The nearly commensurate phase and effect of harmonics on the successive phase transition in $1T$-TaS_2. J. Phys. Soc. Jpn. **43**, 1509–1517 (1977)
46. Nakanishi, K., Shiba, H.: Domain-like incommensurate charge-density-wave states and the first-order incommensurate-commensurate transitions in layered tantalum dichalcogenides. I. $1T$-polytype. J. Phys. Soc. Jpn. **43**, 1839–1847 (1977)